The Philosophy and Science of Language

Ryan M. Nefdt • Carita Klippi
Bart Karstens
Editors

The Philosophy and Science of Language

Interdisciplinary Perspectives

palgrave
macmillan

Editors
Ryan M. Nefdt
Philosophy Department
University of Cape Town
Cape Town, South Africa

Bart Karstens
Science and Technology Studies
Rathenau Institute
The Hague, The Netherlands

Carita Klippi
Tampere University
Tampere, Finland

Laboratoire d'Histoire des Théories
Linguistiques (CNRS, Université
de Paris, Université Sorbonne Nouvelle)
Paris, France

ISBN 978-3-030-55437-8 ISBN 978-3-030-55438-5 (eBook)
https://doi.org/10.1007/978-3-030-55438-5

This Palgrave Macmillan imprint is published by the registered company Springer Nature Switzerland AG.
The registered company address is: Gewerbestrasse 11, 6330 Cham, Switzerland

Preface

While perhaps *all* fields of study have at some point in the course of their existence interacted with other disciplines and exchanged such things as theories, concepts, metaphors, methods, instruments and measurement systems, the extent to which this so-called epistemic transfer has occurred with respect to linguistics appears to be exceptional. In many cases, epistemic and methodological transfer with both the sciences *and* the humanities have been documented but the underlying philosophical issues have not always been adequately addressed. The present volume hopes, that by bringing a diverse set of scholars in a range of fields connected to the study of language, the important philosophical and interdisciplinary questions can be approached and new avenues for research developed.

The aim of this volume, therefore, is to find out how linguistics interacts with neighboring disciplines such as philology, literary studies, philosophy, logic, mathematics, computer science and the cognitive sciences. Questions of the mathematical foundations of the subject, the cognitive implications, and the historical connections take center stage throughout the volume. A number of chapters overlap in their objectives but each offers unique insight into the central questions under discussion.

We have also aimed at showcasing the work of academics from not only different areas of expertise but also varying positions from emerging scholars to more established figures in the their respective fields across three continents. The result is a book that aims above all to interest any student or advanced scholar who is connected in any way to the study of language and more broadly general questions of diachronic and cross-disciplinary interaction.

Cape Town, South Africa — Ryan M. Nefdt
Tampere, Finland — Carita Klippi
The Hague, The Netherlands — Bart Karstens

Contents

Notes on Contributors

Giosuè Baggio, Professor of Psycholinguistics at the Norwegian University of Science and Technology. In addition to publishing in linguistics, psycholinguistics and cognitive neuroscience, he is also the author of *Meaning in the Brain* (MIT Press, 2018).

Els Elffers, Retired Lecturer in Dutch Linguistics at the University of Amsterdam. She is widely published in foundations and history of linguistics and the semantics and pragmatics of Dutch.

Katrin Erk, Professor in the Linguistics Department at the University of Texas at Austin. Her work is in computational semantics and computational lexical semantics.

Esa Itkonen, Emeritus Professor of General Linguistics at the University of Turku. He is a renowned philosopher of linguistics and has published countless articles and chapters on linguistic metatheory over the past decades. He has published a number of influential books including *Grammatical Theory and Metascience* (John Benjamins, 1978), *Causality in linguistic theory* (Indiana University Press, 1983), *Universal History of Linguistics. India, China, Arabia, Europe* (John Benjamins, 1991) and *Analogy as Structure and Process* (John Benjamins, 2015).

General Ontology for Linguistic Description (GOLD) and is currently with the National Science Foundation as a part-time Expert in the Robust Intelligence program.

Outi Vesakoski, Researcher in Biology at the University of Turku and Principal Investigator at the BEDLAN (Biological Evolution and Diversification of Languages) research initiative in Finland. She works mainly within evolutionary biology, which includes a project about evolution of personality in marine isopods *Idotea balthica*, and especially sexual differences therein. She has published a number of articles and book chapters on or related to this work.

Johannes Woschitz, Researcher at the School of Philosophy, Psychology, and Language Sciences at the University of Edinburgh. He is currently involved in theoretical work related to the epistemic status of sociolinguistics. He addresses paradigm changes therein and questions of scientific progress by referring to principles of the philosophy of science, especially scientific realism. He has published various articles on the philosophy of science applied to generative and sociolinguistics.

List of Figures

List of Tables

1
Introduction

Ryan M. Nefdt, Carita Klippi, and Bart Karstens

The philosophy and science of language are emerging multidisciplinary fields of investigation (Scholz et al. 2016; Kempson et al. 2012; Nefdt 2019). Driven by both the advancements in other fields and *sui generis* results, some internal and others external, the study of language has at times been a conduit to the study of the mind, brain, a lens into societal structure, literature and human history. It is therefore only fitting that a

R. M. Nefdt (✉)
Philosophy Department, University of Cape Town, Cape Town, South Africa
e-mail: ryan.nefdt@uct.ac.za

C. Klippi
Tampere University, Tampere, Finland

Laboratoire d'Histoire des Théories Linguistiques (CNRS, Université de Paris, Université Sorbonne Nouvelle), Paris, France
e-mail: carita.klippi@tuni.fi

B. Karstens
Science and Technology Studies, Rathenau Institute, The Hague, The Netherlands
e-mail: b.karstens@rathenau.nl

R. M. Nefdt et al. (eds.), *The Philosophy and Science of Language*,
https://doi.org/10.1007/978-3-030-55438-5_1

full length volume be devoted to the disciplinary connections between linguistics and those fields both past and present. This volume aims to bring together a diverse set of scholars from around the disciplinary landscape to delve into questions related to the history, philosophy, and theoretical interplay between the study of language and fields as distant as logic, physics, biology, classical philology, and cognitive neuroscience.

As stated in the preface, all academic disciplines have at some point exchanged theories, concepts, metaphors, methods, instruments, etc. with other (proto-) disciplines in the course of their existence, the extent to which this so-called epistemic transfer has occurred with respect to the study of language appears to be exceptional. This has of course not gone unnoticed in the historiography of linguistics, and many cases of epistemic transfer with both the sciences and the humanities have been documented. Yet, the vast majority of these studies remain 'isolated', in the sense that they are not compared to other case studies of epistemic transfer, nor are they linked to each other where this would actually be both appropriate and insightful. The goal of the volume is to find out why and how linguistics exhibited the extraordinary capacity to interact frequently, and in many different ways, with other disciplines across the academic spectrum.

This volume of essays is additionally aimed at bringing together scholars from theoretical linguistics, history, classical studies, philosophy, logic, computer science, evolutionary theory and cognitive science in order to approach the study of language across these disciplines and beyond. The theoretical approaches in philosophy include experts from both the analytic and continental traditions.

With each chapter a new avenue of insights and critique are brought forth. From the philosophical side, what sets this volume apart from some recent work on the philosophy of linguistics such as Ludlow (2011) or Tomalin (2006) is the expansion of the linguistic frameworks under discussion. Although generative linguistics features prominently in a number of chapters (Chaps. 2, 3, 6, 9, and 13), sociolinguistics (Chap. 6), structuralism (Chaps. 11 and 13), neurolinguistics (Chap. 11), cognitive linguistics (Chap. 9), and computational linguistics (Chaps. 4 and 13) all receive due philosophical and historical treatment. Additionally, on a broader level, diachronic or historical linguistics is considered in

Chap. 7, Chap. 8 discusses linguistic metatheory, Chap. 5 scientific discourse analysis and Chap. 14 incorporates metaphor studies.

The multiplicity of phenomena often leads scientists of any particular discipline to seek unifying principles for their object of study. The nature of language presents such a multiplicity. A search for such unification could lead to an eternal pursuit of a theory of everything, which would give the ultimate explanation of the linguistic universe. However, a true integral and total linguistics is unrealistic. Instead of tempting a unified theory to apprehend the slippery character of language, interdisciplinary exchanges have provided, one at a time, different solutions to translate the essential substance of language.

The focus on epistemic transfer shifts the center of attention to what unites academic disciplines instead of what divides them. This shift is in tune with a trend in present-day historiography of science to break down barriers between academic fields. The term 'science' is perhaps already inappropriate, if with science we would exclude the humanities. 'History of knowledge' is more in spirit of moving beyond disciplinary categories, although this prompts the question of the difference between 'scientific' knowledge and other forms of knowledge such as practical skills and knowledge by acquaintance.

In various places it has been argued recently that the persistent notion of a divide between the sciences and humanities has to be seriously qualified for at least two reasons (Graff 2015, von Dongen and Paul 2017). First, the line of division has in the past been drawn in different ways and for different reasons. If these demarcations are a local construct we cannot speak of a fundamental divide between the sciences and humanities. Second, it can be demonstrated that the sciences and humanities shared ideas or even have common origins. This volume is a further contribution to the growing body of literature that supports this claim.

Historians of science have developed a number of concepts that can help us to study interdisciplinary interaction. Well-known are Thomas Gieryn's idea of 'boundary work' and Peter Galison's notion of 'trading zone' (Galison 1997). For Gieryn boundaries between disciplines are not permanent but are instead continuously negotiated during "boundary work" (Gieryn 1999). He concludes that disciplinary boundaries are porous: that is, boundary work may involve boundary crossing. Galison has tried to capture the constant redrawing of boundaries between the instrumental, experimental and theoretical subdisciplines of physics, with the trading zone concept. The trading zone is a place of interaction in which a hybrid community seeks ways to communicate, often through new forms of interlanguage. Trading zones can thus be the springboard for new institutional structures and new categories of knowledge. To zoom in closer on what is actually transferred during disciplinary interaction the concept of 'cognitive goods' has recently been proposed (Bod et al. 2019). 'Cognitive goods' is an umbrella term intended to refer to anything that can be exchanged by practitioners i.e. methods, concepts, models, metaphors, formalisms, principles, modes of representation, argumentative and demonstrative techniques, technical instruments, institutional arrangements, and intellectual, theoretical, and epistemic virtues. The cognitive goods concept is meant to serve as a springboard for systematic analysis of 'flows' between disciplines which have occurred throughout the history of science.

What follows is a brief summary of each chapter grouped around four general broad themes, linguistics and the formal sciences, linguistics and the natural sciences, linguistics and the cognitive sciences and finally linguistics and the humanities.

1 Contributions to This Volume

Jaroslav Peregrin's chapter entitled "Syntax and Semantics in Linguistics and Logic" launches an investigation into the very notion of syntactic form within formal linguistics and formal logic respectively. He details the nature of formal properties as it is applied within linguistic and mathematical logic settings while touching on issues of syntax in semantics, cognitive science, and AI. He argues that the concept of syntax has played a crucial role in the development of modern logic and linguistics. Despite this its definition is not unequivocal, especially in cases where no clarification is given as to which of its various senses is being employed at a given time. Peregrin takes pains to distinguish these senses and shows that doing so provides philosophical fecundity in a number of settings, both theoretical and applied in both logic and linguistics.

While Peregrin's chapter offers a high-level theoretical exploration of the very concept of syntax, D. Terence Langendoen in his "Negation of entities and a reanalysis of the logic of reciprocity" showcases the connection between logic and linguistics directly by providing a mathematical linguistic analysis of the syntax of negation and reciprocity in natural language. This is a highly technical work linking advanced concepts in mathematics (Dedekind orderings) to natural language phenomena, e.g. the logic of multiple negation to reciprocity. Langendoen's chapter presents novel results toward a novel conclusion taking formal structures and applying them to linguistic phenomena.

Katrin Erk's chapter "Variations on abstract semantic spaces" moves to an overview of the connections between linguistics and computer science by surveying the vast and technical landscape of lexical semantics. She discusses the various computational means of representing semantic space. Semantic spaces represent the meanings of words and phrases as objects in space. Her chapter concerns abstract semantic spaces which are formalisms that, in one way or another, make use of the idea of meaning as a space. Although a complete account which allows for inference to sentence meaning is still unclear, Erk presents promising computational research possibilities for this eventual extension.

Finally, to close off the linguistics and formal sciences theme, Adrien Mathy mounts an argument that the formalization and the mathematization of linguistics was accompanied by adhering to a specific scientific ethos that linguists actively sought to exemplify. Thus non-epistemological factors played an important role in the acceptance of concepts from foreign fields. Additionally, in his chapter "Mathematical transfers in linguistics: A dynamic between *ethos* and formalization as a process of scientific legitimization", Mathy proposes a case study on the mathematization of linguistics through a publication of Antoine Culioli, *La formalisation en linguistique* (1968). The article aims at showing that disciplinary transfer does not concern only the conceptual apparatus as such, but also the belief system, the author has chosen to call "ethos" (in the footsteps of the French tradition in rhetoric and discourse analysis).

The theme shifts to the connections with the natural sciences in Johannes Woschitz' chapter "Scientific realism and linguistics: Two stories of scientific progress". Here he draws on the scientific realism debate in the philosophy of science to assess progress in two linguistic subdisciplines, namely Chomskyan syntax and the Labovian study of phonological change. He analyses the historical developments of Universal Grammar (UG) and Labov's internal factors as case studies. He then claims that the history shows continuity and similarity in development between these two disciplines, as in both cases epistemologically 'expensive' concepts have been challenged and re-theorized. Woschitz makes use of contemporary work on scientific models, realism, theory comparison and the Kuhnian sociology of science to put forward his picture of scientific continuity.

The next chapter by Unni Leino, Kaj Syrjänen and Outi Vesakoski "Linguistic change and biological evolution" explores the interaction between linguistics and biology, with specific focus on the analogy between linguistic change and biological evolution. They argue that general evolutionary meta-theories give support for finding and using similarities between biology and linguistics. But they caution that such general theories are often too general for the task of adapting specific tools from one field to another which requires detailed analysis of both fields. Nevertheless, they are optimistic that there exists between biology and linguistics, a network of functional similarities which extends from evolutionary biology and historical linguistics to population biology and dialectology/sociolinguistics.

Lastly, under this umbrella theme is Esa Itkonen's chapter "Three models for linguistics: Newtonian Mechanics, Darwinism, Axiomatics". In characteristic style, Itkonen discusses the history of three major influences on linguistics but goes further to ask the value-laden questions of whether or not these three particular influences, classical mechanics, Darwinism evolutionary theory and axiomatic logic have been beneficial, harmful, or irrelevant to its development. He suggests that axiomatics provided a better model for two specific desiderata of linguistic theory, systematization and causal explanation respectively, but outlines its weaknesses with relation to applied settings as well as provides a convincing case for why the other two major influences fail in terms of the aforementioned desiderata.

The next overarching theme concerns the relationship between linguistics and the cognitive sciences. The first chapter by Ryan M. Nefdt "The

Role of Language in the Cognitive Sciences" aims to present a case for why language and linguistics should retain a central role in the emerging Second Generation Cognitive Science. He provides a historical overview of the initial cognitive revolution in the late 1950s and the philosophical as well as historical reasons language occupied a leading position in the nascent interdisciplinary project before using architectural considerations to suggest that language still has much to offer the scientific study of the mind in the twenty-first century.

Els Elffers shifts the timeline back further to the relation between aphasiology and linguistics in the nineteenth century in her "Linguistics and brain science: (dis-) connections in 19th century aphasiology". In her chapter, she presents a nuanced picture of nineteenth century interaction between linguistics and aphasiology. Unlike many of the other chapters in the volume which highlight interdisciplinary cross-pollination, Elffers asks why there was only minimal contact between these two disciplines at a time when there was scientific developments at their intersection. She pays special attention to Steinthal's contribution to aphasiology. His chapter on language disorders has been referred to as a unique and promising example of interdisciplinary neurolinguistics-avant-la-lettre. She argues that although this is an overstatement, Steinthal's text bears witness of psychological sophistication, but is not a programmatic plea for a new linguistically-informed approach to aphasiology. Her narrative is intriguing and aims to provide clues to historical interdisciplinary mysteries at the core of language studies and brain science in the nineteenth century.

Finally, Giosuè Baggio takes us back to trends of epistemic transfer in contemporary neurolinguistics with his chapter entitled "Epistemic transfer between linguistics and neuroscience: Problems and prospects". This chapter focuses on cases of successful, partial, and failed unidirectional epistemic transfer between theoretical linguistics and neuroscience. He distinguishes between three types of transfer, depending on the nature of the linguistic knowledge involved: type-A knowledge, about language as such, essentially invariant across theories or formalisms; type-B knowledge, about alternative formal analyses of basic structures and operations in language; type-C knowledge, about the application of various computational methods to analyzing or modeling behavioral or neural data. He argues that successful epistemic transfer may be achieved, under certain conditions, with type-A and type-C knowledge, but suggests that Type-B transfer has not led so far to new knowledge of the neural correlates and mechanisms of linguistic computation. He concludes that greater theoretical emphasis on algorithmic-level analyses, via a revised notion of linguistic competence and a new model of epistemic transfer, can bring formal linguistics and neuroscience closer together. He also discusses the possible role of a computationalist psycholinguistics as a 'bridge science', that could serve the aim of linking linguistics and neuroscience.

The final theme puts the connections between the human sciences and linguistic under the lens. Anna Novokhatko in her "Linguistics meets hermeneutics: reading early Greek epistemological texts" argues that the interaction between linguistics and philology is reciprocal and benefits both fields tremendously. She cautions that this connection can all too often be forgotten by formal/mathematical approaches, but her analysis of pre-classical Ancient Greek epistemological texts shows that there is hope for synthesis. Much like Langendoen's chapter which provides both logical and linguistic analysis of particular constructions, Novokhatko analyzes the texts of Pindar, Theognis, Archilochus, Heraclitus, Xenophanes and many others. She advocates the position that analysis of formal structure may crucially hinge on content analysis (and vice versa).

Maintaining the historical textual analysis, Jacqueline Léon in her chapter "On the history of models in American linguistics" tracks the emergence of the term "model" through various epochs of modern formal linguistics. She argues that the use of the term stabilized with the Chomskyan paradigm but not before various detours and appropriations from formal logic, model theory, proof theory, and stochastic approaches such as Markov models. Her analysis shows more than a terminological issue in that it provides a glimpse into the evolution of the mathematical concept within linguistic practice.

Carita Klippi argues that metaphors can be understood as interdisciplinary vehicles, but also that they form an integral part of theory formation within scientific disciplines. She focuses on the metaphor of 'life of language' in the second half of nineteenth century linguistics and shows how this metaphor induced historicist and organicist theoretical linguistic conceptions, given rise to *diverging* notions of what kind of science linguistics actually is. Modes of validation of knowledge are not just discipline specific, they may differ within the same discipline depending on diverging interpretations of the same metaphor.

Finally, Kate Hazel Stanton's chapter "Linguistics and Philosophy: Breakup Song" ends the book with a critical look at the future of the interaction between linguistics and philosophy. She reviews past successes and suggests that it was the undeveloped nature of linguistics at the time that accounted for most of them. She argues that despite rhetorical flourish and reflection on erstwhile interactions, there is now a growing methodological chasm between the disciplines. One that will ultimately hinder continued interdisciplinary interchange. Nonetheless, she concludes by considering some promising avenues for fruitful collaboration between the fields in the future.

References

Bod, Rens, Jeroen Van Dongen, Bart Karstens, Sjang Ten Hagen, and Emma Mojet. 2019. The flow of cognitive goods: A historiographical framework for the study of epistemic transfer. *Isis* 110 (3): 483–496.

Galison, Peter. 1997. *Image and logic: A material culture of microphysics*. Chicago: University of Chicago Press.

Gieryn, Thomas. 1999. *Cultural boundaries of science: Credibility on the line*. Chicago: University of Chicago Press.

Graff, Harvey. 2015. *Undisciplining knowledge: Interdisciplinarity in the twentieth century*. Baltimore: Johns Hopkins University Press.

Kempson, Ruth, Timothy Fernando, and Nicholas Asher. 2012. *Philosophy of linguistics*. Oxford: North Holland/Elsevier B.V.

Ludlow, Peter. 2011. *The philosophy of generative linguistics*. Oxford: Oxford University Press.

Nefdt, Ryan. 2019. The philosophy of linguistics: Scientific underpinnings and methodological disputes. *Philosophy Compass* 14 (12): e12636.

Scholz, Barbara C., Geoffrey K. Pullum, and Francis Jeffry Pelletier. 2016. Philosophy of linguistics. In *The Stanford encyclopedia of philosophy*. https://plato.stanford.edu/entries/linguistics

Tomalin, Marcus. 2006. *Linguistics and the formal sciences*. Cambridge: Cambridge University Press.

van Dongen, Jeroen, and Herman Paul. 2017. *Epistemic virtues in the sciences and the humanities*. Cham: Springer Press.

Part I

Linguistics and the Formal Sciences

2

The Complexities of *Syntax*

Jaroslav Peregrin

1 What Is Syntax?

What is *syntax*? The primary sense of the word is usually explained as "the way in which linguistic elements (as words) are put together to form constituents (as phrases or clauses)" and "the part of grammar dealing with this" (this particular wording is taken from the Merriam-Webster Dictionary). This, of course, can be generalized to not necessarily linguistic systems, so that it becomes something as *the way in which elements of a constructional system are put together to form constituents*, where a *constructional system* is anything where some wholes are assembled out of some parts. (Merriam-Webster reflects this, I think not very successfully, by listing a second case of "a connected or orderly system: harmonious arrangement of parts or elements <the *syntax* of classical architecture>").

Work on this paper was supported by the grant No. 20-18675S of the Czech Science Foundation.

J. Peregrin (✉)
Institute of Philosophy, Czech Academy of Sciences, Prague, Czech Republic
e-mail: peregrin@flu.cas.cz

R. M. Nefdt et al. (eds.), *The Philosophy and Science of Language*,
https://doi.org/10.1007/978-3-030-55438-5_2

The term *syntax* comes from the Greek *syntaxis* (σύνταξις—"putting together, order or arrange"), which is a combination of the prefix *syn-* (σύν—"together") with *taxis* (ταξις—"arrangement"). Note that *syntax* in this sense is always a matter of a *system* of items: it is about how items of the system are constructed from other items of the system (e.g. how expressions of a language are constructed from other expressions). It is obvious that an item may count as simple w.r.t. a system (like a word w.r.t. the system of expressions of English), while counting as complex with respect to another system (like the same word w.r.t. the system of all compounds of English letters).

The first treatise devoted to syntax as a linguistic discipline is almost as old as the term itself; it was written by the Greek scholar Apollonius Dyscolus under the telling title *Peri syntáxeos.* However, a systematic study of syntax is not older than some two centuries (Graffi 2001). In his path-breaking *Syntactic Structures* (1957), Chomsky writes:

> Syntax is the study of the principles and processes by which sentences are constructed in particular languages. Syntactic investigation of a given language has as its goal the construction of a grammar; that can be viewed as a device of some sort for producing the sentences of the language under analysis.

In the context of modern linguistics, syntax is one part of linguistics, its other parts being semantics, morphology, phonetics etc. However, outside of the context of linguistics we can often encounter the simplified opposition syntax × semantics.

Probably the widest leakage of this mostly technical concept outside of the narrowly scientific circles is due to the philosopher John Searle: he used it to back up his claim that computers can never think, because, as Searle (1980) puts it, "the computer ... has a syntax but no semantics" (p. 423). This has been, since then, broadly accepted as a revelatory statement; and even many of those who did not quite agree with Searle, took this as an illuminating way of framing the discussion; and the framework is still widely accepted. However, independently of whether we agree with Searle or not, is the concept of syntax as he employs it the same—or similar—to that used by Chomsky and other linguists?

In the 1930's, the concept of syntax also moved to the center of discussions in logic, especially in connection with the path-breaking results of Gödel (1930, 1931). The logician who became famous by his painstaking analyses of the concept is Rudolf Carnap. What he was after was putting the concept of syntax, along with the concept of semantics, on firmer conceptual foundations, for he became aware that without this, it would be impossible to secure firm conceptual foundations for modern logic. In his *Introduction to Semantics* (Carnap 1942),[1] he famously divides the general theory of language, which he calls *semiotics*, into *syntax*, *semantics* and *pragmatics* as follows:

> In an application of language, we may distinguish three chief factors, the speaker, the expression uttered, and the designatum of the expression, i e. that to which the speaker intends to refer by the expression. In semiotics, the general theory of signs and languages, three fields are distinguished. An investigation of a language belongs to pragmatics if explicit reference to a speaker is made, it belongs to semantics, if designata but not speakers are referred to, it belongs to syntax, if neither speakers nor designata but only expressions are dealt with. (p. 8)

Again, it is not quite clear whether this explication of the concept is compatible with the ways it is handled by either Chomsky or Searle.

2 *Syntatic* Versus *Formal* Properties

While syntax, we saw, concentrates on "the way elements are put together to form constituents", the opposition between syntax and semantics may lead to a broader construal of *syntax*, such that it encompasses all properties an expression has "by itself", independently of what it may mean or designate, all its "internal" properties. (Semantic properties are then "external" ones, such that we have conferred on it to let it mean or designate something.) According to Carnap, syntax is a theory of *formal rules*

---

[1] Similar ideas can be found already in his earlier writings—viz., e.g., Carnap (1939). It is not without interest that, as was pointed out to me by A. Klev, Carnap (1932) uses the term *semantics* as a synonym for *logical syntax*. See Tuboly (2017) for a thorough discussion.

governing linguistic forms. In *The Logical Syntax of Language* (Carnap 1934), he explains the term *formal* as follows:

> A theory, a rule, a definition, or the like is to be called *formal* when no reference is made in it either to the meaning of the symbols (for example, the words) or to the sense of the expressions (e.g. the sentences), but simply and solely to the kinds and order of the symbols from which the expressions are constructed. (p. 1)

Hence the concept of *formal* property is much broader than that of *syntactic* property as we have conceived of it up to now. The fact, however, is that the terms "syntactic" is often used, even by Carnap, in the sense of *formal*. Hence an important disambiguation—we must distinguish between the two possible senses of *syntactic* so that we keep using the term *syntactic* in the narrower sense and we engage the term *formal* for the broader sense.

Consider, for the sake of illustration, a "language" constituted by the following list of names:

Ann
Bob
Cynthia
John
Jill
Juliet
Rachel

In the terminology we have introduced, they do not have any (non-trivial[2]) *syntactic* properties (none of the names consists of other names on the list), but they have a lot of *formal* properties. They have, for example, various phonetic properties: some of them are, for instance, mono-syllabic. Or they consist of different numbers of letters.[3] On the

[2] Of course, they all have the trivial property «to be simple, not to be composed of anything».

[3] Remember that the classification of a property as syntactic depends on what we take the system in question to be. If, for example, we took the list as a part of a more exclusive system containing also all the letters of the English alphabet, then the items would have a lot of properties that would be syntactic.

other hand, the names may also possess properties that are not syntactic—they may, for instance, belong to concrete persons (let us suppose they are the names of the inhabitants of a village).

An important (though somewhat trivial) feature of formal properties is that insofar as forms of individual items differ, also any finite set of the items can be formally distinguished from any other. If we look at our list of names, it might seem that not every assortment of items shares a distinctive formal property (only by an improbable coincidence would the names of all and only vegetarians on our list, e.g., begin with the same letter). However, it is always possible to form the requisite property as the disjunction of the distinctive properties of the individual items of the set: for example all and only the items of the set {*Bob*, *John*, *Rachel*} have the formal property «to be of the form of *Bob* or *John* or *Rachel*».

Now call any item on the list a *name*, and suppose that we introduce two new items, *the father of* and *the mother of*, such that if we concatenate any one of them with a name we get a (complex) name. Thus our list becomes potentially infinite, containing such items as *the father of Ann*, *the mother of the father of Bob*, etc.

Ann
Bob
Cynthia
John
Jill
Juliet
Rachel
the father of
the mother of
the father of Ann
the mother of Bob
the father of the mother of Cynthia
...

Now the new, complex items already have not only formal, but also genuinely syntactic properties, such as «to consist of *the father of* and *Ann*» (which is possessed by *the father of Ann*).

3 *Pure* Versus *Descriptive* Syntax

Carnap (1934) writes:

> We have already said that that syntax is concerned solely with the formal properties of expressions. (p. 5)

Thus, Carnap delimits *syntax* as concerning all formal properties, hence in our broad sense. However, at other places in his book this is not so clear, he seems to tend to the narrow construal. Moreover, it would seem that Carnap's concept of *syntax* is even narrower than the one introduced above. Look at how Carnap continues the above quotation:

> We shall now make this assertion more explicit. Assume two languages (*Sprachen*), S_1 and S_2, use different symbols, but in such a way that a one-one correspondence may be established between symbols S_1 and those of S_2 so that any syntactical rule about S_1 becomes a rule of S_2 if, instead of relating it to the symbols S_1, we related it to the symbols of S_2; and conversely. Then although the two languages are not alike, they have the same *formal structure* (we call them isomorphic languages), and syntax is concerned solely with the structure of languages in this sense. (pp. 5–6)

This suggests that Carnap construes syntax as concerning not all the properties that have to do with composition—according to him, it concerns only those of such properties that are invariant to isomorphism. Thus, for example, imagine a system i, i_1, ..., i_n of items such that i consists of i_1, ..., i_n. Then, i has the property «to consist of i_1, ..., i_n», which may seem to be syntactic, but it is not syntactic in Carnap's sense: for if we have another system, j, j_1, ..., j_n, which is isomorphic with the original one in that j consists of j_1, ..., j_n, then any syntactic property i has must be identical with a property j has—hence «to consist of i_1, ..., i_n» cannot be syntactic, for this is a property j certainly does not have.

In fact, however, this concerns only what Carnap later specified as *pure syntax* and what concerns "the possible arrangements, without reference either to the nature of the things which constitute the various elements, or to the question as to which of these possible arrangements of the elements are anywhere actually realized" (p. 7). Besides this, there is also a *descriptive syntax*, which concerns "the syntactical properties and relations of empirically given expressions" (ibid.) Hence, we cannot define the syntax of a particular language in terms of pure syntax, for such a definition requires stating that the abstract categories are instantiated by certain concrete items.

Return to our example language and suppose we make its clone and then we replace the name *Ann* by, say, *Anthony* and we replace the expression *the father of* by, say, *the son of*:

Ann	*Anthony*
Bob	*Bob*
Cynthia	*Cynthia*
John	*John*
Jill	*Jill*
Juliet	*Juliet*
Rachel	*Rachel*
the father of	*the son of*
the mother of	*the mother of*
the father of Ann	*the son of Anthony*
the mother of Bob	*the mother of Bob*
the father of the mother of Cynthia	*the son of the mother of Cynthia*
...	...

Then, if we are within pure syntax, the relation of *the son of Anthony* to its parts must be *the same* as that of *the father of Ann* to its ones (for there is an isomorphism of the two lists which maps one of them on the other). Hence the property «to be the concatenation of *the father of* and *Ann*» is not a syntactic property of *the father of Ann.* A property of this expression that would be genuinely syntactic would have to be «to be the concatenation of … and ---», where the slots marked by "…" and "---" are filled with some "isomorphism invariant" characterizations of the "positions" assumed by *Ann* and *is a father of* in the original list and by *Anthony* and *is a son of* in the modified one. For example, if the items *the father of, the mother of* and *the son of* are called *functors*, then the property might be «to be a concatenation of a functor and a name».

Consider the following rule, which may be part of a definition of a language:

if S and S' are sentences, then $S \wedge S'$ is a sentence

Here, we would assume, "$S \wedge S'$" is an oblique way of referring to the concatenation of the sentence referred to by S, the sign "$\wedge$", and the sentence referred to by S'.[4] But if this rule is to be syntactic in the sense of Carnap's pure syntax, this is not an acceptable reading. As such rules are allowed to refer to exclusively syntactic properties, "$\wedge$" cannot be read as referring to the sign of the very shape it mentions, but rather to any sign that plays the same role in any structure isomorphic to the language in question. In other words, the rules must be read as saying

The concatenation of a sentence, a conjunction, and a sentence is a sentence.

where *conjunction* must be a certain role of a sign *vis-à-vis* purely syntactic rules. What can such a role be? What syntactically characterizes a conjunction is precisely the above rule, namely that when it is put in between two sentences it yields a sentence. However, there are likely to be more such operators in our language, such as X or implication; and all of them will be syntactically characterized just by the same rule. Hence we cannot distinguish a conjunction from a disjunction or an implication—all of them will be of the same syntactic category, which we may call (*sentential*) *connectives*. Hence unless we want also both disjunction and implication to be characterized as "conjunctions" we need to modify the above rule to

The concatenation of a sentence, a connective, and a sentence is a sentence.

What distinguishes conjunction from disjunction or implication? Of course, it is their *meaning*, hence it would seem properties that are paradigmatically not syntactic. However, it is possible to characterize them

[4] It is this obliqueness that Quine (1940), §6, proposed to dispose of by means of his concept of *quasiquotation*: according to him, we should write $\ulcorner S \wedge S' \urcorner$, which "amounts to quoting the constant contextual backgrounds and imagining the unspecified expressions written in the blanks". In our case "the constant contextual background" would be "$\wedge$" and "the unspecified expressions" would be S and S'.

not in the paradigmatically semantic terms such as truth and denotation, but rather in terms of inference and proofs. And this is behind a further generalization of *syntax* effected by Carnap.

4 Syntax *Simpliciter* Versus *Logical* Syntax

In the *Logical Syntax of Language* Carnap defined what he called "logical syntax". This term might easily be taken to imply that logical syntax is a kind of syntax, and hence that this concept is narrower than the general concept of syntax. However, as we will see, the relationship between Carnap's concept of syntax and the intuitive one is rather tricky.

What Carnap says is the following:

> By the *logical syntax* of a language, we mean the formal theory of the linguistic forms of that language – the systematic statement of the formal rules which govern it together with the development of the consequences which follow from these rules. (p. 1)

Carnap further writes that his logical syntax is a matter of two kinds of rules, namely those of *formation* and those of *transformation*. In particular, he writes:

> The difference between syntactical rules in the narrower sense and the logical rules of deduction is only the difference between *formation rules* and *transformation rules*, both of which are completely formulable in syntactical terms. Thus we are justified in designating as 'logical syntax' the system which comprises the rules of formation and transformation. (p. 2)

It is the formation rules which are constitutive of syntax in the usual sense. They determine which wholes can be built from which parts and consequently what is a part of what. The transformation rules are, then, a surplus matter and may be quite arbitrary. The relations and properties they institute are syntactic not in the sense that they would determine the part-whole structure of the system, but in the sense that the formulations of the rules that affect the institution refer only to the syntactic properties of the elements of the system.

The important thing is that the transformation rules are to be "completely formulable in syntactical terms". What exactly does this mean? Are the formulations of the rules allowed to refer only to the purely syntactical properties, or also to the descriptively syntactical, or even to the properties we have dubbed formal?

Consider the first possibility: we have seen that the purely syntactic properties do not allow us to differentiate between conjunction, disjunction and implication. Thus we cannot have a transformation rule of the form

transform S and S' into $S \wedge S'$,

or

transform two sentences into the concatenation of the first sentence, a conjunction, and the second sentence,

for we are not yet in possession of the concept of conjunction; hence we must have

transform two sentences into the concatenation of the first sentence, a connective, and the second sentence,

which is certainly not what we want.

The problem is that we need to stipulate different transformation rules for conjunction, disjunction and implication, respectively, but syntax alone does not make it possible to distinguish between them. Hence to be able to formulate reasonable transformation rules, we need to lean on properties which are not necessarily syntactic, but merely formal.

Now suppose that using those, we formulate the following transformation rules, which characterize conjunction; we use the usual "derive B from A" instead of the literal "transform A into B":

from S and S', derive $S \wedge S'$,
from $S \wedge S'$, derive S,
from $S \wedge S'$, derive S'.

We can say that any sign that satisfies them, in any language, is a *conjunction*; and we can characterize *disjunction* and *implication* analogously. Thus, once we introduce transformation rules (as in effect arbitrary rules formulable in formal terms), we have a much more fine-grained "syntactic" classification of expressions.

Thus the properties that are syntactic if we take into account not only formation, but also transformation rules are not only those that concern the composition of an item, but also all those that are constituted in terms of rules that refer to merely formal—or syntactic?—properties. Of course, a definition of conjunction can be effected not only with the reference to a concrete item, such as "$\wedge$", but also on a purely abstract level, if we stipulate that a connective is a conjunction iff the following rules are in force:

from two sentences, derive the concatenation of the first sentence, the connective, and the second sentence;
from the concatenation of a sentence, the connective, and a sentence, derive the first sentence;
from the concatenation of a sentence, the connective, and a sentence, derive the second sentence.

This can be taken as an *implicit definition* of the concept of *conjunction* as an abstract role that can be instantiated in any language. Seen thus, *conjunction* is a property invariant to isomorphism like the concepts syntactic in Carnap's pure sense; however, defining any *particular* language we would have to refer to a *concrete* item, we would have to stipulate, e.g.

"$\wedge$" is a conjunction.

Hence again as in the case of formation rules, transformational rules, if they are to be formulated for a concrete language, must be allowed to refer to not only syntactic, but also formal properties.

As a more trivial example, return to our extended list, including the formation rule that allows us to concatenate a functor and a name to yield a name. Let us consider the addition of the following two transformation rules

transform *the father of x* into *x*
transform *x* into *the mother of x*

In this way the two functors we have in our list become differentiated. (In contrast to this, names are not differentiated—both the rules treat them indiscriminatingly.) *The father of* becomes what we can call a freely *cancelable functor*—in the beginning of any complex name it can be freely canceled. *The mother of*, on the other hand, becomes what we will call a freely *appendable functor*—it can be freely appended to the beginning of any name. Again, the roles of *cancelable functor* and *appendable functor* are wholly abstract and can be instantiated in many different ways.

5 Carnap's Extended Syntax

Carnap's *Logical Syntax of Language* contains one more generalization, which, though less important for our present purposes, is, nevertheless, remarkable as it indicates that a very relaxed concept of syntax can get quite close to semantics. Carnap considers engaging "generalized transformation rules" (Carnap does not call them thus) of unusual kinds, such as those with infinite numbers of premises.

Normally, a rule is something a human can apply, i.e. check that all the premises hold and hence that she can draw the relevant conclusion; and if the number of its premises is not finite, then this is hardly possible, she would not be able to check all the premises. (As a matter of fact, a rule would not be applicable even if it had a finite, but too enormous number of premises—but in our case it is not applicable even as a matter of principle.) A paradigmatic example of this kind of transformation rule is the so called *omega rule*, which takes us from an infinite number of premises of the shape $P(n)$, where n runs through all numerals for natural numbers, to $\forall x P(x)$.

What is remarkable is that adding this rule to Peano arithmetic makes it complete (the price, of course, is that the class of theorems is no longer recursively enumerable): each sentence becomes either provable or refutable (though, of course, we must keep in mind that the meanings of the terms *provable* and *refutable* in this sentence are now different from the standard one.) Hence this relaxed form of syntax becomes virtually equivalent to semantics—and from the Carnapian perspective we can perhaps say that semantics, though its proponents may claim it goes beyond the limits of syntax and achieves what is unreachable by syntax, *is* this relaxed form of syntax.[5]

6 Embodied Syntax

While the concept of syntax has undergone, within logic, the kind of painstaking vivisections discussed in the previous sections, within linguistics its trajectory has been rather different. The fact is that within linguistics, the concept of syntax is usually assumed to be relatively transparent, not requiring much elucidation. Thus for example, in Chomsky's book *Knowledge of Language* (Chomsky 1986), which is one of his most complex works on the nature of language, we find no attempt at any elucidation, let alone any definition, of *syntax*. However, it is also a fact that Chomsky gave the linguistic understanding of *syntax* a vigorous boost, in that he ascribed to it a new kind of empirical dimension.

Consider the sentence

Bob is stupid

Of which expressions does it consist? Of *Bob*, *is* and *stupid*?; of *Bob* and *is stupid*?; or of *Bob is* and *stupid?* This, of course, depends on the relevant syntactic rules. But what are the relevant syntactic rules? What are the syntactic rules of English or Peano arithmetic? The answer is relatively easy in the latter case: the language of Peano arithmetic is an artificial language, delimited in terms of certain rules, so they are the rules that are relevant in this respect. But the situation is not so clear in the case of a natural language like English. As speakers of English we are able to roughly separate well-formed expressions from non-well-formed ones, but we are usually not able to articulate rules that might underlie this ability. Their formulation is a matter of linguistic theories.

[5] See Peregrin (2014), Chap. 7, for a more detailed discussion of the issues hinted at in this section.

Now one of the ways to construe the task of a linguist who is to articulate a grammar of a language such as English, is that we have the delimitation of the set of well-formed expressions of a given language and the task is to formulate *any* set of rules that leads to this set. There is no *priori* reason why, in the case of English, they cannot be rules that would see *Bob is stupid* as the combination of *Bob is* and *stupid* (though it may turn out that the rules that do this tend to be more cumbersome than those which do not).

But we may also conjecture that there are some specific rules present somewhere in the mind or in the brain of the speakers of English, though they are not consciously aware of them. This leads to a very different construal of the task of a syntactician: she is no longer free to assemble *any* set of rules that leads to the right boundary between well-formed and non-well-formed expressions, she must find *the* set of rules that is responsible for this. This is a sense of syntax made popular especially by Chomsky and his followers.

In fact, initially Chomsky operated with a concept of syntax not so different from Carnap's. Let us repeat what he wrote In *Syntactic Structures* (Chomsky 1957) and which we quoted at the beginning of this article:

> Syntax is the study of the principles and processes by which sentences are constructed in particular languages. Syntactic investigation of a given language has as its goal the construction of a grammar; that can be viewed as a device of some sort for producing the sentences of the language under analysis. (p. 11)

Here, syntactic theory is taken to *construct* a grammar—not describe one that it finds and scrutinizes within the *language faculty* of actual speakers. Hence at this stage, what Chomsky does can be seen as simply studying Carnapian formation rules. However, already in the *Aspects of the Theory of Syntax* (Chomsky 1965), Chomsky characterizes the subject matter of his studies, which include syntax, in the following way:

> A grammar of a language purports to be a description of the ideal speaker-hearer's intrinsic competence. If the grammar is, furthermore, perfectly explicit – in other words, if it does not rely on the intelligence of the understanding reader but rather provides an explicit analysis of his contribution – we may (somewhat redundantly) call it a generative grammar. (p. 4)

Here there is a new element: syntax not only studies how a complex expression of a language can be composed of other expressions, but also how speakers of the language in fact compose them. (Chomsky claims that he is studying the competence of an "ideal speaker", but there is no doubt that this is a method of approximating the competence of real, non-ideal speakers.)

Given a language as a potentially infinite set of expressions, there may be many generative grammars that produce precisely this set. (And in *Syntactic Structures*, finding an *arbitrary* grammar that generates a given set of expressions was one of the basic problems.) However, in the case of a natural language, among its possible grammars there is probably only one that is actual in the sense that it is really "implemented" within the brains of the speakers of the language. And the study of syntax, in the newer Chomskyan sense, is the study of this very structure.

This, then, is one more sense of "syntax", in which syntax is a specific empirical phenomenon, namely something present in the minds/brains of speakers of natural languages.

7 Kinds of Syntactic Properties

Let us take stock. We have distinguished six kinds of properties that may be subsumed under the label *syntactic*:

1. A property of a sign is *formal* iff it is possessed by the sign in virtue of its mere design independently of what the sign means. A formal property of *the father of Bob*, for example, is that its second word is the longest (or also that it consists of *the father of* and *Bob*).
2. A property of a sign is *syntactic in the Carnap's descriptive sense* iff it concerns the way it is composed of other signs of the system in question. Such a property of *the father of Bob* (given that we accept it as an element of the system of English expressions) is that it consists of *the father of* and *Bob*.
3. A property of a sign is *syntactic in Carnap's pure sense* iff it is invariant to isomorphism. Such a property of *the father of Bob* is that it consists of a functor and a name.

4. A property of a sign is *logically syntactic in Carnap's sense* iff it concerns its role w.r.t. transformation rules and is invariant to isomorphism. Such a property of *the father of Bob* is that it consists of a cancelable functor and a name.
5. A property of a sign is *logically syntactic in Carnap's extended sense* iff it concerns its role w.r.t. transformation rules including some generalized ones (e.g. infinitistic), and is invariant to isomorphism.
6. A property of a sign is *syntactic in Chomsky's sense* iff it describes the part of the human mind/brain that is responsible for putting together the linguistic vehicles of our communication.

Here then is a wealth of senses, and failing to distinguish between them, needless to say, is frequently perilous. We are convinced that it is precisely failing to do so which leads to most of the confusions concerning syntax in logic, linguistics and philosophy, for it obscures the role truly played by the concept. To throw more light on this, however, we need to discuss its twin concept of semantics.

8 Semantics

As syntax is usually counterposed to semantics, it will be useful to pay some attention also to the latter concept. So what is semantics? We have already seen Carnap's definition from his *Introduction to Semantics*, according to which "an investigation of a language belongs to semantics, if designata but not speakers are referred to". Hence his concept of semantics is parasitic upon the concept of *designation*. It is only if an item designates something else that it has some semantic properties—it is the properties it has in virtue of its designating the other item.

However, we have seen that the broadest explication of *syntactic* is *formal*, where the formal properties are those the item possesses "by itself", "internally". This concept of a *form* invites being opposed by the concept of a *content*—where a contentual property is something an item has "externally", by dint of us humans having somehow conferred it on the item. However, it would seem that far from every "external" property can be sensibly considered semantic.

External properties, in the typical case, are conferred on expressions by us, by the fact that we come to use them in a certain way. (Note that the properties stemming from it designating something are also of this kind—words do not designate anything by themselves, but only because we make them designate something.) For example, a word can have the property of being the 137th entry in a certain dictionary; which, of course, has only little to do with its semantics. More to the point, a word can have the property of being used in a certain way within reports that it is raining outside, or within orders to shut the door, or within exclamations of pain. Given Carnap's tripartite classification, some of such properties may seem to invite the classification *pragmatic*. However, the truth is that since Carnap's time the distinction between semantics and pragmatics has been considerably blurred.

The point is that many semantic theories dispense with concepts such as designation or representation and see semantics as grounded in usage. Thus, the so called *use theories of meaning* (Peregrin 2011) assume that the meaning of an expression is roughly its role within our language games. In this way, semantics and pragmatics become largely overlapping, and many properties concerning the ways expressions get used are legitimately considered semantic. Hence we have a lot of properties that are external (they are even plausibly called semantic), though they are not properties which an expression would have in virtue of designating anything. As a result, we have properties semantic in a sense broader than Carnap's.

All in all, we have two senses of semantics, a narrower one and a broader one. On the narrower construal, semantic properties are those that an expression has in virtue of designating something; on the broader construal they comprise more of its external properties, especially those which have to do with the employment of the expressions within our "language games".

What is now crucial for us is the question of the existence of an overlap between syntactic and semantic properties. It is clear that, given the intuitive notions of syntax and semantics, such an overlap should *not* exist—*to be syntactic* and *to be semantic* should be mutually exclusive characteristics of properties. But we have already provided some explications of these two properties and demonstrated their complexities. It is clear, to be sure, that there is no overlap between narrowly syntactic and semantic properties—but can we say the same if we broaden the concept of syntax to Carnapian *logical* syntax?

We have seen that there are some broadly—*viz.* logically—syntactic properties which are co-extensional with some semantic properties; but this is not surprising, this may exceptionally happen even to narrowly syntactic properties. But we have also seen that every finite set of expressions can be characterized by a syntactic property; hence *every* semantic property with a finite extension is co-extensional with some broadly syntactic property. Now the question is how it is with properties whose extensions are not finite.

9 Rules Versus Formulations

Return to our example list of names: imagine that all and only those persons on our list whose names begin with *J* (i.e. *John, Jill* and *Juliet*) are vegetarians. Now consider the following two instructions:

(I) *A name is to be written in italics if it begins with 'J'*
(I′) *A name is to be written in italics if its bearer is a vegetarian*

Are they two different rules, or only two formulations of the same rule?

There is a sense in which both express the same rule, namely that the names *John, Jill* and *Juliet* are to be written in italics. There is, to be sure, also a sense in which each of the instructions expresses a different rule. The difference, however, would come to the surface only in a contrafactual situation. If, for example, we extend our list to include the names Bill and Jennifer, such that Bill is a vegetarian, whereas Jennifer is not, then (I) would instruct us to write *Jennifer*, and not *Bill*, in italics, whereas the second one would instruct us to write *Bill*, and not *Jennifer*, in italics. Using traditional terminology, we may say that (I) and (I′) differ in intensions, though they are equivalent with respect to extension.

Now there are many situations in which intensions do not play a nontrivial role. Imagine, for example that the instructions (I) and (I′) are meant to concern merely a concrete fixed list (namely the one in which all and only names of vegetarians begin with *J*). Or imagine a rule concerning numbers or other mathematical objects.[6] In neither of these examples would contrafactual situations play an important role (in the first because going contrafactual would not change the fixed list, and in the second because mathematical objects are considered unchangeable across contrafactual situations). Hence it would seem that we must admit that the same rule may be formulated in different ways; and we should distinguish between a *rule* and *rule formulation.* And it would also seem that the attribute *syntactic*, as formulated above, would pertain to rule formulations rather than to rules—it would seem that one and the same rule may have formulations that are syntactic, while also having formulations that are not syntactic.

Hence rules formulated in the purely syntactic manner, like Carnap's transformation rules, may also have formulations that are semantic (or vice versa). Remember, given our above example, that striking out the names of all vegetarians from our list is *the same thing as* striking out all names beginning with *J*. In this way, some syntactic properties may serve, in certain contexts, as substitutes for semantic properties—or, as we may put it, as their "indicators".[7] (With respect to our example list, the syntactic property of beginning with *J* is the syntactic "indicator" of the semantic property of being a vegetarian.) And in view of the fact that in some context extension is all that counts, it may be that the syntactic properties may even count as *the same* as the semantic ones with which they are co-extensional.

10 Syntax of Artificial Languages of Logic

There are some morals that we may draw from the considerations of the first part of this paper and that are important for understanding the role of syntax in logic. One such moral is that *speaking about syntax we may be speaking about it in the narrow sense, but there is also the broad sense in which not only formation, but also arbitrary transformation rules—viz. rules of derivation—are called syntactic.* Another important moral is that *a*

[6] Cf. Peregrin (2018).

[7] Cf. Peregrin (2020a).

syntactic property can be co-extensional with a semantic one, and if we consider the broadly syntactic properties, then so many of the semantic ones are co-extensional with them that it makes sense to ask whether not all of them are.

The problem which many logicians concentrated on in the first half of the twentieth century was the problem of the delimitation of logical theories. *viz.* sets of sentences in the artificial languages of logic. Can all such theories be delimited by syntactical means, or are there some which cannot, but can be delimited otherwise, especially in terms of semantics?

An artificial language of the kind used by logic has the form of a constructional system of expressions (a set of basic symbols plus a set of formation rules for constructing more complex expressions out of simpler ones), with sentences as a distinguished kind of expressions; together with some rules for deriving some of its sentences from other sentences (transformation rules). Now the problem of the syntactic characterizability of theories assumes the form of the question whether, given a set of sentences of a language, we can always find transformation rules that would produce the very same set.

Note that there is one respect in which this task is similar to the ancient task of the trisection of an angle. The task would be dull were we to pose it in its full generality ("is it possible to trisect an angle?"), for it is almost clear that this can be done *somehow* (more or less precisely). But the task becomes interesting when we stipulate that it must be done using *nothing else* than a compass and a ruler, which may be employed in clearly delimited ways—only then does it become a true yes/no-question.

Now it is only when the concept logical syntax, in the Carnapian sense, is exactly specified that the problem of axiomatic delimitability of theories (which, when we ask about theories delimited semantically becomes the problem of completeness of a logical calculus) becomes deeply interesting. In fact, it was the unprecedentedly clear understanding of the nature of artificial languages and of the distinction between their syntax and semantics that allowed Gödel to launch the revolution in mathematical logic that took place in the twentieth century. He saw the consequences of drawing this distinction meticulously and this enabled him to reach his path-breaking results. In particular, Gödel appreciated the non-trivial problems we must face if we confront the syntactic and semantic properties of a system.

How could semantics determine a set of linguistic items? We saw, in our earlier example, that when we have a list of names of inhabitants of a village, a subset of them may be distinguished as the set of the names of persons with a particular property, e.g. vegetarians. However, we also saw that any subset of a *finite* set of linguistic items can be delimited syntactically (at least when we understand syntactic properties as including all the formal ones), hence here we *cannot* have the case of a set that is delimited semantically, but not syntactically.[8]

What about infinite sets? Is there an infinite set of expressions of a language, which can be delimited semantically, but not syntactically? Unfortunately, there is no village (nor, for that matter, city, state, nor planet) with an infinite number of inhabitants, so we cannot extend the previous case straightforwardly to this one.[9] But it would seem that in an abstract realm there might be an infinite number of objects (e.g. natural numbers) and especially that a domain of objects (even a finite one) may make an infinite number of sentences true. Hence is there a theory that consists of those and only those sentences that are made true by a domain of objects, but which cannot be delimited syntactically?

It is this question the positive answer to which is usually taken to follow from the discovery of Gödel (1931). What Gödel showed was that if a theory includes sentences that express some *prima facie* obvious principles of arithmetic and it is closed to some *prima facie* obvious transformation rules, it cannot contain the negation of every sentence it does not contain. And if we consider the interpretation of the theory with respect to the standard model and if we assume that hence each of its sentences is true or false, the existence of a theory that is delimited semantically, but not syntactically, follows. Thus, semantics appears to "reach beyond" syntax.[10]

---

[8]We have seen that if we have a *finite* set of items *formally distinct* from each other, then the set itself is formally characterizable (such a characterization can be produced as the disjunction of the distinctive properties of the individual items); hence for every property of elements of a *finite* domain there is a formal property that is co-extensional with it. Thus, any finite set of sentences can be delimited syntactically.

[9]It is not without interest that one of the original senses of *finitism* concerned the finiteness of the universe, for its proponents took for granted that any usable language cannot but have such kind of limited universe (Frost-Arnold 2013).

[10]Unless, of course, we accept the "generalized" transformation rules of Carnap, which may render syntax in fact equivalent to semantics.

11 The Importance of the *Clear* Distinction Between Syntax and Semantics

It is not easy to see how path-breaking the clarification of the concept of syntax, which underlay Gödel's discovery, really was at the time, for today the understanding pioneered by Gödel has become commonplace. By way of illustration, let us look at two of the early logicians who wrestled with problems similar to those dealt with by Gödel but with lesser understanding of the distinction between syntax and semantics. The first example is taken from Coffa (1991); it concerns Carnap's proof (from his manuscript written around 1929 and published as Carnap (2000)) that an axiomatic system that has no model is inconsistent in the sense of entailing a contradiction. Coffa points out that Carnap's proof goes as follows: from the fact that a formula *f* does not have a model ("-(E)f") it follows that it entails its own negation ("f ⇒ -f"); and as the formula trivially entails itself ("f ⇒ f"), it follows that it is inconsistent ("f ⇒ f & -f").

Carnap's theorem thus concerns the relationship of two properties a formal system can have: "emptiness", defined as not having any model, and "inconsistency", defined as entailing a contradiction. The problem is that if we understand "entailment" in the semantic sense (roughly as "A entails B iff every model of A is a model of B", as Tarski (1936) would later explicate it), the relationship claimed by Carnap is trivial: if a system is empty, then it entails everything just *eo ipso*, and so it also trivially entails a contradiction.

It takes the careful distinguishing of syntax and semantics, effected by Gödel, for it to become clear that such a claim *may* have also a very non-trivial sense, *if* we interpret the "entailment" in the syntactic sense, *viz.* as derivability. In such a case, the emptiness of the system does not entail its inconsistency trivially. In fact, in the case of first-order predicate calculus, this claim amounts to the famous completeness theorem proved by Gödel (1930): what Gödel proved was that any consistent first-order theory has a model, which yields Carnap's theorem by contraposition.

Another example: Prior to Gödel, some mathematicians and logicians mused about the unformalizability or "incompleteness" of formal mathematics—they tried to prove that not every mathematical claim could be formalized and thus that "formal mathematics" does not exhaust the whole of mathematics. One such was Finsler (1926), who claimed to establish that "in the case of the axiom system for the real numbers ... there exists a proposition that, though its truth is from the logical point of view unambiguously decided by the axiom system, can be neither proved nor refuted by purely formal means". (It is interesting that Finsler does not give a concrete axiomatic system, hence his claim is presumably that this holds for *any* axiom system for the real numbers.)

Though Finsler's considerations, eventually resting on the discrepancy between the uncountability of real numbers and countability of any language, are not without interest, again it took Gödel's insight into the distinction between syntax and semantics to prove the incompleteness of certain logical systems in the well-known and extremely revealing sense presented by Gödel (1931). Heijenoort (1967) diagnoses the situation in the following way:

> [A]ccording to Finsler, Gödel has not exhibited a formally undecidable proposition at all: Gödel's true undecidable proposition is formally decidable and is simply a true and provable proposition that turns out to be unprovable in some limited system, namely, Gödel's system P. This clearly shows the limits of Finsler's anticipation of Gödel's results. While Gödel puts the notion of formal system at the very center of his investigations, Finsler attempts to lay bare the fallacy hidden in any paradox and to remove this fallacy by means adapted to the specific case, without a general reconstruction of language. (p. 440)

Hence, by sharpening the Carnapian concept of syntax and contrasting it properly with semantics, we gain an extraordinarily useful tool. However, the tool may also be misused, as sometimes happens.

12 Syntax in the Brain

In view of many linguists, the concepts of syntax and semantics are not so enigmatic to necessitate anything as the Carnapian painstaking dissections. We have already seen that one of Chomsky's most extensive works about language, *Knowledge of Language*, contains no attempt to delimit the concept in a clear and unambiguous way; and we should add that the concept of semantics is paid even less attention in the book.

I think this is because for Chomsky the concept of syntax is too obvious to deserve any such musings. (Needless to add, saying that the *concept* is transparent is not saying that the actual syntax of natural language is either transparent or simple.) It is just something we can observe in action all the time and hence know all too well—after all, we usually do not spend hours scrutinizing what sunshine or walking is. And Chomsky's former fundamental conviction is that to get hold of the nature of language we must see it as a system with an essentially recursive structure.

I think this slightly biases the tension between the concepts of syntax and semantics which otherwise hold each other in a dialectic clinch. Chomsky's view is that language *is* syntax, in a sense; and semantics is something of a by-product. But we should not forget that, in a clear sense, any language is here because of its semantics: the recursive structure of linguistic items is here not because we need it as such, but because it lets us produce an unlimited number of expressions each with distinctive semantic properties. Of course, as we already indicated, there is no general agreement on the nature of the semantic properties; but there seems to be little doubt that it is these that make linguistic expressions useful.

Admittedly, one construal of semantics consists in seeing language as a matter of our inner representations of the outer world. Then we can say, together with Chomsky, that language is primarily in the brain (as not only syntax, but also representations are in the brain) and we can perhaps also say that meanings amount to "logical forms"—insofar as these can be construed as representations. (Unfortunately, Chomsky does not tell us much about what a logical form being associated with an expression allows us to use the expression for.)

A reaction to this bias is represented by those who propose importing the tools of logic into linguistics—to build a "formal semantics" for natural language.[11] Here, we find ourselves looking at natural language through Carnapian glasses: we have the rules of syntax (especially the formation ones—whether we also need the transformation ones is less

---

[11] The founding father of this movement is usually taken to be Montague (1974); for a compendium of the achievements of the movement see van Benthem and ter Meulen (1996).

clear) and we have "designations", the mapping of expressions of which is reconstructed within model theory.

This might do away with the Chomskyan bias and restore the delicate balance between syntax and semantics. However, while in the case of the artificial languages of logic we construct the semantics, which may be done by defining a model-theoretic interpretation (but also in other ways; see, e.g. the recent boom in proof-theoretic semantics[12]). In the case of natural language, however, we *reconstruct* the semantics, and it is not immediately obvious that the model-theoretic tools are well suited for this purpose. (The point is that the model-theoretic reconstruction is usually taken as a vindication of the notions of expressions as representations of their meanings,[13] which—as I argued elsewhere[14]—should not go without saying.)

Anyway, we see one more grave ambiguity of the term "syntax" at work here. Syntax may be considered either as merely a theoretician's tool for systematizing the empirical phenomenon of well-formedness in a natural language, or as a part of the subject matter of her empirical investigation—as a mechanism within the brains of those producing the well-formed expressions which the theoretician is trying to capture.

13 Syntax and Artificial Intelligence

We saw that the very clear understanding of the concept of *syntax* has led to some radical progress in logical theory. However, it has come to play an important role also in philosophy, as a result of the way Searle employed it to distinguish the kind of understanding that is characteristic of us humans from its mere imitations.

Searle's well-known pronouncements were made in the context of the philosophical discussions centering around the question *Can machines think?*. The *locus classicus* of this discussion is Turing (1950), but Searle

[12] See Francez (2015).

[13] As Button and Walsh (2018) put it: "model theory seems to be providing us with a perfectly precise, formal way to understand certain aspects of linguistic representation."

[14] See Peregrin (1995).

(1980) gave it a new, important impetus by claiming that "the computer ... has a syntax but no semantics" (p. 423). This has been, since then, broadly accepted as a revelatory statement; and even many of those who did not quite agree with Searle, took this as an illuminating way of framing the discussion; and the framework is still widely accepted.[15]

At first sight, this may sound plausible: a computer can surely recognize formal properties of symbols, but it cannot understand them, it cannot recognize their semantic properties. In this way, the concept of syntax has come to assume a prominent place also within the philosophical conceptual toolbox: these are the concepts in terms of which we characterize genuine thinking and understanding, as contrasted to their imitations and simulacra.

However, given our conceptual analysis, what does it take to "have (only) syntax"? The syntactic (or better, formal) properties of an expression are those which one can recognize if one is confronted with an expression as such. However, if one is able to learn how the expression is used by its competent users (and there is no reason to think that this is something beyond the ken of a computer), one may learn its semantic properties. (And note that this claim does not presuppose the use-theory of meaning; *no* semantic property can be conferred on an expression other than by the way it is employed by the speakers.) This is not yet to take a stance on whether computers or other entities can or cannot think; it is just pointing out that addressing this in terms of "having syntax/semantics" is awkward.

But let us waive this; let us assume that "having only syntax" simply means being restricted to syntactic rules. Does it follow that an entity which "has only syntax" cannot have semantics? Even this is not necessarily the case. To see this, let us consider what precise concept of *syntax* is here in play. One possibility is that it is syntax in the narrow sense of the word, i.e. concerning properties bestowed on expressions by the *formation* rules (plus, possibly, their other formal properties). Then it is quite clear that such properties cannot yield semantics—and if an entity cannot accommodate properties other than of this kind, then Searle would indeed be right that it is bound to fall short of semantics.

[15] See, for example Bozşahin (2018).

However, what if we construe *syntax* in the broad sense, in which it concerns also *transformation* – *viz.* inferential—rules? It would seem that there is no reason why a computer could not accommodate them. (And indeed computers routinely carry out derivations, proofs, etc.) So a computer can have syntax not only in the narrow sense of the word, but also in the broad one. And once we have syntax in the broad sense, it is no longer so obvious that it cannot yield us semantics.

Suppose we teach the computer to derive a statement of the form $A \wedge B$ from the statements A and B, and to derive both A and B from $A \wedge B$. What part of understanding of the meaning of "$\wedge$" will it still be lacking? (Notice that if we hold that an inference includes the possibility of its antecedent being true and its consequent, at the same time, being false, these inferences are enough to recover the classical truth table.[16]) If none, then there does not seem to be a reason to say that a computer cannot have a semantics, for it *can* have the semantics of "$\wedge$".[17]

An opponent might argue that the point was that a computer cannot have the semantics of empirical words like "dog" or "house" or "run". Then, firstly, a question is why the thesis is formulated so as they can have no semantics whatsoever. (Note that no semantics for empirical words does not even rule out the possibility of ever having the semantics for a complete language—for example the language of Peano arithmetic does not contain any empirical words.) And secondly, there is the question why it cannot acquire the semantics of "dog" or "house" or "run" in a similar way to which it can acquire that of "$\wedge$". Because to understand empirical terms it must have contact with the world? But a computer can have contact with the world; just as it can be taught that it can infer "Fido is a dog" from "Fido is a dog and Dumbo is an elephant", so it can be taught to infer "This is a dog" from a certain picture, acquired by its inbuilt camera, of what is in front of it.

True, these issues are much more complex than this brief discussion can encompass.[18] But this is precisely the point: we cannot expect to

---

[16] See Peregrin (2010) for an analysis of the relationship between inference and truth-based meaning.

[17] Cf. also Rapaport (2000).

[18] It is not the purpose of the present paper to argue for an inferentialism, the view that meaning is an inferential role (see Peregrin (2014)). The point is merely that its rejection is not something that can be taken for granted.

answer questions like *can computers have semantics, or only syntax?* until we are completely clear about what they mean, hence until we are completely clear about the concepts they contain. And one of the concepts crucially involved is that of *syntax*, which is much trickier than would justify the all too easy way in which it is usually employed in these discussions.

14 Conclusion

The concept of *syntax* plays a crucial role in the development of modern logic and linguistics; and it has overspilled into philosophical discussions. It is, however, far from unequivocal, particularly when no clarification is given as to in which of its various senses it is being employed. Aside of the narrow concept, which reflects the part-whole structure of a system, there is a much broader concept (brought to the fore especially by Carnap) which is, however, often confused with the narrow one. The Carnapian broadening has turned out to be very fruitful, especially in the context of studying the properties of formal languages of logic; and we have learned, thanks to Gödel, that if we are meticulous about its nature, it can point us to truly path-breaking results. Moreover, there is a notion of syntax as a purely theoretical matter, as a means of reconstruction of empirical phenomena, which contrasts with the notion of syntax as an empirically describable mechanism to be found in the brain. (Needless to say these two notions of syntax lead to very different verdicts with respect to how syntax can be studied.) The fact, however, is that the concept is often employed, especially in philosophy, without due care, leading to dangerous confusions and engendering treacherous illusions of explanation.

References

Bozşahin, Cem. 2018. Computers aren't syntax all the way down or content all the way up. *Minds and Machines* 28: 543–567.

Button, Tim, and Sean Walsh. 2018. *Philosophy and model theory*. Oxford: Oxford University Press.

Carnap, Rudolf. 1932. Erwiderung auf die vorstehenden Aufsätze von E. Zilsel und K. Duncker. *Erkenntnis* 3: 177–188.
———. 1934. *Logische Syntax der Sprache*. Vienna: Springer. Quoted from the English translation *he logical syntax of language*, London: Routledge, 2000.
———. 1939. *Foundation of logic and mathematics* (International Encyclopedia of Unified Sciences 1). Chicago: Chicago University Press.
———. 1942. *Introduction to semantics*. Cambridge, MA: Harvard University Press.
———. 2000. *Untersuchungen zur allgemeinen Axiomatik*. Darmstadt: Wissenschaftliche Buchgesellschaft; written cca 1929.
Chomsky, Noam. 1957. *Syntactic structures*. The Hague: Mouton.
———. 1965. *Aspects of the theory of syntax*. Cambridge, MA: MIT Press.
———. 1986. *Knowledge of language*. Westport: Praeger.
Coffa, Alberto. 1991. *The semantic tradition from Kant to Carnap*. Cambridge: Cambridge University Press.
Finsler, Paul. 1926. Formale Beweise und die Entscheidbarkeit. *Mathematische Zeitschrift* 25: 676–682; Quoted from English translation Formal proofs and undecidability in van Heijenoort (1967), 438–445.
Francez, Nissim. 2015. *Proof-theoretic semantics*. London: College Publications.
Frost-Arnold, Greg. 2013. *Carnap, Tarski, and Quine at Harvard: Conversations on logic, mathematics, and science*. La Sale: Open Court.
Gödel, Kurt. 1930. Die Vollständigkeit der Axiome des logischen Funktionenkalküls. *Monatshefte Für Mathematik Und Physik* 37: 349–360.
———. 1931. Über formal unentscheidbare Sätze der Principia Mathematica und verwandter Systeme I. *Monatshefte Für Mathematik Und Physik* 38: 173–198.
Graffi, Giorgio. 2001. *200 years of syntax: A critical survey*. Amsterdam: Benjamins.
Heijenoort, Jean. van, ed. 1967. *From Frege to Gödel: A source book from mathematical logic*. Cambridge, MA: Harvard University Press.
Montague, Richard. 1974. *Formal philosophy: Selected papers of R. Montague*. New Haven: Yale University Press.
Peregrin, Jaroslav. 1995. *Doing worlds with words*. Dordrecht: Kluwer.
———. 2010. Inferentializing semantics. *Journal of Philosophical Logic* 39: 255–274.
———. 2011. The use-theory of meaning and the rules of our language games. In *Making semantics pragmatic*, ed. K. Turner, 183–204. Bingley: Emerald.
———. 2014. *Inferentialism: Why rules matter*. Basingstoke: Palgrave.

———. 2018. Intensionality in mathematics. In *Truth, existence and explanation*, ed. Mario Piazza and Gabriele Pulcini, 57–70. Cham: Springer.

———. 2020a. Carnap's inferentialism. In *Vienna Circle in Czechoslovakia*, ed. R. Schuster, 97–109, Dordrecht: Springer.

———. 2020b. *Philosophy of Logical Systems*. New York: Routledge.

Quine, Willard van Orman. 1940. *Mathematical logic*. Cambridge, MA: Harvard University Press.

Rapaport, William. 2000. How to pass a Turing test. *Journal of Logic, Language, and Information* 9: 467–490.

Searle, John. 1980. Minds, brains & programs. *Behavioral and Brain Sciences* 3: 417–457.

Tarski, Alfred. 1936. O pojeciu wynikania logicznego. *Przeglad Filozoficzny* 39, 58–68; English translation On the concept of following logically. *History and Philosophy of Logic* 23 (2000): 155–196.

Tuboly, Adam. 2017. From "Syntax" to "Semantik"—Carnap's inferentialism and its prospects. *Polish Journal of Philosophy* 11: 57–78.

Turing, Alan. 1950. Computing machinery and intelligence. *Mind* 59: 433–460.

Van Benthem, Johan, and Alice ter Meulen (eds.). 1996. *Handbook of logic and language*. Amsterdam: North Holland.

3

Negation in Dedekind Orderings and the Logic of Reciprocity

D. Terence Langendoen

1 Dedekind Orderings

The mathematician Richard Dedekind was the first to explore the nature of partially ordered sets (posets) Δ_n that can be defined over a base set A_n of n members (Dedekind 1931 [1897]) as follows. First let D_n be an intermediate poset of $2^n - 1$ members consisting of all the members of a classical (Boolean) poset defined by an ordering $\geq$ over A_n less the supremum $\top$. Then, let Δ_n be the largest poset of the power set D_n less the empty set, whose members are

D. Terence Langendoen (✉)
Department of Linguistics, University of Arizona, Tucson, AZ, USA
e-mail: langendt@arizona.edu

R. M. Nefdt et al. (eds.), *The Philosophy and Science of Language*,
https://doi.org/10.1007/978-3-030-55438-5_3

unordered with respect to $\geq$ as in 1.[1] The ordering of Δ_n is derived from that of D_n as in 2.[2]

1. For all $\delta \in \Delta_n$ and all $d, e \in \delta$: $d \not\gtreqless e$.
2. For all $\delta, \varepsilon \in \Delta_n$: $\delta \geq^* \varepsilon$ if and only if for all $d \in \delta$ there is an e in ε such that $d \geq e$.

The sequence $\#\Delta_n$ consisting of the number of members of Δ_n is known as the Dedekind numbers, and the problem of calculating their values is known as Dedekind's problem.[3] It is appropriate to name the orderings Δ_n for Dedekind, despite the fact that his name is also associated with a different notion of ordering, the Dedekind-MacNeille completion of a poset (Davey and Priestley 2002: 166).

A Dedekind ordering Δ_n is instantiated by a choice of A_n, D_n and $\geq$. First, following Dedekind's lead, let A_n^f be a set of $n > 0$ prime numbers, and let D_n^f be the poset consisting of the $2^n - 1$ multiples of A_n^f excluding unity, which has no prime factor. The factor-preserving ordering of D_n^f is $\geq^f$ defined in 3, and the ordering of Δ_n^f is the derived ordering $\geq^{f*}$ in 4. For example, let $A_2^f = \{2, 3\}$. Then $D_2^f = \{2, 3, 6\}$ and $\Delta_2^f = \{\{2\}, \{3\}, \{6\}, \{2, 3\}\}$. Henceforth for convenience in representing the members of Δ_n, the outer brackets are omitted, and the internal comma separator is replaced by the vertical bar, so that

[1] The members of Δ_n are the nonempty antichains of the power set of D_n. Davey and Priestley (2002) refer to Δ_n itself as a free distributive lattice with n generators. Generation of the members of Δ_n is discussed in the next section.

[2] The multiple antecedent definition of $\geq^*$ is given here in i.

(i) For all $\delta_1, \ldots, \delta_n, \varepsilon \in \Delta_n$: $\delta_1, \ldots, \delta_n \geq^* \varepsilon$ if and only if for all $d \in \bigcup_{i=1}^{n} \delta_i$ there is an $e \in \varepsilon$ such that $d \geq e$.

[3] Dedekind (1897 [1933: 147]) stated the values of what he called Gruppe 𝔓 for $3 \leq n \leq 4$: "Die Anzahl in dieser Gruppe 𝔓 enthaltenen Elemente scheint mit der Anzahl n der gegebenen Elemente sehr rasch zu wachsen; sie ist = 18 im Falle n = 3, und (wenn ich nicht irre) = 166 im Falle n = 4; einen allgemeinen Ausdruck für diese Anzahl zu finden, habe ich noch nicht versucht." In fact $\#\Delta_5$ = 7,579 was not calculated until several years after the reprinting of Dedekind (1897) had appeared (Church 1940), and its exact values are currently known only for $n \leq 8$ only, with $\#\Delta_8$ = 56,130,437,228,687,557,907,786. (Wiedermann 1991)

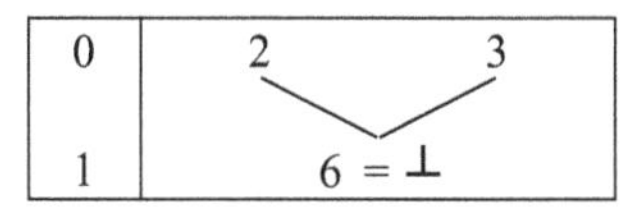

Fig. 3.1 $\geq^f$ ordering of D_2^f

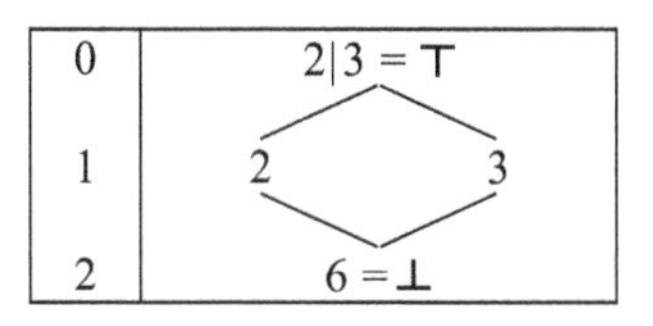

Fig. 3.2 $\geq^{f*}$ ordering of Δ_2^f

$\Delta_2^f = \{2, 3, 6, 2|3\}$.[4] The orderings of D_2^f and Δ_2^f are shown as lattices in Figs. 3.1 and 3.2, in which ⊤ is the supremum, ⊥ is the infimum, and the numbers in the column on the left are the lattices' row numbers, starting with 0 in the top row.

3. For all $d,e \in D_n^f$: $d \geq^f e$ if and only if every factor of e is a factor of d.
4. For all $\delta,\varepsilon \in \Delta_n^f$: $\delta \geq^{f*} \varepsilon$ if and only if for all $d \in \delta$ there is an e in ε such that $d \geq^f e$.

Next let A_n^s be a set of n members, and D_n^s the power set of A_n^s less the empty set. To obtain Δ_2^s, replace 3 and 4 with the orderings in 5 and 6. For example, let $A_2^s = \{0,1\}$. Then $D_2^s = \{\{0\}, \{1\}, \{0, 1\}\}$ and $\Delta_2^s = \{\{0\}, \{1\}, \{0,1\}, \{\{0\}|\{1\}\}\}$. The orderings of D_2^s and Δ_2^s are obtained from those of D_2^f and Δ_2^f in Figs. 3.1 and 3.2 by substituting {0} for 2, {1} for {3} and {0, 1} for 6.

5. For all $d,e \in D_n^s$: $d \geq^s e$ if and only if $d \supseteq e$.
6. For all $\delta,\varepsilon \in \Delta_n^s$: $\delta \geq^{s*} \varepsilon$ if and only if for all $d \in \delta$ there is a e in ε such that $d \supseteq e$.

[4] Thus a numeral representing a member of D_n^f stands for a number, whereas one representing a member of Δ_n^f stands for a singleton set containing of that number.

Finally let A^i_n be a set of n entities, and D^i_n the set of size $2^n - 1$ of mereological sums of those entities as in the dual of the calculus of individuals of Leonard and Goodman (1940). To obtain Δ^i_n, replace 3 and 4 with the orderings in 7 and 8. For example, let $A^i_2 = \{a, b\}$, where a and b are dogs named Tripod and Towzer.[5] Then $D^i_2 = \{a, b, ab\}$, in which $ab = a \oplus b$, where $\oplus$ is the sum operator of the dual of the calculus of individuals,[6] and $\Delta^i_2 = \{a, b, ab, a|b\}$. The orderings of D^i_2 and Δ^i_2 are obtained from those of D^f_2 and Δ^f_2 in Figs. 3.1 and 3.2 by substituting a for 2, b for 3, and ab for 6.

7. For all $d,e \in D^i_n$: $d \geq^i e$ if and only if every part of e is a part of d.
8. For all $\delta,\varepsilon \in \Delta^i_n$: $\delta \geq^{i*} \varepsilon$ if and only if for all $d \in \delta$ there is an e in ε such that $d \geq^i e$.

2 Conjunction and Disjunction in Dedekind Orderings

In D^s_n, set union serves as the conjunction (meet) operator over a non-null base set A^s_n, and in D^f_n and D^i_n, comparable union operators (multiplication of the union of factors and sum) do so over non-null base sets of prime numbers A^f_n and atomic individuals A^i_n.[7] In virtue of the general definition of D_n, conjunction alone generates its members from

[5] The dogs' names are taken from the discussion in Geach (1962: 67) of the interpretation of the phrase 'Tripod or Towzer' when talking about one's family's dogs. About this phrase, he writes "We must not … too readily assume that we understand a disjunction of proper names." In this paper it is to be understood as choices between the entities named by each disjunct.

[6] The ordering of the calculus of individuals may be defined as in ii, with the entailment relation of 7 reversed.

(ii) For all $d,e \in D^s_n$: $d \leq^i e$ if and only if if and only if every part of d is a part of e.

[7] In D^s_n set intersection serves as the disjunction (join) operator, and comparable intersection operators (multiplication of common prime factors and mereological product) do so in D^f_n and D^i_n. See Koslow (1992: Theorems 22.1 and 22.2) for proof that sum and product are equivalent to disjunction and conjunction in the calculus of individuals, hence that they have the opposite relation in D^i_n; see n. 6.

the base set without regard to the specific nature of the union operator, as in 9. In Δ_n^s on the other hand, set union serves as the disjunction (join) operator over D_n^s, and by virtue of the distributivity of conjunction over disjunction for sets of sets, the conjunction operator in D_n^s also serves as the conjunction (meet) operator in Δ_n^s. To get this same effect for all other Δ_n, D_n can be converted to a set of sets D_n^* as in 10, and Δ_n generated from D_n^* using both conjunction and disjunction as in 11. For example in D_2^f, the non-classical 3-member poset in Fig. 3.1, $6 = 2 \wedge 3$, and in Δ_2^f, the classical 4-member poset in Fig. 3.2, $\{6\} = \{2\} \wedge \{3\}$ (also notated as $6 = 2 \wedge 3$) and $\{\{2\},\{3\}\} = \{2\} \vee \{3\}$ (notated as $2|3 = 2 \vee 3$).[8]

9. Recursive definition of D_n

   (a) Base set: $A_n = \{a_1, \ldots, a_n\} \subseteq D_n$
   (b) Recursive step: If $x, y \in D_n$, then $x \wedge y \in D_n$
   (c) Closure: Nothing else is in D_n.

10. $D_n^{s*} = D_n^s$, and for all other D_n: $D_n^* = \{\{x\} \mid x \in D_n\}$
11. Recursive definition of Δ_n

   (a) Base set: $D_n^* \subseteq \Delta_n$
   (b) Recursive steps: If $x, y \in \Delta_n$ then $x \wedge y \in \Delta_n$ and $x \vee y \in \Delta_n$
   (c) Closure: Nothing else is in Δ_n.

This process can be simulated by a one-step recursive definition of Δ_n^b directly over a base set A_n^b of n bit strings of length $\sum_{k=1}^{n-1}\binom{n}{k} = 2^n - 2$ with the 1-preserving ordering in 12, so that conjunction and disjunction receive their familiar definitions in 13 and 14, and the recursive definition is in 15, for which the real work is done by defining membership in A_n^b. For each $a_k \in A_n^b (1 \le k \le n)$, each bit $a_k^i (1 \le i \le 2^n - 2)$ can be thought of as representing $d^i \in D_n^b - \{\perp\}$, and having the value 1 if and only if $a_k \ge^b d^i$.[9] For example, let $A_2^b = \{10, 01\}$, where the first bit of each string

[8] Note that the definition in 9–11 is semantic, not syntactic. For example in Δ_2^f 2|3 and 2|3|6 are both generated, but treated as equivalent.

[9] See Kisielewicz (1988) on determining the exact value of $\#\Delta_n$ from the bit string representations of its members.

represents the entity *a* and the second bit the entity *b*. Then $\Delta_2^b = \{00, 10, 01, 11\}$, in which 00 = 10 ∧ 01 and 11 = 10 ∨ 01.

12. For all $s, t \in \Delta_n^b : s \geq^b t$ if and only if for each bit pair s^i, t^i, $(1 \leq i \leq 2^n - 2)$, if $s^i = 1$ then $t^i = 1$.
13. Conjunction in Δ_n^b. For all $s,t \in \Delta_n^b : s \wedge t = u$, where $u^i = 1$ if and only if $s^i = t^i = 1$, $(1 \leq i \leq 2^n - 2)$.
14. Disjunction in Δ_n^b. For all $s,t \in \Delta_n^b : s \vee t = u$, where $u_i = 0$ if and only if $s_i = t_i = 0$, $(1 \leq i \leq 2^n - 2)$.
15. Recursive definition of Δ_n^b for the ordering in 12

    (a) Base set: $A_n^b = \{a_1, \ldots, a_n\} \subseteq \Delta_n^b$
    (b) Recursive steps: If $x, y \in \Delta_n^b$ then $x \wedge y \in \Delta_n^b$ and $x \vee y \in \Delta_n^b$
    (c) Closure: Nothing else is in Δ_n^b.

Negation in Δ_2^b also receives its familiar definition in 16, so that ¬10 = 01, ¬ 01 = 10, ¬ 00 = 11, and ¬11 = 00.

16. Negation in Δ_2^b. For all $s \in \Delta_2^b : \neg s = t$, where $t^i = 1$ if and only if $s^i = 0$.

However for Δ_n^b $(n > 2)$, 16 is correct for the infimum and supremum only. For example, let $A_3 = \{a, b, c\}$, where *a* and *b* are the dogs Tripod and Towzer as before, and *c* is a cat named Taffy, so that D_3 in 17 is the 7-membered set that results from conjunction over A_3, and together with D_3^b, is ordered as in Fig. 3.3.[10] Then Δ_3 in 18 is the 18-membered set that results from conjunction and disjunction over D_3^*; its members are represented here in disjunctive normal form (DNF).[11] Its ordering is shown in Fig. 3.4 together with that of Δ_3^b generated from $A_3^b = \{100110, 010101, 001011\}$.[12] The individual bits of the members

[10] Starting with this paragraph, wherever mereological (whole-part) orderings of the type defined in 8 are considered, the superscript '*i*' is omitted.

[11] Because conjunction and disjunction distribute over each other in Δ_n, the members of Δ_3 can also be represented in conjunctive normal form (CNF), i.e. with conjunction having scope over disjunction. For example, *ab*|*ac* in DNF is equivalent to *a*(*b*|*c*) in CNF, and *a*|*bc* in DNF is equivalent to (*a*|*b*)(*a*|*c*) in CNF.

[12] The element *ab*|*ac*|*bc* in the middle row 3 of Fig. 3.4 is identified as a + 'middle' member of Δ_3 because it entails all the elements in the rows above it and is entailed by all the elements in the rows below it. For *n* > 3, Δ_n has more than 1 middle member, each of which entails all of the elements

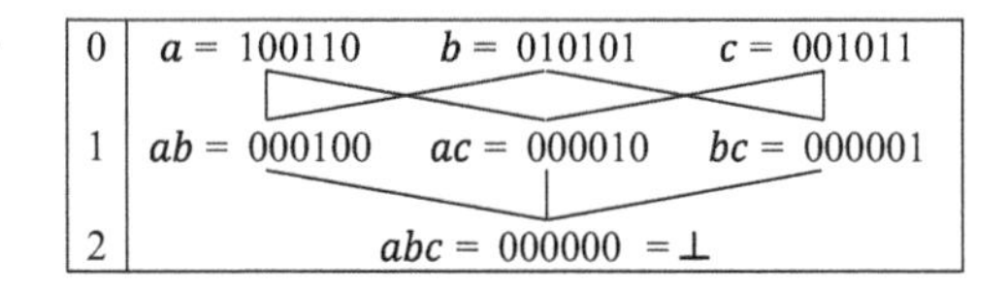

Fig. 3.3 Ordering of D_3 and D_3^b

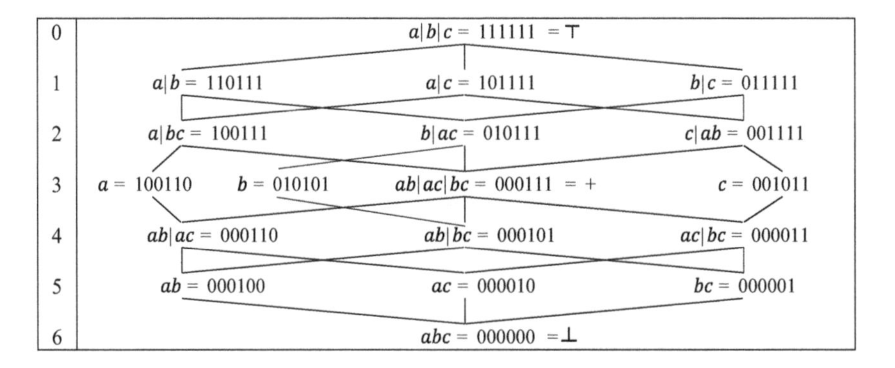

Fig. 3.4 Ordering of Δ_3 and Δ_3^b

of Δ_3^b can be determined by their association r with the members of $D_3^b - \bot$ as in 19. The definition of negation in 16, if extended to Δ_3^b, gives an incorrect result for every member except for $\bot$ and $\top$. For example, $\neg 100110 = 000001$, not 011001, which is not a member of Δ_3^b. The correct analysis of negation in Δ_n and Δ_n^b $(n > 2)$ is taken up in the next section.

17. $D_3 = \{a, b, c, ab, ac, bc, abc\}$
18. $\Delta_3 = \{a, b, c, ab, ac, bc, abc, a|b, a|c, b|c, a|bc, b|ac, c|ab, ab|ac, ab|bc, ac|bc, a|b|c, ab|ac|bc\}$
19. For all $s_3 \in \Delta_3^b$, let $r(s_3^1) = a$, $r(s_3^2) = b$, $r(s_3^3) = c$, $r(s_3^4) = ab$, $r(s_3^5) = ac$, and $r(s_3^6) = bc$. Then $s_3^i = 1$ if and only if $r(s_3) \geq r(s_3^i)$ for all i $(1 \leq i \leq 6)$.

---

in row 1 and is entailed by all of the elements in row $2^n - 3$. For example in Δ_4 defined over $A_4 = \{a, b, c, z\}$, there are 12 middle members: $+ = \{ab|ac|bz, ab|az|bc, ac|ab|cz, ac|az|bc, az|ab|cz, az|ac|bz, bc|ab|cz, bc|bz|ac, bz|ab|cz, bz|bc|az, cz|ac|bz, cz|bc|az\}$.

Corresponding to each member of Δ_3 are English expressions that can be said to uniquely identify it in contexts in which no other individuals occur, as in Table 3.1, in which the members are listed from top to bottom and left to right as shown in Fig. 3.4, and for which one or more such expressions are provided.[13] From this demonstration it can reasonably be concluded that English has the expressive power to identify all the members of any Δ_n, and that it has various grammatical devices for doing so in addition to the explicit conjunction and disjunction of proper names of the members of A_n.

Table 3.1 Members of Δ_3 and some of their identifying English expressions

Δ_3	Identifying English expressions
$a\|b\|c$	Tripod, Towzer or Taffy; one of the pets
$a\|b$	Tripod or Towzer; one of the dogs
$a\|c$	Tripod or Taffy
$b\|c$	Towzer or Taffy
$a\|bc$	Tripod, or Towzer and Taffy
$b\|ac$	Towzer, or Tripod and Taffy
$c\|ab$	Taffy or Tripod and Towzer; the cat or the dogs
a	Tripod
b	Towzer
$ab\|ac\|bc$	Tripod and Towzer, or Tripod and Taffy, or Towzer and Taffy; two of the pets
c	Taffy; the cat
$ab\|ac$	Tripod and Towzer, or Tripod and Taffy; Tripod, and Towzer or Taffy
$ab\|bc$	Tripod and Towzer, or Towzer and Taffy; Towzer, and Tripod or Taffy
$ac\|bc$	Tripod and Taffy, or Towzer and Taffy; the cat and one of the dogs
ab	Tripod and Towzer; the dogs; both dogs
ac	Tripod and Taffy; Tripod and the cat
bc	Towzer and Taffy; Towzer and the cat
abc	Tripod, Towzer and Taffy; all of the pets

[13] This is not to say, for example, that 'Tripod and Towzer' is the name of *ab* or that 'Tripod or Towzer' is the name of $a|b$ in Δ_3, but only that such compound expressions may be used to refer to them unambiguously in appropriate contexts, and that subsequent use of pronouns may be coreferential with them, as in 'Tripod and Towzer must really have been hungry, the way they ate that bone.' and 'Tripod or Towzer growled at Taffy and I really hate it when she does that.'

3 Singly Negated Mappings of Dedekind Orderings

Two kinds of negation can be defined for Δ_n, the standard one that satisfies the law of contradiction in 20, symbolized '$\neg$',[14] and the other that satisfies the law of excluded middle in 21, symbolized '$\ulcorner$'.[15] Because $\ulcorner$ serves as the negation operator in $\hat{\Delta}_n$, the dual of Δ_n defined in 22, we call it, following Koslow (1992: 99), the dual of negation, or d-negation. The corresponding definitions for Δ_n^b are in 23 and 24.[16]

20. Negation in Δ_n: For all $\delta \in \Delta_n$:

    (a) $\delta \wedge \neg\delta = \bot$
    (b) $\neg\delta$ is the weakest member of Δ_n to satisfy 20.a: For all $\zeta \in \Delta_n$, if $\delta \wedge \zeta = \bot$ then $\zeta \geq^* \neg\delta$.

21. D-negation in Δ_n: For all $\delta \in \Delta_n$:

    (a) $\delta \vee \ulcorner\delta = \top$
    (b) $\ulcorner\delta$ is the strongest member of Δ_n to satisfy 21.a: For all $\zeta \in \Delta_n$, if $\ulcorner\delta \vee \zeta = \top$ then $\ulcorner\delta \geq^* \zeta$.

22. For all $\delta, \varepsilon \in \hat{\Delta}_n : \delta \leq^* \varepsilon$ if and only if $\delta \geq^* \varepsilon$ in Δ_n.

---

[14] "Law of contradiction" is a misnomer for Dedekind and other non-propositional orderings because there is nothing contradictory about their infimum.

[15] D-negation can also be defined as a "weakest" operator on Δ_n (Koslow 1992: Theorem 12.11). A definition specifically for Δ_n is in iii.

(iii) D-negation. For all $\delta \in \Delta_n$:

(a) If there is $\zeta \neq \top \in \Delta_n$ such that $\delta >^* \zeta$, then for all such ζ: $\ulcorner\delta \ngeq^* \zeta$; else for all $\theta \in \Delta_n$ such that $\theta \ngeq^* \delta$, $\ulcorner\delta \geq^* \theta$.

(b) $\ulcorner\delta$ is the weakest member of Δ_n to satisfy i.a. For all $\gamma \in \Delta_n$: If it is the case that if there is $\zeta \neq \top \in \Delta_n$ such that $\delta >^* \zeta$, then for all such ζ, $\gamma \ngeq^* \zeta$; else for all $\theta \in \Delta_n$ such that $\theta \ngeq^* \delta$, $\gamma \geq^* \theta$, then $\gamma \geq^* \ulcorner\delta$.

[16] I have not confirmed the correctness of these definitions for $n > 3$.

23. Negation in Δ_n^b: For all $s \in \Delta_n^b$: If for all i ($1 \le i \le 2^n - 2$) $s_n^i = 1$ then $\neg s_n^i = 0$, else $\neg s_n^i = \left(s_n^i =\right)0$ if and only if $1 \le i \le n$ and $s_n^j = 1\left(2^n - (n+1) \le j \le 2^n - 2\right)$.
24. D-negation in Δ_n^b: For all $s \in \Delta_n^b$: If for all i ($1 \le i \le 2^n - 2$) $s_n^i = 0$ then $s_n^i = 1$, else $s_n^i = \left(s_n^i =\right)1$ if and only if $2^n - (n + 1) \le i \le 2^n - 2$ and $s_n^j = 0$ $(1 \le j \le n)$.

Δ_2 and Δ_2^b are classical orderings, because negation and d-negation are identical functions from Δ_2 and Δ_2^b **onto** Δ_2 and Δ_2^b respectively (Koslow 1992: Theorem 12.6), satisfying 20, 21, 22 and 23, as shown in Fig. 3.5.

However Δ_n and Δ_n^b ($n > 2$) are non-classical orderings because both negation and d-negation are many to 1 functions from Δ_n and Δ_n^b **into** Δ_n and Δ_n^b respectively, as shown in Table 3.2 for Δ_3 and Δ_3^b, in which the boldfaced bits in the members of $\neg\Delta_3$ and $⌐\Delta_3$ are unchanged by negation and d-negation in accordance with 23 and 24. Identifying English expressions for the members of $\neg\Delta_3$ and $⌐\Delta_3$ can be found in Table 3.1, but are irrelevant in Table 3.2; for example, no one would ever intend to use either 'not Towzer and Tripod' or 'not Towzer or Tripod' to refer to Taffy even though they could, assuming they had in mind and understood their audience also had in mind the properties of $\neg\Delta_3$ in the first case and of $⌐\Delta_3$ in the second.

The image of Δ_3 under negation is $\neg\Delta_3$, which is the classical subordering of Δ_3 in Fig. 3.6; its row numbers and those of Figs. 3.7–3.9 correspond to those of Fig. 3.4. It consists of the supremum and the singleton members of Δ_3, whose members in turn are the individual members of D_3. Similarly, the image of Δ_3 under d-negation is the classical subordering of $⌐\Delta_3$ in Fig. 3.7, which partially overlaps $\neg\Delta_3$. It consists of the infimum and the members of Δ_3 that are the non-empty sets of members of A_n. Let us call $\neg\Delta_3$ and $⌐\Delta_3$ the singly negated mappings of Δ_3. The properties of these mappings hold for all Δ_n.

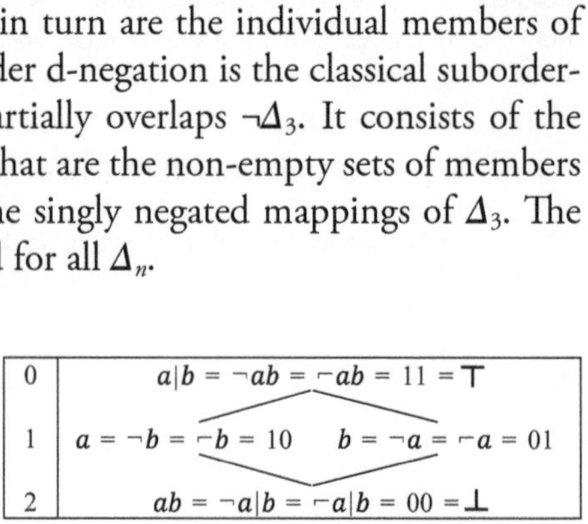

Fig. 3.5 Negation and d-negation in Δ_2 and Δ_2^b

0	$a\|b = \neg ab = ⌐ab = 11 = \top$
1	$a = \neg b = ⌐b = 10$ $\quad$ $b = \neg a = ⌐a = 01$
2	$ab = \neg a\|b = ⌐a\|b = 00 = \bot$

Table 3.2 Negation and d-negation in Δ_3 and Δ_3^b

Δ_3	$\neg\Delta_3$	$\ulcorner\Delta_3$	Δ_3^b	$\neg\Delta_3^b$	Δ_3^b
a|b|c	abc	abc	111111	000000	000000
a|b	abc	c	110111	000000	001011
a|c	abc	b	101111	000000	010101
b|c	abc	a	011111	000000	100110
a|bc	abc	b|c	100111	000000	011111
b|ac	abc	a|c	010111	000000	101111
c|ab	abc	a|b	001111	000000	110111
a	bc	b|c	100110	000001	011111
b	ac	a|c	010101	000010	101111
c	ab	a|b	001011	000100	110111
ab|ac|bc	abc	a|b|c	000111	000000	111111
ab|ac	bc	a|b|c	000110	000001	111111
ab|bc	ac	a|b|c	000101	000010	111111
ac|bc	ab	a|b|c	000011	000100	111111
ab	c	a|b|c	000100	001011	111111
ac	b	a|b|c	000010	010101	111111
bc	a	a|b|c	000001	100110	111111
abc	a|b|c	a|b|c	000000	111111	111111

Fig. 3.6 Ordering of $\neg\Delta_3$

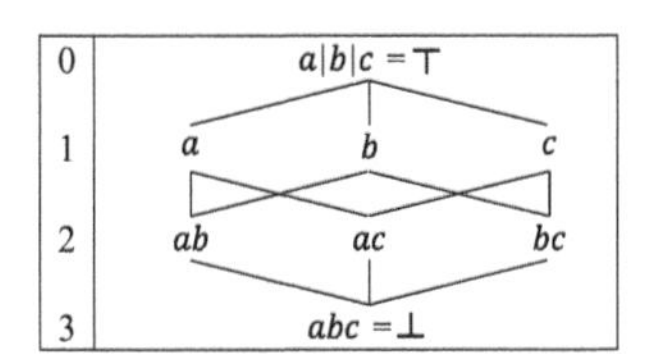

Fig. 3.7 Ordering of $\ulcorner\Delta_3$

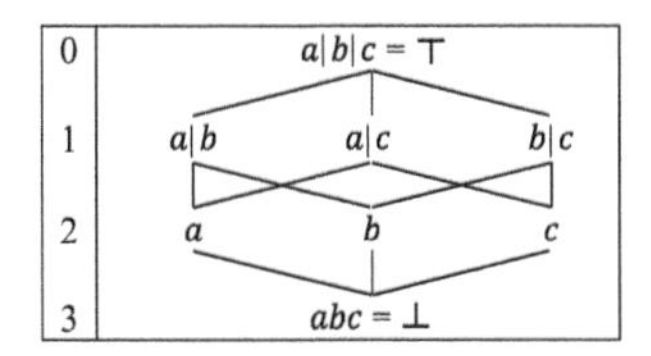

By examining Table 3.3 below, one can determine that for all $\delta \in \Delta_3$, $\neg\delta \geq^* \ulcorner\delta$.[17] This can be expressed more compactly as $\neg\Delta_3 \geq^o \ulcorner\Delta_3$, where $\geq^o$

[17] In Table 3.3, the numbers in the $\neg\Delta_3$ and $\ulcorner\Delta_3$ columns are the rows to which the negation and d-negation operators map the elements in each row of Δ_3. Call those numbers $\# \neg i$ and $\# \ulcorner i$. Then $\neg\Delta_3 \geq^o \ulcorner\Delta_3$ if and only if for all i, $\# \neg i \geq \# \ulcorner i$ (where $\geq$ signifies ordinary numerical 'greater than or equal to'). The circumflex over 1, 3, and 5 indicates that the mapping reverses the order of the

Table 3.3 Members of Γ_3^1 showing that $\neg\Delta_3 \geq^o \ulcorner\Delta_3$

Δ_3 members	row	$\neg\Delta_3$	$\ulcorner\Delta_3$
a|b|c = ⊤	0	6	6
<a|b,a|c, b|c>	1	6	$\hat{3}$
<a|bc,b|ac, c|ab>	2	6	$\hat{1}$
<a, b, c>	3	$\hat{5}$	$\hat{1}$
ab|ac|bc	3+	6	0
<ab|ac,ab|bc, ac|bc>	4	$\hat{5}$	0
<ab, ac, bc>	5	$\hat{3}$	0
abc = ⊥	6	0	0

is an ordering of orderings derived from ≥* as defined in 25, in which Γ_n is a poset of the negation and d-negation mappings of Δ_n distinguished by their prefix strings μ, ν of negation and d-negation operators.[18] Let $\Gamma_n^l \subseteq \Gamma_n$ be the set of members of Γ_n whose shortest prefix strings are of length l. Then $\Gamma_2^1 = \{\neg\Delta_2 = \neg\Delta_2\}$ so that $\#\Gamma_2^1 = 1$, and for $l > 1$, $\Gamma_2^l = \varnothing$ so that $\#\Gamma_2^l = 0$.[19] For $n > 2$, $\Gamma_n^1 = \{\neg\Delta_n, \neg\Delta_n\}$, so that $\#\Gamma_n^1 = 2$. In the next section, the members of Γ_n^l and the values $\#\Gamma_n^l$ for $n > 2$ and $l > 1$ are presented.

25. For all $\mu\Delta_n$, $\nu\Delta_n \in \Gamma_n$: $\mu\Delta_n \geq^o \nu\Delta_n$ if and only if for every $\delta \in \Delta_n$: $\mu\delta \geq^* \nu\delta$.

4 Multiply Negated Mappings of Dedekind Orderings

The set of doubly negated Dedekind orderings $\Gamma_n^2, (n > 2)$, is given in 26, so that $\#\Gamma_n^2 = 4$ for each such n, as can be determined from Table 3.4 for Γ_3^2. Their ordering together with that of Δ_n is the linear one in 27

elements in the Δ_3 members column for those rows. In subsequent tables, the Δ_3 members column is omitted, and the row column is labeled "Δ_3".

[18] That is, $\mu, \nu \in (\neg|\ulcorner)^+$ in the notation for regular expressions. The phrase "negation and d-negation" is shortened below to "negation" when used as a modifier of "mapping" or "operator".

[19] These are instances of the well-known results for all classical orderings O that $\neg^2 O = \ulcorner^2 O = O$, and that $\neg^3 O = \ulcorner^3 O = \neg\, O = \ulcorner\, O$.

Table 3.4 Ordering of $\Gamma_3^2 \cup \Delta_3$

$\neg⌐$	$⌐^2$	Δ_3	$\neg^2$	$⌐\neg$
0	0	**0**	0	0
5	1	**1**	0	0
6	3	**2**	0	0
6	3	**3**	3	0
6	6	**3+**	0	0
6	6	**4**	3	0
6	6	**5**	5	1
6	6	**6**	6	6

Fig. 3.8 Ordering of $⌐\neg \Delta_3$

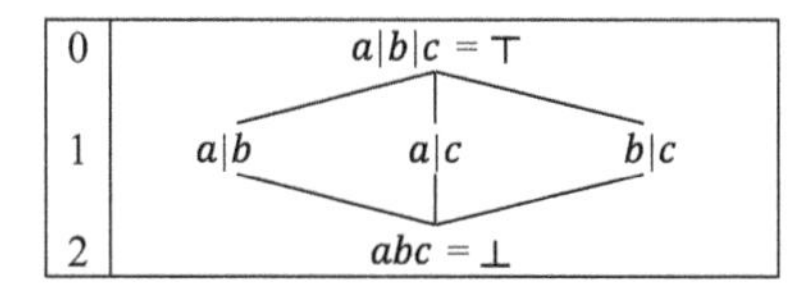

Fig. 3.9 Ordering of $\neg⌐ \Delta_3$

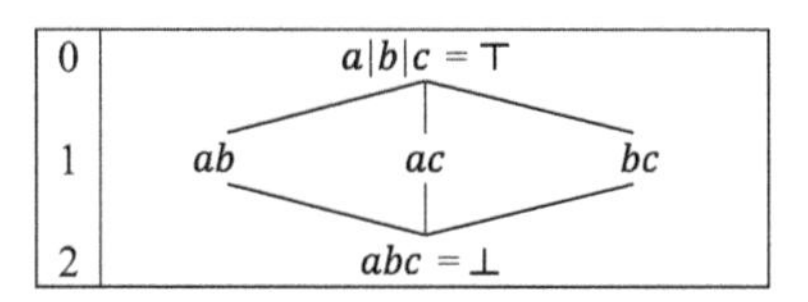

(Koslow 1992: Theorem 17.7), which can be termed the law of double negation for Dedekind orderings. The parallel-negation mappings are from $\neg\Delta_n$ onto $\neg\Delta_n$ and from $⌐\Delta_n$ onto $⌐\Delta_n$, whereas the cross-negation ones are from $\neg\Delta_n$ into $⌐\Delta_n$, whose image is a non-classical ordering like $⌐\neg \Delta_3$ in Fig. 3.8, and from $⌐\Delta_n$ into $\neg\Delta_n$, whose image is also a non-classical ordering like $\neg⌐ \Delta_3$ in Fig. 3.9.[20]

[20] None of the members of row 1 of Fig. 3.8 nor of row 5 of Fig. 3.9 have negations. For example, neither $\neg a|b$ nor $⌐a|b$ exist in $⌐\neg \Delta_3$, because in both cases $a|c$ and $b|c$ are candidates, but neither is stronger than the other. In $⌐\neg \Delta_n$, the members of row 1 are all and only all the members of $⌐a_i$ $(1 \leq i \leq n)$, and in $\neg⌐\Delta_n$, the members of row $2^n - 3$ are all and only all the members of $\neg a_i$ $(1 \leq i \leq n)$. Nevertheless, these members can be mapped by $\{\neg⌐\ \neg, ⌐^2\neg, \neg^2⌐, ⌐\neg\ ⌐\}\Delta_3$. For example, $\neg⌐\neg\ a|b = ⌐^2\neg\ a|b = \bot$, $\neg^2⌐\ a|b = c$, and $⌐\neg\ ⌐a|b = \top$; see Table 3.5.

26. Double negations for Dedekind orderings:
$$\Gamma_n^2 = \{\neg^2\Delta_n, \ulcorner^2\Delta_n, \neg\ulcorner\Delta_n, \ulcorner\neg\Delta_n\}$$
27. Ordering of Γ_n^2: $\neg\ulcorner\,\Delta_n \geq^o \ulcorner^2\Delta_n \geq^o \Delta_n \geq^o \neg^2\Delta_n \geq^o \ulcorner\neg\,\Delta_n$

The set of strictly triply negated Dedekind orderings Γ_n^3 for $n > 2$ is given in 28, so that $\#\Gamma_n^3 = 6$ for each such n. The ordering of $\Gamma_n^1 \cup \Gamma_n^3$, the full set of triply negated Dedekind orderings for $n > 2$, including $\neg\Delta_n = \neg^3\Delta_n$ and $\ulcorner\Delta_n = \ulcorner^3\Delta_n$, can be determined from Table 3.5 for Γ_3^3, and is shown as the lattice in Fig. 3.10.

Table 3.5 Ordering of the members of $\Gamma_3^3 \cup \Gamma_3^1$ (tabular form)

Δ_3	$\neg\ulcorner\neg$	$\ulcorner^2\neg$	$\neg$	$\neg\ulcorner^2$	$\ulcorner\neg^2$	$\ulcorner$	$\neg^2\ulcorner$	$\ulcorner\neg\ulcorner$
0	6	6	6	6	6	6	6	6
1	6	6	6	6	6	$\hat{3}$	$\hat{3}$	0
2	6	6	6	6	$\hat{5}$	$\hat{1}$	0	0
3	6	6	$\hat{5}$	$\hat{1}$	$\hat{5}$	$\hat{1}$	0	0
3+	6	6	6	6	0	0	0	0
4	6	6	$\hat{5}$	$\hat{1}$	0	0	0	0
5	6	$\hat{3}$	$\hat{3}$	0	0	0	0	0
6	0	0	0	0	0	0	0	0

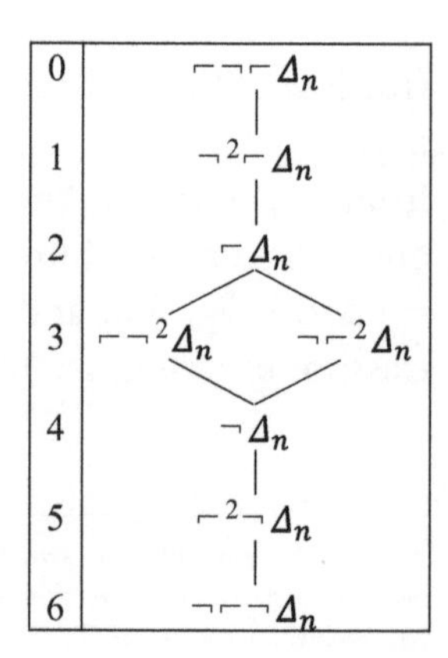

Fig. 3.10 Ordering of $\Gamma_3^3 \cup \Gamma_3^1$ (lattice form)

28. Strictly triple negations:

$$\Gamma_n^3 = \{\neg\neg\Delta_n, \neg\Delta_n, \neg^2\Delta_n, {}^2\neg\Delta_n, \neg^2\Delta_n, \neg^2\Delta_n\}$$

The observation that there are nonclassical orderings for which $\#\Gamma^l$ continues to grow for small values of l led Koslow (1992: 148) to observe "We are left with an open question: Are there infinitely many or only finitely many nonequivalent combinations of [¬] and [⌐]?" Using the notation of this paper, Koslow's question can be formulated as in 29.a. With that answer we also have answers to 29.b: How many are there, and what are those combinations? The remainder of this section deals first with the questions in 29 for Dedekind orderings only, and second with the questions for all non-classical orderings.

29. Koslow's open question extended:

    (a) For all non-classical orderings, is there a length v at which $\#\Gamma^{v+1}$ drops to 0?
    (b) If so what is $\sum_{l=1}^{v} \Gamma^l$, and what is $\Gamma = \bigcup_{l=1}^{v} \Gamma^l$?

The set of strictly quadruply negated Dedekind orderings Γ_n^4 for $n > 2$ is given in 30, so that $\#\Gamma_n^4 = 8$, and $\#\left(\Gamma_n^4 \cup \Gamma_n^2\right) = 12$, the size of the full set of quadruply negated Dedekind orderings The ordering of $\Gamma_n^4 \cup \Gamma_n^2 \cup \Delta_n$ can be determined from Table 3.6 for Δ_3, and is shown as the lattice in Fig. 3.11.

Table 3.6 Ordering of $\Gamma_3^4 \cup \Gamma_3^2 \cup \Delta_3$ (tabular form)

(¬⌐)^2	¬⌐	⌐¬^{2}⌐	⌐^{2}	¬^{2}⌐^{2}	⌐¬⌐^{2}	Δ_3	¬⌐¬^{2}	⌐^{2}¬^{2}	¬^{2}	¬⌐^{2}¬	⌐¬	(⌐¬)^2
0	0	0	0	0	0	**0**	0	0	0	0	0	0
6	5	1	1	0	0	**1**	0	0	0	0	0	0
6	6	6	3	3	0	**2**	0	0	0	0	0	0
6	6	6	3	3	0	**3**	6	3	3	0	0	0
6	6	6	6	6	6	**3+**	0	0	0	0	0	0
6	6	6	6	6	6	**4**	6	3	3	0	0	0
6	6	6	6	6	6	**5**	6	6	5	5	1	0
6	6	6	6	6	6	**6**	6	6	6	6	6	6

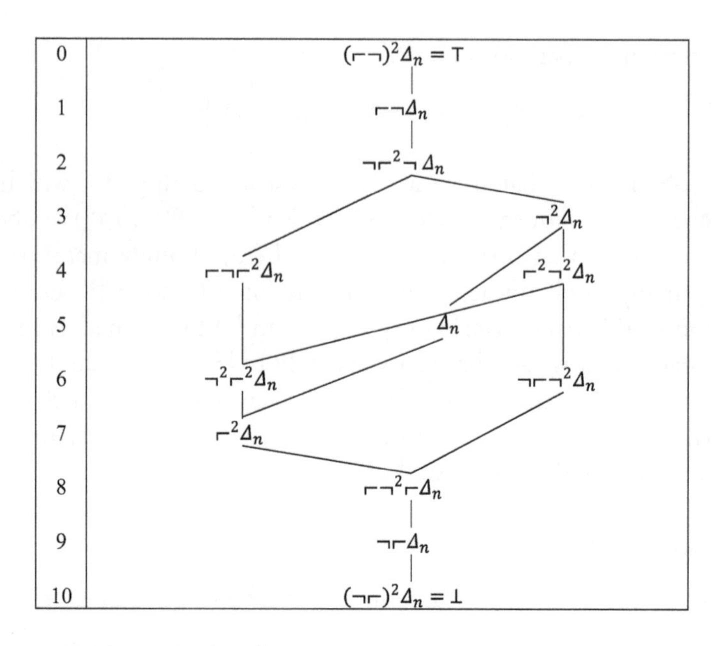

Fig. 3.11 Ordering of $\Gamma_n^4 \cup \Gamma_n^2 \cup \Delta_n$ (lattice form)

30. Strictly quadruple negations:

$$\Gamma_n^4 = \{(\neg)^2 \Delta_n, (\neg)^2 \Delta_n, \neg^2\neg\Delta_n, \neg^2\Delta_n, \neg^2\ulcorner^2\Delta_n, \ulcorner^2\neg^2\Delta_n,$$
$$\neg\ulcorner\neg^2\Delta_n, \ulcorner\neg\ulcorner^2\Delta_n\}$$

The set of strictly quintuply negated Dedekind orderings Γ_n^5 for $n > 2$ is given in 31, so that $\#\Gamma_n^5 = 6$, and $\#\left(\Gamma_n^5 \cup \Gamma_n^3 \cup \Gamma_n^1\right) = 14$, the size of the full set of quintuply negated Dedekind orderings. The ordering of $\Gamma_n^5 \cup \Gamma_n^3 \cup \Gamma_n^1$ can be determined from Table 3.7 for Δ_3, and is shown as the lattice in Fig. 3.12.

31. Strictly quintuple negations:

$$\Gamma_n^5 = \{\neg\neg^2\Delta_n, \neg^2\neg\Delta_n, \neg^2\neg^2\Delta_n, \neg^{2\,2}\Delta_n, \neg^2\ulcorner\neg^2\Delta_n, \ulcorner^2\neg\ulcorner^2\Delta_n\}$$

The set of strictly sextuply negated Dedekind orderings Γ_n^6 for $n > 2$ is given in 32, so that $\#\Gamma_n^6 = 2$, and $\#\left(\Gamma_n^6 \cup \Gamma_n^4 \cup \Gamma_n^2\right) = 14$, the size of the full set of sextuply negated Dedekind orderings. The ordering of

Table 3.7 Ordering of $\Gamma_3^5 \cup \Gamma_3^3 \cup \Gamma_3^1$ (tabular form)

Δ_3	$\neg\ulcorner\neg$	$\ulcorner^2\neg$	$\neg$	$\ulcorner\neg\ulcorner^2\neg$	$\neg\ulcorner^2\neg^2$	$\neg^2\ulcorner\neg^2$	$\ulcorner\neg^2$	$\neg\ulcorner^2$	$\ulcorner^2\neg\ulcorner^2$	$\ulcorner\neg^2\ulcorner^2$	$\neg\ulcorner\neg^2\ulcorner$	$\ulcorner$	$\neg^2\ulcorner$	$\ulcorner\neg\ulcorner$
0	6	6	6	6	6	6	6	6	6	6	6	6	6	6
1	6	6	6	6	6	6	6	6	6	6	$\hat{5}$	$\hat{3}$	$\hat{3}$	0
2	6	6	6	6	6	6	6	$\hat{5}$	6	$\hat{1}$	0	$\hat{1}$	0	0
3	6	6	$\hat{5}$	6	$\hat{5}$	0	$\hat{1}$	$\hat{5}$	6	$\hat{1}$	0	$\hat{1}$	0	0
3+	6	6	6	6	6	6	6	0	0	0	0	0	0	0
4	6	6	$\hat{5}$	6	$\hat{5}$	0	$\hat{1}$	0	0	0	0	0	0	0
5	6	$\hat{3}$	$\hat{3}$	$\hat{1}$	0	0	0	0	0	0	0	0	0	0
6	0	0	0	0	0	0	0	0	0	0	0	0	0	0

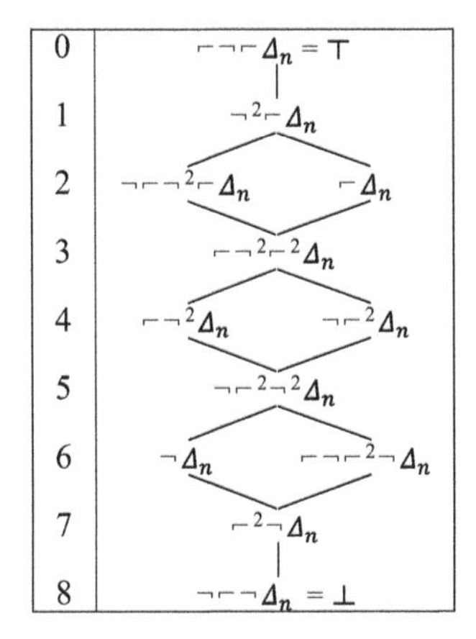

Fig. 3.12 Ordering of the members of $\Gamma_n^5 \cup \Gamma_n^3 \cup \Gamma_n^1$ (lattice form)

$\Gamma_n^6 \cup \Gamma_n^4 \cup \Gamma_n^2 \cup \Delta_n$ can be determined from Table 3.8 for Δ_3, and is shown as the lattice in Fig. 3.13.

32. Strictly sextuple negations: $\Gamma_n^6 = \left\{\left(\neg^2\right)^2, \left(^2\neg\right)^2\right\}$

The reader can confirm that the set of strictly septuply negated orderings $\Gamma_n^7 = \varnothing$, and consequently that $\#\Gamma_n^l = 0$ for $l = 7$. Therefore, for all non-classical Dedekind orderings, the answers to the questions in 29 are those in 33.

Table 3.8 Ordering of $\Gamma_3^6 \cup \Gamma_3^4 \cup \Gamma_3^2 \cup \Delta_3$ (tabular form)

$(¬⌐)^2$	$¬⌐$	$⌐¬^2⌐$	$⌐^2$	$(⌐^2¬)^2$	$¬^2⌐^2$	$¬⌐¬^2$	Δ_3	$⌐¬⌐^2$	$⌐^2¬^2$	$(⌐^2¬)^2$	$¬^2$	$¬⌐^2¬$	$⌐¬$	$(⌐¬)^2$
0	0	0	0	0	0	0	**0**	0	0	0	0	0	0	0
6	5	1	1	0	0	0	**1**	0	0	0	0	0	0	0
6	6	6	3	0	3	0	**2**	0	0	0	0	0	0	0
6	6	6	3	6	3	0	**3**	6	3	0	3	0	0	0
6	6	6	6	6	6	6	**3+**	0	0	0	0	0	0	0
6	6	6	6	6	6	6	**4**	6	3	6	3	0	0	0
6	6	6	6	6	6	6	**5**	0	6	6	5	5	1	0
6	6	6	6	6	6	6	**6**	6	6	6	6	6	6	6

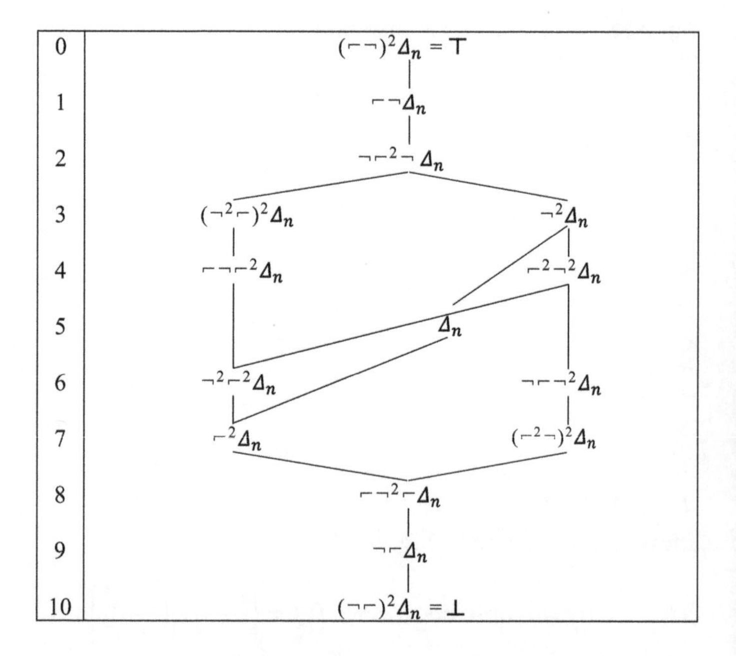

Fig. 3.13 Ordering of $\Gamma_n^6 \cup \Gamma_n^4 \cup \Gamma_n^2 \cup \Delta_n$ (lattice form)

33. For all non-classical Dedekind orderings Δ_n:

    (a) The length v at which $\#\Gamma_n^{v+1}$ drops to 0 is 6.
    (b) The number of distinct negation orderings $\sum_{l=1}^{v} \#\Gamma_n^l$ is 28.
    (c) The set of the 28 orderings $\Gamma_n = \cup_{l=1}^{v} \Gamma_n^l$ is shown in the chart in Fig. 3.14.

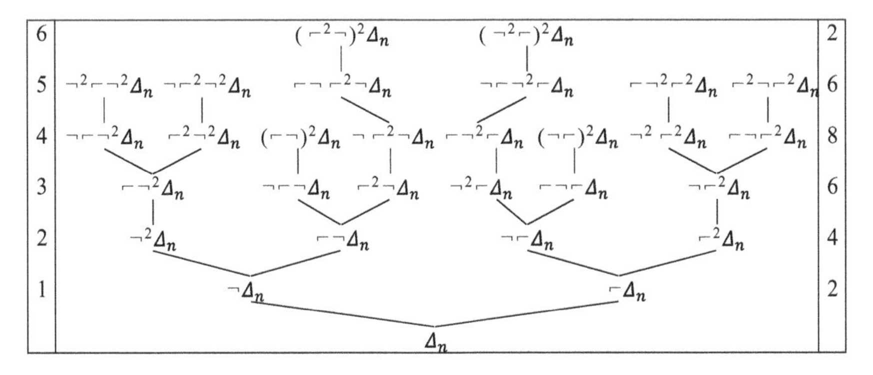

Fig. 3.14 $\Gamma_n \cup \Delta_z$ organized by left-edge growth of prefix μ; length of prefix for given row is in leftmost column; number of distinct mappings for that row is in rightmost column

The answers in 33 do not change materially for the complete class of nonclassical orderings. For any such ordering *O*, the images of negation and d-negation are classical, just as they are for Dedekind orderings, so at most the subsequent mappings follow the same course as a Dedekind ordering with homomorphic images of negation and d-negation.[21] To analyze the behavior of non-classical non-Dedekind ordering under negation, let Ω be the poset of the negation mappings of *O*, and 25 and 33 be modified to read as in 34 and 35.

34. For all $\mu O, \nu O \in \Omega$: $\mu O \geq^{o} \nu O$ if and only if for every $o \in O : \mu o \geq \nu o$.
35. For some non-classical orderings *O*:

    (a) The length v at which $\#\Omega^{v+1}$ drops to 0 is 6.
    (b) The number of distinct negation orderings $\sum_{l=1}^{v} \#\Omega^{l}$ is 28.
    (c) The set of the 28 orderings $\Omega = \cup_{l=1}^{v} \Omega^{l}$ is in the chart in Fig. 3.14 with *O* replacing Δ_n.

$O_{3a} = \neg\Delta_3 \cup \ulcorner\Delta_3 \cup \{ab|ac|bc\}$, whose 12 members are listed in 36 and whose ordering is shown in Fig. 3.15, exemplifies the smallest

[21] Each $\neg\Delta_n$ and $\ulcorner\Delta_n$ is a classical ordering with *n* generators in row 1 of its lattice under conjunction, with disjunction needed only for the supremum, and every classical ordering can be generated in that way.

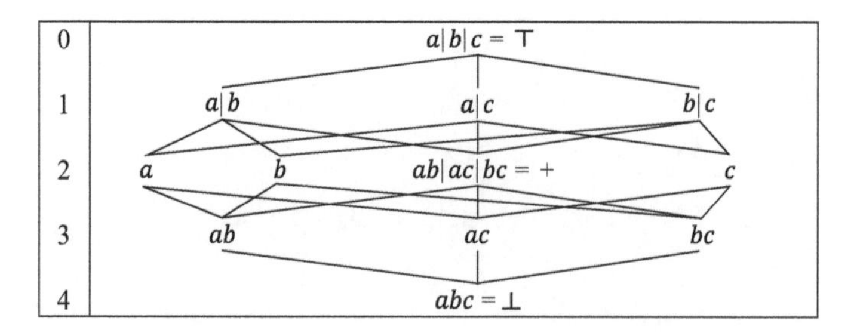

Fig. 3.15 Ordering of O_{3a}

non-Dedekind ordering that satisfies 35. The ordering of the 28 members of Ω_{3a} is shown in Table 3.9 for the 14 odd-number combinations of negation and d-negation, and Table 3.10 for the 14 even-number combinations plus O_{3a}.[22] Figures 3.12 and 3.13 are the lattice representations of the orderings of the members of Ω_{3a} by substituting O_{3a} for Δ_3 throughout.

36. $$O_{3a} = \{a, b, c, ab, ac, bc, a|b, a|c, b|c, abc, ab|ac|bc, a|b|c\}$$

The conditions for an ordering O_n ($n > 2$) to satisfy 35 are listed in 37, with O_{3a} an example of the smallest one for which $k = 2$. Dedekind orderings, on the other hand, are those for which $k = 2^{n-1} - 1$ and for which the n members of row k designated in 37.c generate O_n (i.e. Δ_n) using conjunction and disjunction. None of the non-Dedekind orderings that satisfy 35 can be generated in this way because conjunction and disjunction in them are partial, not total, functions mapping $O_n \times O_n$ to O_n. For example in O_{3a}, $a \vee ab|ac|bc$ is undefined because both $a|b$ and $a|c$ are candidates, and neither is stronger than the other. Likewise $a \wedge ab|ac|bc$ is undefined because both ab and ac are candidates, and neither is weaker than the other.[23]

[22] Tables 3.9 and 3.10 can be derived from Tables 3.7 and 3.8 by removing rows 2 and 4 and renumbering.

[23] Alternatively one can allow conjunction and disjunction to have 2 values in these situations, but then they are no longer functions but relations.

Table 3.9 Ordering of $\Omega_{3a}^{5} \cup \Omega_{3a}^{3} \cup \Omega_{3a}^{1}$ (tabular form)

O_{3a}	$¬⌐¬$	$⌐^{2}¬$	$¬$	$⌐¬⌐^{2}¬$	$¬⌐^{2}¬^{2}$	$¬^{2}⌐¬^{2}$	$⌐¬^{2}$	$¬⌐^{2}$	$⌐^{2}¬⌐^{2}$	$⌐¬^{2}⌐^{2}$	$¬⌐¬^{2}⌐$	$⌐$	$¬^{2}⌐$	$⌐¬⌐$
0	4	4	4	4	4	4	4	4	4	4	4	4	4	4
1	4	4	4	4	4	4	4	4	4	4	$\hat{3}$	$\hat{2}$	$\hat{2}$	0
2	4	4	$\hat{3}$	4	$\hat{3}$	0	$\hat{1}$	$\hat{3}$	4	$\hat{1}$	0	$\hat{1}$	0	0
2+	4	4	4	4	4	4	4	0	0	0	0	0	0	0
3	4	$\hat{2}$	$\hat{2}$	$\hat{1}$	0	0	0	0	0	0	0	0	0	0
4	0	0	0	0	0	0	0	0	0	0	0	0	0	0

Table 3.10 Ordering of $\Omega_{3a}^{6} \cup \Omega_{3a}^{4} \cup \Omega_{3a}^{2} \cup O_{3a}$ (tabular form)

$(¬⌐)^{2}$	$¬⌐$	$⌐¬^{2}⌐$	$⌐^{2}$	$(¬^{2}⌐)^{2}$	$¬^{2}⌐^{2}$	$¬⌐¬^{2}$	O_{3a}	$⌐¬⌐^{2}$	$⌐^{2}¬^{2}$	$(⌐^{2}¬)^{2}$	$¬^{2}$	$¬⌐^{2}¬$	$⌐¬$	$(⌐¬)^{2}$
0	0	0	0	0	0	0	**0**	0	0	0	0	0	0	0
4	3	1	1	0	0	0	**1**	0	0	0	0	0	0	0
4	4	4	2	4	2	0	**2**	4	2	0	2	0	0	0
4	4	4	4	4	4	4	**2+**	0	0	0	0	0	0	0
4	4	4	4	4	4	4	**3**	0	4	4	3	3	1	0
4	4	4	4	4	4	4	**4**	4	4	4	4	4	4	4

37. An ordering O_n satisfies 35 if and only if:
    (a) the lattice for O has $2k + 1$ ($2 \leq k \leq 2^{n-1} - 1$) rows with ⊤ in row 0, ⊥ in row $2k$, and at least one + in row k,[24]
    (b) rows 1 and $2k - 1$ have n members, and
    (c) row k includes n members that are the negations of the members of row $2k$ and the d-negations of the members of row 1.

Conjunction and disjunction are generative functions in the 11-membered $O_{3b} = O_{3a} - \{ab|ac|bc\}$, but $\#\Omega_{3b} = 24$, not 28, as can be seen by removing the row designated 2+ for the middle member of O_{3a} from Tables 3.9 and 3.10.[25] In the modified Table 3.9, $¬⌐^{2}¬^{2}O_{3b} = ¬⌐^{2}O_{3b}$ and $⌐¬^{2}⌐^{2}O_{3b} = ⌐¬^{2}O_{3b}$, yielding the 12 odd-numbered negation mappings of Ω_{3b}. In the modified Table 3.10, $(¬^{2}⌐)^{2}O_{3b} = ⌐¬⌐^{2}O_{3b}$ and

[24] For the requirements on being a + (middle) member, see n. 12.

[25] However, the definitions of conjunction and disjunction for the bit string model O_{3b}^{b} of O_{3b} have to be modified from the familiar form in 13 and 14 to require the conjunction of 2 '1' bits to be '0' in certain cases, and the disjunction of 2 '0' bits to be '1' in certain others.

$(⌐^{2}\neg)^{2}O_{3b} = ⌐\neg^{2}O_{3b}$, yielding the 12 even-numbered negation mappings of Ω_{3b}, which are comparable to those for Δ_3 in Table 3.6.

Conjunction and disjunction are also generative functions in the 9-membered $O_{3c} = O_{3b} - \{ac, a|c\}$ with the diamond-pattern structure shown in Fig. 3.16, together with its length 4 bit-string model O^{b}_{3c}, in which the generator $b = 1001$ is the medium member, and the other 2 generators, $a = 1100$ and $c = 0011$, are each other's negation and d-negation.[26] The ordering of the 4 members of, $\Omega_{3c} = \{\neg O_{3c}, ⌐O_{3c}, \neg^{2}O_{3c}, ⌐^{2}O_{3c}\}$, is shown in Table 3.11.

The 8-member ordering $O_{2a} = O_{3c} - \{b\}$ in Fig. 3.17, in which the middle member of O_{3c} has been removed, has the same 4 negation mappings, but as Table 3.12 shows, the properties of Ω_{2a} are quite

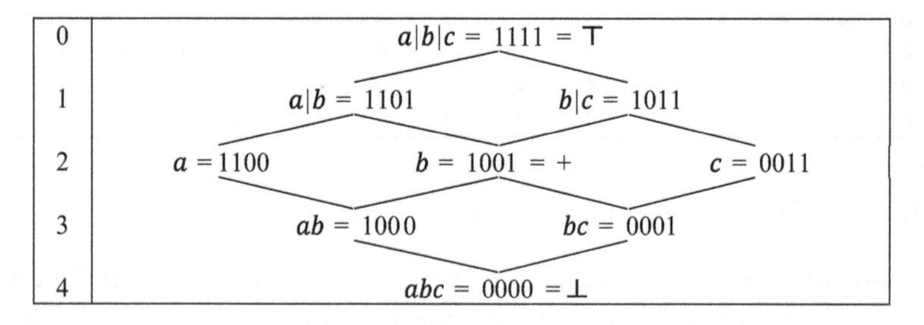

Fig. 3.16 Ordering of O_{3c} and O^{b}_{3c}

Table 3.11 Ordering of Ω_{3c} (tabular form)

Ω_{3c}	$\neg$	$⌐$	$⌐^{2} = \neg⌐$	$\neg^{2} = ⌐\neg$
0	4	4	0	0
1	4	$\hat{2}$	2	0
2	$\hat{2}$	$\hat{2}$	2	2
2+	4	0	4	0
3	$\hat{2}$	0	4	2
4	0	0	4	4

[26] In O^{b}_{3c}, conjunction and disjunction have their familiar definitions in 13 and 14.

Fig. 3.17 Ordering of O_{2a} and O^b_{2a}

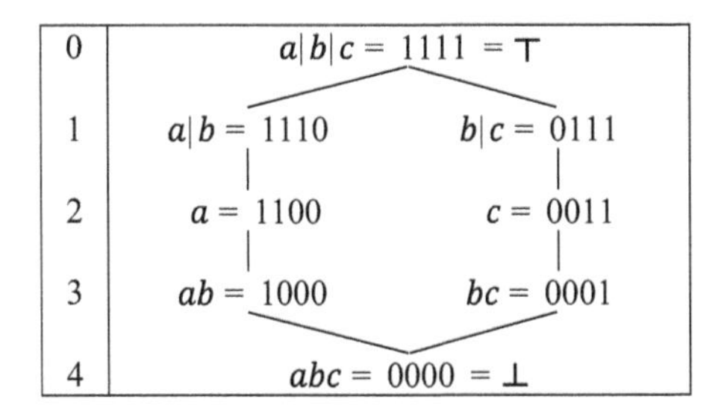

Table 3.12 Ordering of Ω_{2a} (tabular form)

Ω_{2a}	$\ulcorner$	$\neg$	$\ulcorner^2 = \ulcorner\neg$	$\neg^2 = \neg\ulcorner$
0	4	4	0	0
1	$\hat{3}$	$\hat{1}$	3	1
2	$\hat{3}$	$\hat{1}$	3	1
3	$\hat{3}$	$\hat{1}$	3	1
4	0	0	4	4

distinctive[27] In addition, the 2 remaining members in row 2 are no longer generators, and its 4-bit string model (also shown in Fig. 3.17) is crafted to fit the ordering it models without much help from the conjunction and disjunction operators.

Finally, the 7-member ordering $O_{2b} = O_{3c} - \{a, c\}$, in which the middle member of O_{3c} is retained but its other generators are removed, is shown in Fig. 3.18 together with that O^b_{2b}. Its set of negation mappings Ω_{2b} has the 10 members shown in Table 3.13; the 6 odd-number negation mappings are linearly ordered as in 38.[28]

38. Ordering of $\Omega^3_{2b} \cup \Omega^1_{2b}$: $\neg\ulcorner\neg O_{2b} \geq^o \neg O_{2b} \geq^o \ulcorner\neg^2 O_{2b} \geq^o \neg\ulcorner^2 O_{2b} \geq^o \ulcorner O_{2b}$ $\geq^o \ulcorner\neg\ulcorner O_{2b}$

[27] In particular, Ω_{2a} presents a countermodel to Koslow (1992: Theorem 17.2) which asserts that $\neg O \geq \ulcorner O$ for all O meeting certain conditions that Ω_{2a} satisfies.

[28] O_{2b} is isomorphic to an ordering that Koslow (1992: 147) analyzed as having the 6 strictly triply negated mappings like those of Δ_3 in Table 3.5 and Fig. 3.10, and of O_{3a}. In fact it has just the 4 listed in Table 3.13 for Ω^3_{2b}.

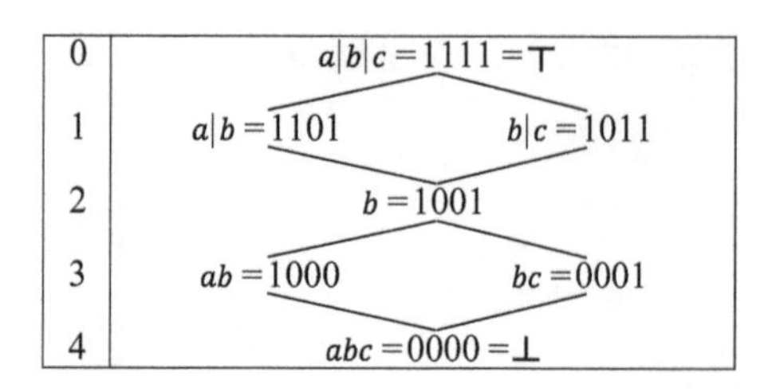

Fig. 3.18 Ordering of O_{2b} and O^{b}_{2b}

Table 3.13 Ordering of $\Omega^2_{2b} \cup O_{2b}$ and $\Omega^3_{2b} \cup \Omega^1_{2b}$ (tabular form)

$\neg ⌐$	$⌐^2$	O_{2b}	$\neg^2$	$⌐\neg$	$\neg ⌐ \neg = ⌐^2\neg$	$\neg$	$⌐\neg^2$	$\neg ⌐^2$	$⌐$	$⌐\neg⌐ = \neg^2⌐$
0	0	**0**	0	0	4	4	4	4	4	4
4	1	**1**	0	0	4	4	4	4	$\hat{1}$	0
4	4	**2+**	0	0	4	4	4	0	0	0
4	4	**3**	3	0	4	$\hat{3}$	0	0	0	0
4	4	**4**	4	4	0	0	0	0	0	0

5 Reference to Negation and d-Negation of Entities

Although as noted in the previous section, negation and d-negation of an entity in Δ_n are not referred to directly in English, they can be referred to indirectly by reference to an antecedent. For example, let Δ_3 be as described in the previous section except that Taffy is a dog, not a cat, and let 39 be interpreted in that context. Then 39.a refers to $\neg a = bc$, which 'Towzer and Taffy' also refers to, and 39.b refers to $⌐a = b|c$, which 'Towzer or Taffy' also refers to.

39. Tripod is defending herself against
    (a) the other dogs.
    (b) another dog.

By not naming the antecedent, we can obtain compact statements of a range of possible situations as in 40; compare the more prolix counterparts 41 and 42–44.

40. One of the dogs is defending herself against

    (a) the others.
    (b) one of the others.

41. Tripod, Towzer or Taffy is defending herself against

    (a) Towzer and Taffy, Tripod and Taffy, or Towzer and Taffy, respectively.
    (b) Towzer or Taffy, Tripod or Taffy, or Towzer or Taffy, respectively.

42. Tripod is defending herself against

    (a) Towzer and Taffy,
    (b) Towzer or Taffy, or

43. Towzer is defending herself against

    (a) Tripod and Taffy,
    (b) Tripod or Taffy, or

44. Taffy is defending herself against

    (a) Tripod and Towzer.
    (b) Tripod or Towzer.

A full study of the properties of expressions like those in 39.a-b and 40.a-b that can be understood as referring to the negation or the d-negation of their antecedents would be too large to report on here. In the remainder of this paper, I focus on a reanalysis of some aspects of the logic of reciprocity presented in Langendoen (1978).

6 The Role of Negation and d-Negation in a Reformulation of the Logic of Reciprocity

Among the definitions of reciprocity proposed in Langendoen (1978), two have come to be viewed as more or less standard in the literature, weak and strong reciprocity defined as in 45 and 46, where *A* is a plural antecedent of an expressed or implicit reciprocal anaphor π such as 'each other' in English, and R is a two-place relation.

45. Weak reciprocity: $R(A, \pi)$ is understood as $(\forall x \in A)(\exists y, z \in A)(x \neq y \wedge x \neq z \wedge R(x, y) \wedge R(z, x)$.
46. Strong reciprocity: $R(A, \pi)$ is understood as $(\forall x, y \in A)(x \neq y \rightarrow R(x, y)$.

For example, let A_3 denote the members of the set of candidates $\{a, b, c\}$ (named Tripod, Towzer and Taffy respectively) for a certain elective office, and consider the interpretation of sentence 47 in the circumstances described in 48. The reader can verify that 47 is understood in accordance with 45 given 48.a, and in accordance with 46 given 48.b-c.[29]

47. Tripod, Towzer and Taffy voted for each other.
48. Circumstances for 47.
    (a) There is one open position to be filled. To be counted, each ballot cast must indicate which one of the three candidates listed thereon is chosen.
    (b) There are two open positions to be filled. To be counted, each ballot cast must indicate which two of the three candidates listed thereon are chosen.
    (c) None of the candidates voted for herself.

The definitions of weak and strong reciprocity in 45 and 46 can be translated into those in 49 and 50, which consider the antecedent to refer to the infimum $\perp_n$ of a Dedekind ordering Δ_n, $(n > 1)$ and use negation (for strong reciprocity) and d-negation (for weak reciprocity) operators on the term in one of the argument positions in place of the inequality statements on variables. These reformulations have the advantage of being easily integrated with the corresponding definition of reflexivity in

[29] Without circumstance 48.c, 47 could also be understood in accordance with 45, with each candidate voting for another candidate for one position, and by implication for herself for the other.

51, in which φ is a reflexive anaphor (e.g. 'themselves') and $n > 0$.[30] They also provide an explanation for why reciprocal constructions, like reflexive ones, are grammatically place-reducing, because the anaphor can be incorporated into the predicate resulting in an $k - 1$-place predicate over the antecedent and the remaining arguments (if any), as in 52 and 53 for $k = 2$.[31]

49. Weak reciprocity (revised): $R(\bot_n, \pi)$ is understood as $R(\delta, \ulcorner\delta)$ for all $\delta \in \Delta_n$ such that $\{ \delta, \ulcorner \delta\}$ is an antichain.
50. Strong reciprocity (revised): $R(\bot_n, \pi)$ is understood as $R(\delta, \neg\delta)$ for all $\delta \in \Delta_n$ such that $\{ \delta, \neg \delta\}$ is an antichain.
51. Reflexivity: $R(\bot_n, \varphi)$ is understood as $R(\bot_n, \bot_n)$.
52. $R\varphi(\bot_n)$ is reflexive if and only if $R(\bot_n, \varphi) = R(\bot_n, \bot_n)$.
53. $R\pi(\bot_n)$ is reciprocal if and only if for all $\delta \in \Delta_n$ such that $\{\delta, \ulcorner \delta\}$ is an antichain, $R(\bot_n, \pi) = R(\delta, \ulcorner\delta)$, or for all $\delta \in \Delta_n$ such that $\{\delta, \neg \delta\}$ is an antichain, $R(\bot_n, \pi) = R(\delta, \neg\delta)$.

References

Church, Randolph. 1940. Numerical analysis of certain free distributive structures. *Duke Mathematics Journal* 6: 732–734.

Davey, Brian A., and Hilary A. Priestley. 2002. *Introduction to lattices and order*. 2nd ed. Cambridge: Cambridge University Press.

[30] On unifying the notions of reflexivity and reciprocity, see Langendoen and Magloire (2003: 257). For $n > 1$ in 51, the reflexive relation holds for all parts of $\bot_n$ if R is dissective in the sense of Goodman (1952), or is understood distributively. For example 'weigh' is non-dissective, so that iv does not entail v, unless iv is understood distributively, i.e. as equivalent to vi.

(iv) Tripod and Towzer weighed themselves.
(v) Tripod weighed herself.
(vi) Tripod weighed herself and Towzer weighed herself.

[31] The reformulation of strong reciprocity also subsumes the strong reciprocity for subsets condition (Langendoen 1978: 190), which is needed to account for the interpretation of certain $R\pi$ predicates like 'be similar', first noted by Leonard and Goodman (1940).

Dedekind, Richard. 1897. Über Zerlegungen von Zahlen durch ihre größten gemeinsamen Thaler. In *Fest-Schrift der herzoglichen Technischen Hochschule Carolo-Wilhelmina dargeboten den naturwissenschaftlichen Teilnehmern an der 69. Versammlung deutscher Naturforscher und Ärzte*, ed. Heinrich Beckurts, 1–40. Braunschweig: Vieweg und Sohn. Reprinted in Robert Fricke, Emmy Noerther, and Oysten Öre (eds.). 1931. *Gesammelte Mathematische Werke*, vol. II, 103–147. Braunschweig: Vieweg und Sohn.

Geach, Peter T. 1962. *Reference and generality*. Ithaca: Cornell University Press.

Goodman, Nelson. 1952. *The structure of appearance*. Cambridge, MA: Harvard University Press.

Kisielewicz, Andrzej. 1988. A solution of Dedekind's problem on the number of isotone Boolean functions. *Journal für die reine und angewandte Mathematik* 386: 139–144.

Koslow, Arnold. 1992. *A structuralist theory of logic*. Cambridge: Cambridge University Press.

Langendoen, D. Terence. 1978. The logic of reciprocity. *Linguistic Inquiry* 9: 177–197.

Langendoen, D. Terence, and Joël Magloire. 2003. The logic of reflexivity and reciprocity. In *Anaphora: A reference guide*, ed. Andrew Barss, 237–263. Malden/Oxford: Blackwell Publishing.

Leonard, Henry, and Nelson Goodman. 1940. The calculus of individuals and its uses. *Journal of Symbolic Logic* 5: 45–55.

Wiedermann, Doug. 1991. A computation of the eighth Dedekind number. *Order* 8 (1): 5–6.

4

Variations on Abstract Semantic Spaces

Katrin Erk

1 Introduction

Distributional models, also called semantic spaces or embeddings, represent the meanings of words and phrases as points or regions in some "meaning space." Semantic space representations can be generated automatically from data, for example by collecting the context words that appear around a word of interest, and mapping the word to a point in space that represents those contexts. Semantic spaces have long been used in language technology, in information retrieval (Salton et al. 1975) and computational lexical semantics (Landauer and Dumais 1997). They

I am deeply grateful to Aurélie Herbelot and Gabriella Chronis, who read an earlier version of the paper and gave me much appreciated feedback. Many thanks also to Jacob Andreas, Jason Baldridge, John Beavers, Ann Copestake, Jonathan Davis, Guy Emerson, Hans Kamp, Jessy Li, Louise McNally, and Steve Wechsler for great discussions and helpful comments. All remaining errors are of course my own.

K. Erk (✉)
Linguistics Department, University of Texas at Austin, Austin, TX, USA
e-mail: katrin.erk@utexas.edu

R. M. Nefdt et al. (eds.), *The Philosophy and Science of Language*,
https://doi.org/10.1007/978-3-030-55438-5_4

have become vastly more important in deep learning, and are now core components of most language technology applications. But semantic spaces are also interesting for theoretical linguistics. They are interesting as a source of data, an automatically generated compressed representation of the contexts in which a word has been observed; the recent overview of Boleda (2020) discusses this use of semantic spaces in depth. But they are also interesting as a source of ideas for formal semantics, they show us how else we could describe meaning.

This text is about the latter perspective on semantic spaces. I am going to distinguish semantic spaces *in practice*, or *practical*, *implemented* semantic spaces, from *abstract semantic spaces*. By abstract semantic spaces, I mean approaches that build on the idea of a space – meaning space, or conceptual space – as a way of representing meaning. There is no uniform definition of an abstract semantic space, rather a collection of approaches that differ in which properties of practical semantic spaces they adopt, and that differ in the phenomena they most care about. Some approaches consider lexical phenomena, such as Asher et al. (2016) and McNally and Boleda (2017); others focus on compositionality and the construction of semantic space characterizations of larger phrases from their components, for example Baroni and Zamparelli (2010), Grefenstette and Sadrzadeh (2011), and Socher et al. (2011), and in particular Baroni et al. (2014). In this text I will concentrate on lexical semantics: how abstract semantic spaces can characterize word meanings, what aspects of word meanings they nudge us to look at, and how they integrate with sentence meaning beyond word meanings.

Semantic spaces have several characteristics that make them interesting for lexical semantics. First, they are rich, fine-grained representations of lexical meaning, which can possibly let us make progress with understanding and describing the intricate shifts in meaning brought about by context. Second, they are what McNally (2017) calls "ersatz conceptual representations", they can be taken to represent conceptual knowledge. Word representations in semantic spaces have been shown many times to ap- proximate human behavioral data well across a number of experiments (Lenci 2008). Third, operations on semantic spaces are typically graded: Similarity between points in space comes in degrees, and classifiers that operate on semantic spaces typically produce weighted or probabilistic outputs. So the formalism does not force us to always make hard category distinctions, which can be difficult to make (Gehrke and McNally 2019; Erk 2010). Fourth, practical semantic spaces do not distinguish between linguistic meaning and world knowledge (Gehrke and McNally 2019), as they are learned from collections of utterances that reflect all the knowledge of their writers. They do not distinguish between semantics and pragmatics either, as Potts (2019) points out[1]: "[learned vectors] will reflect many aspects of language use: biases in word frequency, preferences for certain readings, pragmatic refinements of lexical items, and so forth." This can be a problem in practical semantic spaces, when language technology systems pick up on prejudices reflected in the training data (Bolukbasi et al. 2016), but at the theoretical level it can be an opportunity to study language in terms of its use conditions (Potts 2019).

Not all of the characteristics above apply to all semantic spaces, and in fact they reveal two different perspectives on what abstract semantic spaces "truly" might be: We can view them either as describing fine-grained lexical and conceptual knowledge about the world, or – this second perspective is articulated most clearly in Potts (2019) – a record of observed usage, pragmatics and all. These two perspectives are not incompatible, they just constitute two different facets of meaning that semantic spaces nudge us to think about.

There are already several formalisms that I would call abstract semantic spaces, that take ideas from semantic spaces into formal semantics. Some have design choices that differ radically from today's practical semantic spaces, some define mechanisms for integrating semantic spaces with other formalisms for describing meaning. In this text,

I want to take stock of where abstract semantic spaces are now, and speculate about where else we could take them. I start with a brief introduction to the prevalent flavors of practical semantic spaces in Sect. 2 – so we know what we are abstracting from. Section 3 discusses some examples of abstract semantic spaces, with an emphasis on what can be different from practical semantic spaces. Section 4 looks at formalisms

[1] He writes about deep learning, but almost everything he says applies to pre-deep-learning distributional models as well.

that use semantic spaces for lexical phenomena, and at the different ways in which these word meaning characterizations can be integrated into sentence meanings. Section 5 looks at the inferences a semantic space makes available, and how these inferences manifest outside of the semantic space. Finally Sect. 6 is about the learnability of semantic spaces. It has often be listed as a strength of semantic spaces that they can be learned from data, but practical semantic spaces are usually learned in a way that is not particularly cognitively plausible.

2 Some Semantic Spaces in Practice

In this section I give a very short sketch of three prominent types of semantic spaces. I do not have room here for a longer introduction, but see Turney and Pantel (2010); Boleda (2020) for an overview of count-based spaces, Jurafsky and Martin (2019) for more information on prediction-based embeddings, and the excellent blog post of Alammar (2018) for contextualized word embeddings.

Count-Based Spaces The fundamental idea of distributional models is that words that have similar meanings tend to occur in similar contexts, so by computing an aggregate representation of a word's contexts, we can approximate its meaning. Table 4.1 shows an example featuring the target word *violin*, with a tiny corpus of three sentences. I have (rather arbitrarily) defined "context" to be three words to the left and three words to the right of the target, not crossing sentence boundaries, italicized in the table. Computing an aggregate representation of the contexts can then be as simple as counting how often each word appears in the context region around the target, as illustrated in the table of counts below the "corpus." (Context words that did not appear in the context of the target, like *bow* in our example, are also listed but here get a count of 0.) To move from this table of counts to a semantic space, we interpret the counts as coordinates in a space whose dimensions are the context words from the table, so there is a *sonata* dimension on which *violin* as a value of 1, a *piano* dimension with a value of 2, and so on. This, then, allows us to compute the similarity of two words, say violin and cello, in spatial terms. The

Table 4.1 Illustration of a count-based semantic space: a small corpus (top left), the resulting table of context word counts (bottom), and a sketch of vectors and vector similarity in a semantic space

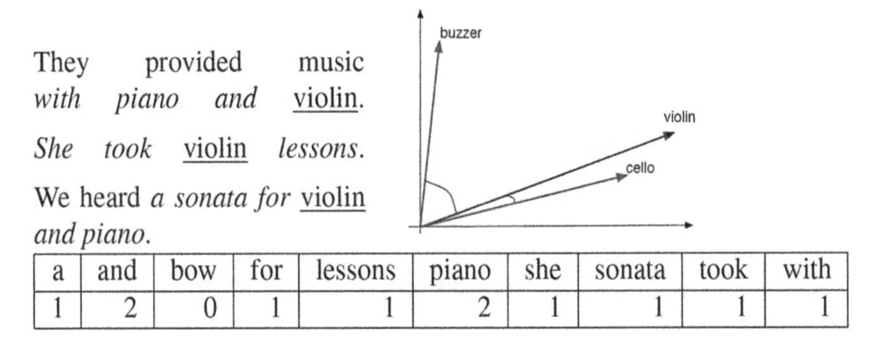

They provided music *with piano and* violin.

She took violin *lessons.*

We heard *a sonata for* violin *and piano.*

a	and	bow	for	lessons	piano	she	sonata	took	with
1	2	0	1	1	2	1	1	1	1

most widely used measure of similarity in semantic spaces is the cosine of the angle of the vectors, illustrated to the right of Table 4.1. The vectors constitute a summary of the whole table of counts, so the sketch can be read as saying that by their cosine values, *violin* and *cello* appear in more similar contexts, and are more semantically similar, than *violin* and *buzzer*.

The example in Table 4.1 uses textual contexts, but embeddings can also be computed from other kinds of contexts, in particular images that appear close to a word of interest. Contexts can also be multimodal, mixing text and images. Bruni et al. (2012) is an example of both visual and multimodal contexts.

Prediction-Based Word Embeddings Word vectors can also be obtained by a quite different method, as a side product of classification. Word2vec skip-gram with negative sampling (Mikolov et al. 2013) is a widely used type of prediction-based word embeddings, where the classification task is to guess, given a target word and a potential context word, whether or not they actually did co-occur in the text. A neural model for this task can be described as a function that is parametrized by a large number of weights. The function is by observing training data (word pairs labeled by whether they did or did not co-occur) and repeatedly tweaking the weights to make the function's output better match the training data. For skip-gram, the function has weights for each target word, and weights for each specific context word. And the target word weights, tweaked to best predict co-occurrence over a large number of training data-points, can be viewed as the vector of the target. Similarity between vectors can as before be measured as the cosine of their angle, and under particular conditions the resulting space is even formally equivalent to a count-based space (Levy and Goldberg 2014). It is by now a common move to build classifiers not for the sake of the classification results but in order to obtain weight vectors that can be used as word embeddings – which raises the question what classification tasks will yield the best word embeddings as a side product (Wang et al. 2019).

Contextualized Word Embeddings The past year has seen a stunning improvement in many natural language processing task through contextualized language models, models that compute an embedding for each word token in context (rather than for each word across all its occurrences, as older approaches did).[2] As in the prediction-based models in the previous paragraph, the embeddings are a side product of classification, here typically the prediction of the next word in a sequence, or the prediction of a masked word within a sentence. The currently most widely used such model is called BERT (Devlin et al. 2019). As part of its classification task, a BERT model learns to update the embedding of a word based on the embeddings of the other words in the same passage. Not all surrounding words are equally important for this update, so the model learns to assign different weights to context words, called *attention* weights, that govern how much influence they have on the update. A BERT model does not just one such update but several layers of updates, and learns not just one set of attention weights but several such sets for each layer, so it has a massive number of parameters.

[2] This is not the first attempt to compute contextualized embeddings. Previous approaches computed them from count-based word embeddings, for example Erk and Padó (2008) and Thater et al. (2010).

More Flexible Word Vectors, More Flexible Uses For a long time, the main use of word vectors was to compute similarity ratings. This was sometimes leveled as a criticism against semantic spaces, as similarity does not distinguish between different semantic relations between words (Murphy 2002). But similarity is no longer the only, or even the main operation on semantic spaces. Similarity ratings are still widely used on both count-based and prediction-based vectors, as mentioned above, and there has been some success on using them on contextualized word embeddings (Garí Soler et al. 2019). But today the prevalent use of word vectors is as part of some classifier, and classifiers can be used to read much more, and more nuanced, information from word vectors than similarity ratings can. There are classifiers that test for specific semantic relations like hypernymy (Baroni et al. 2012; Roller and Erk 2016) and classifiers that predict conceptual properties (Herbelot and Vecchi 2015). Also, the vectors themselves, if they are prediction-based, can be computed to contain particular information, for example on lexical entailment (Kruszewski et al. 2015). Contextualized word embeddings have not seen much use in lexical semantics so far, so it remains to be seen what can be done with them. But one potentially interesting direction is that learning of attention weights (for the influence of different context words, as discussed above) can be guided by linguistic knowledge (Strubell et al. 2018).

3 Some Abstract Semantic Spaces

What I call "abstract semantic spaces" are approaches that use the idea of meaning spaces or conceptual spaces in formalisms for representing meaning, though they can be quite different from standard practical semantic spaces in some respects. In this section I showcase some examples.

Gärdenfors' Conceptual Spaces Gärdenfors (2000) describes properties and concepts as shapes in a *conceptual space* whose dimensions are quality dimensions such as the hue, saturation and brightness of colors. One core assumption that Gärdenfors makes is that properties are convex regions in conceptual space. The convexity requirement solves the "riddle of induction", due to Goodman (1955), of why humans do not form properties such as "grue", the property of being green before the year 2000 and blue afterwards: "grue" would not be a convex area in conceptual space, so hu- mans never hypothesize it as a property. Convexity also yields a simple mechanism for property learning: If a cognizer has observed a number of exemplars, each of them be- longing to one property out of a set of competing properties (for example, blue, red and yellow), then a convex region for each property can be derived by Voronoi tesselation of the space, deriving prototypes (centroids) of each property and then classifying each point in space as belonging to the nearest prototype.

The quality dimensions are grouped into *domains*, for example the dimensions that together describe color perception would form a domain, and dimensions related to sound would form another. A domain is "a set of integral dimensions that are separable from other dimensions" (p. 26): an object cannot have a value on one color dimension without having a value on all color dimensions, but it can have color values without also having sound coordinates. The distinction between properties and concepts is then that a property lives in a single domain, while a concept is often associated with regions across a number of domains. Gärdenfors (2014) expands the inventory of dimensions to include representations of actions and evens: Actions are characterized through patterns of forces, and events are characterized through pairs of vectors, a force vector for the action, and a result vector that describes the change brought about in a patient.

Gärdenfors' conceptual space is maybe the clearest example of the first perspective of semantic spaces from the introduction: as a spatial characterization of knowledge about language and the world. Some of Gärdenfors' ideas have found their way into practical semantic spaces, most directly in McMahan and Stone (2015), who learn a probabilistic representation of color terms as convex regions in color space (this removes one of the main drawbacks I see in conceptual spaces, namely that all category boundaries are strict), but also the idea of having words be regions rather than points in space (Vilnis and McCallum 2015; Vilnis et al. 2018). But there are certainly other interesting ideas that could be adopted into other approaches, in particular the spatial characterizations of actions and events in Gärdenfors (2014).

Herbelot and Copestake: A Semantic Space That Is a Model Structure Herbelot and Copestake (2013) point out that a model structure from model theory can be considered as a semantic space that has been learned from an *ideal distribution*, a corpus containing all true statements about all entities in a world. In the resulting space, the "vector" associated with a predicate symbol would consist of all the individuals (or tuples of individuals) of which the predicate symbol is true –its extension. Conversely, the "vector" for an individual (or a tuple of individuals) would consist of all the true statements that can be made about it. While this space is a model structure, it can still be endowed with the same kinds of operations as other semantic spaces, in particular a similarity metric between predicate symbols and a similarity metric between individuals.

An *actual distribution* is then a partial world, the portion of an ideal distribution that an individual cognizer has observed. A semantic space computed from such an actual distribution can be viewed as an exemplar model, that is, a conceptual representation based on remembered exemplars (Nosofsky 1986; Murphy 2002). So learning, as learning from exemplars, is quite similar in a Herbelot and Copestake space and in Gärdenfors' conceptual space. And in fact Gärdenfors, too, locates individuals in space (Gärdenfors 2000, p. 135): "all points [. . .] can be taken to represent *possible objects*," and a possible world can be defined as a subset of the points in conceptual space.

The most important message of Herbelot and Copestake (2013), in my opinion, is that semantic spaces can contain information about individual exemplars or referents after all, and support reasoning about them in all ways that exemplar models would.

Many people have suggested that semantic spaces are suitable only for information at the kind level, not the individual level (Lenci 2008, p. 22), (McNally and Boleda 2017; Boleda 2020). But a space in the style of Herbelot and Copestake could certainly be constructed in practice, and would support reasoning about referents.

4 Word Meaning in Space, and Its Integration into Sentence Meaning

In this section, we look at ways in which semantic spaces have been used, or can be used, to address lexical phenomena, in particular fine-grained effects of meaning in context. The second theme of this section is how fine-grained representations of lexical meaning, such as semantic space representations, can be connected with the meaning representation for the entire sentence. To use terminology introduced by Pelletier (2017), the approaches we present in this section all have two tiers, one that is a semantic space (or that can be implemented through a semantic space), and one which is logical form, connected through some nexus. Many important questions for such theories are still in flux: How should the tiers be combined, and what should the nexus be? What kinds of inferences does the semantic space tier make available, and how does it affect the other tier?

Asher's Type Composition Logic The theory of Asher (2011) associates each term of the logic with a fine-grained type; Asher et al. (2016) later build on this by implementing the types through a semantic space. Asher's types are conceptual in nature, and have two main functions. The first is to describe meaning shifts in component expressions when they are combined into a larger whole. The second function is to check for semantic anomaly: at the external content level, a proposition can be true or false – this is what distinguishes (1-a) and (1-b) below; but it is the internal content level that detects semantic anomaly: (1-b) may be false but at least imaginable, while (1-c) is hard to even wrap ones head around.[3]

(1)

a. Tigers are animals.
b. Tigers are robots.
c. #Tigers are financial institutions. Example (1.2.) in Asher (2011, p. 5)

[3]Vecchi et al. (2011) study semantic anomaly in semantic spaces from a computational perspective.

In combinations of modifiers and nouns, the meaning of the head noun can in some cases subtly shift the meaning of the modifier, as in the case of *red*, shown in (2): A *red apple* has red skin only and is usually white on the inside, while a *red shirt* needs to be mostly red. A *red pen* writes red but can be blue on the outside. Asher addresses this by giving *red* a polymorphic type, such that the type of an occurrence of *red* in a particular context can be made to depend on the type of its argument. Kamp (1975) already noted this problem of meaning shifts in modifiers, and Kamp and Partee (1995, p. 161) suggested that it is a general principle:

> (HPP) Head primacy principle: In a modifier-head structure, the head is interpreted relative to the context of the whole constituent, and the modifier is interpreted relative to the local context created from the former context by the interpretation of the head.

But they note that there are exceptions like the *stone lion* in (3), where head primacy seems to be overridden, as it is doubtful whether a stone lion qualifies as an actual lion. Asher discusses the case of the stone lion too (p. 301), and proposes a mechanism for material modifiers that acts differently on types that already allow for the material in question (the material of jars can be stone, and stone jars are still jars) than on types that do not (the material of lions is not usually stone).

(2)
  a. red apple
  b. red shirt
  c. red pen
     cf. Asher's example (2.36), p. 51

(3) stone lion (vs. actual lion; Asher's example (11.2a), p. 302)

The *red shirt* and the *stone lion* are examples of the general problem of *concept combination*, which is discussed from a linguistic point of view in Kamp and Partee (1995) but also has an extensive literature in psychology, see for example Barsalou (2017) and Hampton (2017). The types that Asher defines for the modifiers take an initial step toward addressing the problem, but barely scratch the surface: How exactly can the internal meaning of pen determine the meaning shift of *red*? How can the internal meaning of *stone* or *lion* specify that stone lions have lion shapes, but do not usually have lungs and are also not dangerous? And how about less lexicalized cases like the *striped apple* (an example discussed in Kamp and Partee (1995) and originally introduced by Osherson and Smith (1981)), where the meaning of striped needs to be shifted to something like *striped in the not very striped way that apples can be, definitely not striped like a shirt*? Would this, too, be hardcoded in the lexicon? And if it is not, how would it be inferred? Concept combination has many other odd phenomena in need of explanation. For example, Hampton (1988) discovered cases of overextension in conjunctive concepts, for example, a *blackboard* would be judged to be a member of the category *school furniture* but not of the category *furniture*. And Wu and Barsalou (2009) found that mental simulation also seems to play a role: For a concept like *rolled-up lawn*, participants generated properties like *roots* and *dirt*, which they did not for either *lawn* or, say, *rolled-up snake*.

At least some of these problems could in principle be addressed by semantic spaces that have fine-grained feature representations of concepts. For example, Hampton (1987) has a proposal for concept combination at the feature level where features that are more important to a component concept are retained in the compound concept, which might solve the *stone lion*. But existing practical semantic spaces may not be up to the task yet. Asher et al. (2016) performed experiments with practical semantic spaces where they derived context-specific representations of modifiers and nouns and tested whether the resulting vectors yielded the right entailments or if nearest neighbors of in-context vectors at least could be interpreted as being semantically coherent with the contextually appropriate sense. Their results were mixed, though the more lenient semantic coherence evaluation showed more promising results. Given how quickly deep learning systems improve in performance right now, it is possible that today's semantic spaces could be up to the task, in particular spaces learned specifically for fine-grained entailment. But at the least, *abstract* semantic spaces can provide a framework where we can think about what kinds of spaces would be needed to make progress on concept combination.

Asher's types are proof-theoretic, their meaning is not given through an extension or intension, but through rules that dictate their behavior. A type is associated with inference rules that govern the combinability and combination of types, and with introduction rules that say under what condition an object can be categorized as realizing this type or concept. A proof theoretic perspective is also interesting for semantic spaces, in particular because some proof-theoretic approaches view meaning as an algorithm for testing justifications, or for constructing mental models (Moschovakis 1994, 2006; Muskens 2004; van Lambalgen and Hamm 2003). This can be a useful perspective to take on inferences that semantic spaces afford – we will get back to this below.

Above I have introduced the terminology of two-tier formalisms and a nexus from Pelletier (2017). Pelletier uses this terminology specifically for theories that combine an external (mind independent) and an internal (mind dependent) view of meaning. The logical tier in the type composition logic of Asher (2011) is definitely mind in- dependent; Asher argues for the importance of external, mind independent aspects of meaning, famously defended by Putnam (1975). Asher characterizes the types as mind dependent, but the information he describes them as containing is mainly taxonomic, the kind of information one would expect all cognizers to share, and Asher speculates that the types "track" external meaning (p. 37) unless something goes wrong with the cognizer.

McNally and Colleagues: Kinds as a Nexus McNally and colleagues (McNally and Boleda 2017; McNally 2017; Gehrke and McNally 2019) also use semantic spaces for lexical phenomena, in particular relational adjectives and idioms. McNally (2017) motivates the use of semantic spaces in formal semantics as follows:

> First, the use of these representations allows for the integration into formally-oriented semantic analysis of techniques for handling the problems of polysemy in modification and other phenomena involving the lexicon that are poorly handled by traditional formal semantic tools. This integration can improve the empirical coverage of existing formal semantic theories and yield models that are better suited to natural language processing.

> Second, distributional models arguably come closer to capturing the intuition that common nouns and adjectives name concepts, and thus establishes a point of connection to conceptual approaches to meaning.

McNally and colleagues put an emphasis on the ability of meaning vectors to contain a superposition of senses: For polysemous words, practical semantic spaces learn a representation that is a mixture of the contexts observed with the different senses, which can be viewed as a mixture of the senses. Composition of the superposed word vector with the vector of a context word, for example using simple vector addition, yields a vector that is (more or less) disambiguated to a context-specific sense (Erk and Padó 2008; Mitchell and Lapata 2010).

As the nexus between tiers, they use kinds. They build on the idea of Zamparelli (2000) that nouns denote kinds, but characterize kinds as vectors. An entity can *realize* a kind. If an entity denoted by the discourse referent u is an apple, this would be described, in the logic, as $\exists u\,(\mathbf{Realize}(u, \overrightarrow{\text{apple}})$

Here, $\overrightarrow{\text{apple}}$ is a vector, and it is the kind *apple*, so the vectors from the semantic space appear directly in the logical form.

Emerson: A Probabilistic Nexus The basic assumption of Emerson (2018) is that speakers can learn how other speakers usually label objects, for example, how likely they would be to categorize a particular object (whose conceptual representation is given a point in semantic space) as a *zucchini*, *gourd* or *squash*. This probabilistic categorization function then becomes the nexus that links a semantic space (which contains conceptual representations of objects in the world) to logical form (with the categorization labels being predicate symbols). In more detail, he assumes a logical form in the shape of a semantic dependency graph (Copestake 2009), where nodes stand for discourse referents and are labeled with predicate symbols, and edges indicate se- mantic relations. Every node is paired with a point in semantic space, where the points in space are jointly optimized for all nodes in the graph: Each point in space should best match its node, that is, it should be likely to be categorized with the predicate symbol labeling the node. And points in space for neighboring graph nodes are connected by learned selectional constraints. This approach is reminiscent of BERT (the contextualized embedding model discussed in Sect. 2), where the context-specific embedding for each word is computed in based on its neighbors, except that Emerson's model computes embeddings relative only to semantic relation neighbors, while BERT considers all words in the passage as neighbors.

5 Inferences from Semantic Spaces

The internal, mind dependent meaning of a word, say *cat* in *There is a cat in the yard*, should allow the cognizer to make inferences, for example that there is an animal in the yard, or that it might not be safe to let the canary out. But what does it mean to make inferences from an internal meaning, what kinds of inferences can be made, and what is the mechanism? This question is particularly urgent for approaches like those in Sect. 4, which combine abstract semantic space characterizations of word meaning with some other representation of sentence meaning: What is it that can be inferred from the sentence meaning representation when we have lexical representations in a semantic space, can we infer things that we cannot infer otherwise? This question is, however, not specific to abstract semantic spaces, it arises for any framework that combines multiple different frameworks for describing sentence meaning.

Strict Inferences, Probabilistic Inferences, and What They Mean For Asher (2011), the inferences that can be made from internal content (types) are analytical entailments, supported by the subsumption hierarchy of types. For example, CAT is a subtype of ANIMAL. So if we know that that Kim is of type CAT, we also know that Kim is of type ANIMAL. From this it can be inferred, projected from internal content to logical form, that *animal* (*Kim*) is true. This direct entailment of external content from the internal types is possible because, as mentioned above, Asher assumes that the internal types "track" the external, denotation meaning. But this assumption is also limiting, as it restricts inferences to propositions that are definitely true (as opposed to only likely) and that all competent speakers can be expected to share.

In a probabilistic framework, we would like to be able to say that the cognizer can draw inferences not only about what is definitely true, but also about what is likely to be the case. For example, Emerson (2018), as described in Sect. 4, learns a categorization probability: Given a point in semantic space that characterizes an object, how likely is an arbitrary speaker to label this object as, say, *gourd*? Then once an entity is associated with a point in semantic space, additional labels (predicate symbols) for the entity can be inferred from the point in space; How else would a competent speaker be likely to label this entity? It does not become entirely clear, however, what the status of such inferences would be. If the cognizer infers that a label of *squash* is also likely, this cannot mean that the object is in the (external, mind-independent) extension of *squash* – the inference is, after all, a labeling estimate made by an individual cognizer based on that cognizer's experience, and it is an estimate of what other speakers would call the object, not of what mind independent categories the object falls in.

In Erk (2016), I tried to explain learning from textual context: If I have never seen an alligator, but I have read the word in text frequently enough, how do I form an approximation of what an alligator is? I formalized the mental state of the cognizer as an update system (Veltman 1996). An update system describes the cognizer's state as a set of worlds that, from the information that the cognizer has observed, could possibly be the actual world. I further assumed a probability distribution over the worlds in the cognizer's state – the cognizer considers some worlds to be more likely than others to be the actual world. I associated kinds (or concepts) with semantic space representations, and assumed that the cognizer observes that kinds that have similar properties tend to appear in similar textual distributional contexts. Then if I have never seen an alligator, but I know what properties crocodiles have, and I have observed that alligators tend to appear in similar textual contexts as crocodiles, I will consider those worlds more likely in which alligators share many properties with crocodiles. This approach is somewhat clunky – it assumes that is possible to get a probability distribution over worlds, which is questionable (Cooper et al. 2015), and it can only make use of properties that hold of *all* exemplars of a kind –, but it points the way to a possible interpretation of probabilistic inferences from semantic spaces: We can view them as part of *mental* models, imagined situations. Here, then, is the connection to van Lambalgen and Hamm (2003) that I promised in the previous section. Van Lambalgen and Hamm take a proof-theoretic perspective on meaning, specifically meaning as an algorithm that can construct mental models. They view meaning as an algorithm that constructs a *minimal* model of the utterance at each point in time. Maybe it is possible to view semantic space characterizations of meaning as algorithms that probabilistically construct mental models of an utterance, but not necessarily minimal ones. A mental model would be more likely to be constructed if it is based on more likely inferences from semantic space.

Reasoning About Affect Practical semantic spaces are being used for many purposes, with one prominent purpose being *sentiment analysis*, the analysis of affect. Affect can be voiced through expressives like *damn* and *bastard*, which Potts (2007) studied, where the affect is not part of the at-issue content, but affect can also be at issue. Fillmore (1985) points out that *thrifty* and *stingy* both designate people who do not spend much money, but they differ in affect, or as Fillmore puts it, in their frame. One test for non-at-issue content is behavior under denial: In (4), the denial cannot be used to negate the sentiment about the car. But in the case of *thrifty* versus *stingy*, both the amount of money spent and the sentiment can be denied without any problem, as shown in (5). (Fillmore describes these two cases as within-frame negation in (5-a) and cross-frame negation in (5-b))

(4)

a. That damn car is broken.
b. #No, that is not so. (to negate the affect about the car)

(5)

a. Mary isn't stingy, she is generous.
b. Mary isn't stingy, she is thrifty.
adapted from Fillmore (1985) examples 16–19

The pair *thrifty/stingy* is not a one-off, there are more adjective pairs that state the same or almost the same property but with different affect, and where either the property or the affect can be denied. (6) shows a few, with translations in German where I have a stronger speaker intuition.

(6)
a. thrifty/stingy
b. generous/profligate (German großzügig/verschwenderisch)
c. taciturn/secretive (German schweigsam/verschlossen)
d. communicative/garrulous (German gesprächig/geschwätzig)

Practical semantic spaces strongly pick up on this difference in affect. Table 4.2 shows nearest neighbors (via cosine) for the words *thrifty*, *stingy*, *thriftiness* and *stinginess*, computed using gensim (Řehůřek and Sojka 2010) from 300-dimensional GloVE vectors (Pennington et al. 2014), that were trained on Wikipedia and Gigaword.[4] The nearest neighbors for *thrifty* are useless, as they are dominated by a brand of that name, but both *stingy* and *stinginess* clearly pick up on a general unpleasantness of character in a way that *thriftiness* does not.

Humans clearly draw inferences from affect. Affect has an influence on affordances, that is, on how the cognizer can interact with objects: *good* items are useful, *bad* items are not (Ellis (1993) considers categories like *good* and *bad* as more basic than descriptive words). But what would a calculus of affect look like? If we wanted to explore this question, then semantic spaces, which strongly encode affect, would be a good starting point.

Table 4.2 Ten nearest neighbors of *thrifty, stingy, thriftiness* and *stinginess* in to a GLoVe model trained on Wikipedia and Gigaword

Thrifty	**Stingy**
Payless, avis, frugal, a-car, hertz, drugstore, parsimonious, rental, shoesource, dansk	Miserly, reticent, lackadaisical, generous, hypocritical, standoffish, stingier, flighty, timid, parsimonious
Thriftiness	**Stinginess**
Frugality, gutsiness, combativeness, straightforwardness, brashness, precocity, conservationism, obsessiveness, exemplifying, nature-based	Coldness, partiality, ardor, evasiveness, wrongheadedness, nonchalance, bossiness, ingratitude, rapacity, combativeness

[4] word2vec and GloVE are often used interchangeably. I have chosen GloVE because it was trained on Wikipedia, which often yields particularly good meaning representations (Chris Manning, p.c.).

6 Learning from Both Examples and Explanations

The question of learning is important for any formalism that describes mind dependent meaning: How would a speaker learn to form representations like this?[5] For abstract semantic spaces, it is relatively straightforward to say how they would be learned from occurrences (examples). In fact, semantic spaces have been used for exploring human concept learning; Lenci (2008, p. 16) gives an overview of this research. But there is a particular problem of learning that is quite hard for semantic spaces: how to learn from both examples and explanations.

Children observe evidence about word meanings that comes in the form of examples, and also in the form of definitions or explanations. (7) shows a fairly typical example from a children's book, *Blood in the Library* (Dahl 2012), where a superhero named the Librarian battles super villains like, in this case, the Eraser. The book contains words that the reader may not be familiar with, like *suffer* and *fate* in (7-a). A glossary at the end of the book defines the words, as shown in (7-b) and (7-c).

(7)
  a. "Watch your fingers," says the Eraser. "Or you'll suffer the same fate."
  b. fate: what will happen to someone
  c. suffer: experience something bad, hard, or painful

It is not only in books that children get explanations of word meanings. (8) lists some explanations that I observed myself giving to a five-year-old child (translated from the original German; the original German word that was explained is given in parentheses).

(8)
  a. doubt (zweifeln): to think that something is not the case
  b. huffy (eingeschnappt): upset, annoyed, disgruntled
  c. learn by heart (auswendig lernen): read or hear something often enough that you can recite it from memory. This typically involves something that can be written down, like a story or piece of music
  d. mock (verspotten): nyah-nyah-nyah (imitates mocking sound)

---

[5] See Liang and Potts (2015) for a discussion of learning logical form from data.

The examples in (7) and (8) are quite diverse. Some take the form of definitions, for example (7-b) and (8-a), others look more like characterizations of a typical scenario, in particular (8-d) but also (8-c). The explanations also differ in structure: (8-b) is simply a list of (near-)synonyms, while (8-c) is a complex event unfolding over time. In what follows, I use the term "explanations" to refer to all of these cases: definitions or rules as well as characterizations of prototypical settings, synonym collections as well as complex event descriptions.

Learning Semantic Spaces from Explanations Semantic spaces can be learned from rule-based information, not just from examples. Bernardy et al. (2018) learn semantic space representations to best match a small set of premise statements – an example from the paper is shown in (9).

(9)

a. Most animals do not fly.
b. Most birds fly.
c. Every bird is an animal.

Individuals are represented as points in space, and predicates as "dividing lines" that separate extensions from non-extensions, with all parameters learned jointly to best fit the set of premises. This is feasible because the overall number of constraints on the formation of the semantic space is small; in the example above there are only three. So spaces can be learned from rule-based information. Learning a space from both examples and explanations is much harder; we next see how it can be done.

Explanations as Constraints on Example-Based Learning The KBANN algorithm (Towell et al. 1990) is an early approach that integrates logical rules with neural net- work learning. Given a set of propositional rules, Towell and colleagues construct a neural network to mirror the structure of the rules, with each atom of the logic be- coming a node in the network. They then train the network with examples to learn the importance and reliability of the rules. The main characteristics of KBANN that I would like to point out are: (1) Example-based learning is guided by and constrained by a given set of rules that are available during example-based learning. (2) The rules are taken to only hold approximately, not universally, and (3) the integration of rules and examples is not immediate but is a slow process.

Ganchev et al. (2010) is a probabilistic approach that shares these three characteristics. During parameter learning, the model is guided toward parameters that on the one hand characterize the training data well (that is, they make the training data likely), and that on the other hand make the model's output labels mostly, approximately, obey the rules (more precisely, the expectation of the output labels only violates the rules by some small margin). Hu et al. (2016) is a neural approach that builds on the work of Ganchev and colleagues. They have two neural networks, one regularized with a method analogous to Ganchev, but restricted in its structure; the second, which is trained to mimic the output of the first, can vary widely in its structure.

At first glance, these approaches look discouraging: Their architectures are complex, rule integration is slow and laborious, and rules are never fully obeyed, the systems only learn to mostly stay in parameter regions that obey the rule constraints. But whenever we have both rules and examples for a concept, it makes sense that the integration would require some effort from the learner. And when a rule contradicts observed examples, it makes sense that the learner would not just reject the examples and follow the rule, but would learn to what extent the rule should be softened.

When explanations are used as constraints, they are mostly formulated as logical rules. Andreas et al. (2017) points to another option: Formulate them as natural language instructions, and let the model learn how to make use of them. Andreas and colleagues train learners to navigate through a map (and other tasks) with the help of natural language instructions (sequences of steps the learner should take). How the learner uses the instructions is left to learning. It is not clear at this point how general this idea will turn out to be, but it could in principle let the system learn to use explanations in more flexible ways.

Explanations as Instructions for Constructing Meanings Suppose we have examples for some concepts, but for other concepts we only have explanations. Then we would like to treat an explanation as an instruction for constructing a concept of a similar kind as the example-learned concepts. The most straightforward example from this group comes from computer vision. Farhadi et al. (2009) and Lampert et al. (2009) learn to recognize visual attributes, such as "black," "stripes," or "has wheel." Then even if the system has never seen an elephant during training, it can (if all goes well) identify elephants just from knowing that they are large, gray, and have long trunks. That is, the explanation that elephants are large, gray, and have long trunks can be seen as a manual for assembling an elephant recognizer based on its attributes.

Herbelot and Baroni (2017) also use explanations as "construction manuals." They learn an initial word embedding from a definition, in such a way that the embedding can later be expanded by learning from examples. So what is important here is that the concept built from an explanation is constructed so it can be expanded: The explanation forms an initial prototype or a highly informative proto-example that can be combined with examples.

Both Herbelot and Baroni and the visual attribute papers use simple explanations that are basically bags of attributes or bags of words. How about more complex explanations, like (7-b) or especially (8-c) above, what would it mean to construct a usable concept, a prototype for them? The most straightforward answer is that this is what compositional distributional approaches like Baroni et al. (2014) do, especially Clark et al. (2013), who compositionally construct semantic space representations for definitions and test them for their (cosine) similarity to their definienda. Another interesting direction is Andreas et al. (2016) and Gupta and Lewis (2018), who compositionally construct "actionable" embeddings of statements that can be used to perform a task, in their case question answering.

7 Conclusion

Abstract semantic spaces are not a precisely defined class, they are approaches that in some way work with the idea of meaning as a space. Because they tend to have fine-grained, high-dimensional word representations and because of the way they are learned from data in practice, they encourage us to engage with difficult lexical phenomena, in particular meaning in context. When we view them as conceptual, mind dependent representations of meaning, we have to face the question of what kinds of inferences these representations afford and how they connect with overall sentence meaning. There is no single clear answer to this question yet, but some intriguing directions. In the context of practical, implemented semantic spaces there are several machine learning approaches that try to combine learning from examples and learning from explanations. What these approaches show us most of all is how difficult it is to combine these two ways of learning, but they also indicate some new ways we can think about learning from explanations.

References

Alammar, Jay. 2018. The illustrated BERT, ELMo, and co. https://jalammar.github.io/illustrated-bert/

Andreas, Jacob, Marcus Rohrbach, Trevor Darrell, and Dan Klein. 2016. Learning to compose neural net- works for question answering. In *Proceedings of NAACL*, San Diego, California.

Andreas, Jacob, Dan Klein, and Sergey Levine. 2017. Modular multitask reinforcement learning with policy sketches. In *Proceedings of ICML*, Sydney, Australia.

Asher, Nicholas. 2011. *Lexical meaning in context: A web of words*. Cambridge: Cambridge University Press.

Asher, Nicholas, Tim van de Cruys, Antoine Bride, and Marta Abrusán. 2016. Integrating type theory and distributional semantics: A case study on adjective-noun compositions. *Computational Linguistics* 42 (4): 703–725.

Baroni, Marco, and Roberto Zamparelli. 2010. Nouns are vectors, adjectives are matrices: Representing adjective-noun constructions in semantic space. In *Proceedings of EMNLP*. Stroudsburg, PA, USA: Association for Computational Linguistics.

Baroni, Marco, Raffaella Bernardi, Ngoc-Quyh Do, and Chung-chieh Shan. 2012. Entailment above the word level in distributional semantics. In *Proceedings of EACL*, Avignon, France.

Baroni, Marco, Bernardi Raffaella, and Roberto Zamparelli. 2014. Frege in space: A program for compositional distributional semantics. *Linguistic Issues in Language Technology* 9 (6): 5–110.

Barsalou, Lawrence. 2017. Cognitively plausible theories of concept combination. In *Compositionality and concepts in linguistics and psychology*, Volume 3 of language, cognition, and mind, ed. J. Hampton and Y. Winter, 9–30. Cham: Springer Open.

Bernardy, Jean-Philippe, Rasmus Blanck, Stergios Chatzikyriakidis, and Shalom Lappin. 2018. A compositional Bayesian semantics for natural language. In *First international workshop on lan- guage cognition and computational models*, 1–10, Santa Fe, NM.

Boleda, Gemma. 2020. Distributional semantics and linguistic theory. *Annual Review of Linguistics*. https://doi.org/10.1146/annurev-linguistics-011619-030303.

Bolukbasi, Tolga, Kai-Wei Chang, James Zou, Venkatesh Saligrama, and Tauman Kalai. 2016. Man is to computer programmer as woman is to homemaker? Debiasing word embeddings. In *Proceedings of NeurIPS*. San Diego, CA, USA: Neural Information Processing Systems, Inc.

Bruni, Elea, Gemma Boleda, Marco Baroni, and Nam Khanh Tran. 2012. Distributional semantics in technicolor. In *Proceedings of ACL*, Jeju Island, Korea.

Clark, Stephen, Bob Coecke, and Mehrnoosh Sadrzadeh. 2013. The Frobenius anatomy of relative pronouns. In *Proceedings of the 13th meeting on the mathematics of language* (MoL 13), Sofia, Bulgaria.

Cooper, Robin, Simon Dobnik, Shalom Lappin, and Staffan Larsson. 2015. Probabilistic type theory and natural language semantics. *Linguistic Issues in Language Technology* 10: 1–43.

Copestake, Aurélie. 2009. Slacker semantics: Why superficiality, dependency and avoidance of commitment can be the right way to go. In *Proceedings of EACL*, Athens, Greece.

Dahl, M. 2012. *Blood in the library. Return to the library of doom*. North Mankato: Stone Arch Books.

Devlin, Jacob, Ming-Wei Chang, Kenton Lee, and Kristina Toutanova. 2019. BERT: Pre-training of deep bidirectional transformers for language understanding. In *Proceedings of NAACL*, Minneapolis, MN.

Ellis, John. 1993. *Language, thought, and logic*. Evanston: Northwestern University Press.

Emerson, Guy. 2018. *Functional distributional semantics: Learning linguistically informed representations from a precisely annotated corpus*. PhD thesis, University of Cambridge.

Erk, Katrin. 2010. What is word meaning, really? (and how can distributional models help us de- scribe it?). In *Proceedings of the 2010 workshop on geometrical models of natural language semantics*, Uppsala, Sweden.

———. 2016. What do you know about an alligator when you know the company it keeps? *Semantics and Pragmatics* 9 (17): 1–63.

Erk, Katrin, and S. Padó. 2008. A structured vector space model for word meaning in context. In *Proceedings of EMNLP*, Honolulu, Hawaii.

Farhadi, Ali, Ian Endres, Derek Hoiem, and David Forsyth. 2009. Describing objects by their attributes. In *2009 IEEE conference on computer vision and pattern recognition*, 1778–1785.

Fillmore, Charles. 1985. Frames and the semantics of understanding. *Quaderni di Semantica* 6: 222–254.

Ganchev, Kuzman, Joao Graca, Jennifer Gillenwater, and Ben Taskar. 2010. Posterior regularization for structured latent variable models. *Journal of Machine Learning Research* 11: 2001–2049.

Gärdenfors, Peter. 2000. *Conceptual spaces: The geometry of thought*. Cambridge, MA: MIT Press.

———. 2014. *The geometry of meaning: Semantics based on conceptual spaces*. Cambridge, MA: MIT Press.

Garí Soler, Aina, Marianna Apidianaki, and Alexandre Allauzen. 2019. Word usage similarity estimation with sentence representations and automatic substitutes. In *Proceedings of *SEM*, Minneapolis, MN.

Gehrke, Berit, and Louise McNally. 2019. Idioms and the syntax/semantics interface of descriptive content vs. reference. *Linguistics* 57 (4): 769–814.

Goodman, Nelson. 1955. *Fact, fiction, and forecast*. Cambridge, MA: Harvard University Press.

Grefenstette, Edward, and Mehrnoosh Sadrzadeh. 2011. Experimental support for a categorical compo- sitional distributional model of meaning. In *Proceedings of EMNLP*, Edinburgh, Scotland.

Gupta, Nitish, and Mike Lewis. 2018. Neural compositional denotational semantics for question answering. In *Proceedings of EMNLP*, Brussels, Belgium.

Hampton, James. 1987. Inheritance of attributes in natural concept conjunctions. *Memory and Cognition* 15 (1): 55–71.

———. 1988. Overextension of conjunctive concepts: Evidence for a unitary model of concept typicality and class inclusion. *Journal of Experimental Psychology: Learning, Memory, and Cognition* 14 (1): 12–32.

———. 2017. Compositionality and concepts. In *Compositionality and concepts in linguistics and psychology*, Volume 3 of language, cognition, and mind, ed. J. Hampton and Y. Winter, 95–122. Cham: Springer Open.

Herbelot, Aurélie, and Marco Baroni. 2017. High-risk learning: Acquiring new word vectors from tiny data. In *Proceedings of EMNLP*, Copenhagen, Denmark.

Herbelot, Aurélie, and Ann Copestake. 2013. Lexicalised compositionality. https://www.cl.cam.ac.uk/~aac10/papers/lc3-0web.pdf

Herbelot, Aurélie, and Eve Vecchi. 2015. Building a shared world: mapping distributional to model-theoretic semantic spaces. In *Proceedings of EMNLP*, Lisbon, Portugal.

Hu, Zhiting, Xuezhe Ma, Zhengzhong Liu, Eduardm Hovy, and Eric Xing. 2016. Harnessing deep neural networks with logic rules. In *Proceedings of ACL*, Berlin, Germany.

Jurafsky, Dan, and James Martin. 2019. *Speech and language processing, chapter 6: Vector semantics and embeddings*. 3rd ed. Prentice Hall. Upper Saddle River: New Jersey.

Kamp, Hans. 1975. Two theories about adjectives. In *Formal semantics for natural language*, ed. E. Keenan. Cambridge University Press: Cambridge, UK.

Kamp, H., and B. Partee. 1995. Prototype theory and compositionality. *Cognition* 57 (2): 129–191. Cambridge University Press, Cambridge, UK.

Kruszewski, German, Denis Paperno, and Marco Baroni. 2015. Deriving Boolean structures from distributional vectors. *Transactions of the Association for Computational Linguistics* 3: 375–388.

Lampert, Christoph, Hannes Nickisch, and Stefan Harmeling. 2009. Learning to detect unseen object classes by between-class attribute transfer. In *Proceedings of CVPR*.

Landauer, Thomas, and Susan Dumais. 1997. A solution to Plato's problem: The latent semantic analysis theory of acquisition, induction, and representation of knowledge. *Psychological Review* 104: 211–240.

Lenci, Alessandro. 2008. Distributional semantics in linguistic and cognitive research. *Rivista di Linguistica* 20 (1): 1–31.

Levy, Omar, and Yoav Goldberg. 2014. Neural word embedding as implicit matrix factorization. In *Proceedings of NeurIPS*. San Diego, CA, USA: Neural Information Processing Systems, Inc.

Liang, Percy, and Christopher Potts. 2015. Bringing machine learning and compositional semantics together. *Annual Review of Linguistics* 1 (1): 355–376.

McMahan, Brian, and Matthew Stone. 2015. A Bayesian model of grounded color semantics. *Trans- actions of the Association for Computational Linguistics* 3: 103–115.

McNally, Louise. 2017. Kinds, descriptions of kinds, concepts, and distributions. In *Bridging formal and conceptual semantics. Selected papers of BRIDGE-14*, ed. K. Balogh and W. Petersen, 39–61. Düsseldorf, Germany: Düsseldorf University Press.

McNally, Louise, and Gemma Boleda. 2017. Conceptual versus referential affordance in concept com- position. In *Compositionality and concepts in linguistics and psychology*, ed. J. Hampton and Y. Winter, vol. 3. Cham: Springer.

Mikolov, Tomas, Ilya Sutskever, Kai Chen, Gregory Corrado, and Jeffrey Dean. 2013. Distributed representations of words and phrases and their compositionality. In *Proceedings of NeurIPS*, Lake Tahoe, Nevada.

Mitchell, Jeff, and Mirella Lapata. 2010. Composition in distributional models of semantics. *Cognitive Science* 34 (8): 1388–1429.

Moschovakis, Yiannis. 1994. Sense and denotation as algorithm and value. In *Logic colloquium '90*, Volume 2 of lecture notes in logic, 210–249. Cambridge, UK: Cambridge University Press.

———. 2006. A logical calculus of meaning and synonymy. *Linguistics and Philosophy* 29: 27–89.

Murphy, Gregory. 2002. *The big book of concepts*. Cambridge, MA: MIT Press.

Muskens, Reinhard. 2004. Sense and the computation of reference. *Linguistics and Philosophy* 28 (4): 473–504.

Nosofsky, Robert. 1986. Attention, similarity, and the identification-categorization relationship. *Journal of Experimental Psychology: General* 115 (1): 39–57.

Osherson, Daniel, and Edward Smith. 1981. On the adequacy of prototype theory as a theory of concepts. *Cognition* 9 (1): 35–58.

Pelletier, Jeffrey. 2017. Compositionality and concepts – A perspective from formal semantics and philosophy of language. In *Compositionality and concepts in linguistics and psychology*, Volume 3 of language, cognition, and mind, ed. J. Hampton and Y. Winter, 31–94. Cham: Springer Open.

Pennington, Jeffrey, Richard Socher, and Christopher Manning. 2014. Glove: Global vectors for word representation. In *Proceedings of EMNLP*, 1532–1543. Boca Raton, London, New York: CRC Press.

Potts, Christopher. 2007. The expressive dimension. *Theoretical Linguistics* 33 (2): 165–198.

———. 2019. A case for deep learning in semantics: Response to pater. *Language* 2019. https://doi.org/10.1353/lan.2019.0003.

Putnam, Hilary. 1975. The meaning of 'meaning'. In *Language, mind and knowledge*, Volume 7 of Minnesota studies in the philosophy of science, ed. K. Gunderson, 131–193. Minneapolis: University of Minnesota.

Řehůřek, Radim, and Petr Sojka. 2010. Software framework for topic modelling with large corpora. In *Proceedings of the LREC 2010 workshop on new challenges for NLP frameworks*, 45–50, Valletta, Malta.

Roller, Stephen, and Katrin Erk. 2016. Relations such as hypernymy: Identifying and exploiting hearst patterns in distributional vectors for lexical entailment. In *Proceedings of EMNLP*. Boca Raton, London, New York: CRC Press.

Salton, Gerard, Andrew Wong, and Chungshu Yang. 1975. A vector space model for automatic indexing. *Communincations of the ACM* 18 (11): 613–620.

Socher, Richard, Eric Huang, Jeffrey Pennington, Andrew Ng, and Christopher Manning. 2011. Dynamic pooling and unfolding recursive autoencoders for paraphrase detection. In *Proceedings of NIPS*, ed. J. Shawe-Taylor, R. Zemel, P. Bartlett, F. Pereira, and K. Weinberger. San Diego, CA, USA: Neural Information Processing Systems, Inc.

Strubell, Emma, Patrick Verga, Daniel Andor, David Weiss, and Andrew McCallum. 2018. Linguistically-informed self-attention for semantic role labeling. In *Proceedings of EMNLP*, Brussels, Belgium.

Thater, Stefan, Hagen Fürstenau, and Manfred Pinkal. 2010. Contextualizing semantic representations using syntactically enriched vector models. In *Proceedings of ACL*, Uppsala, Sweden.

Towell, Geoffrey, Jude Shavlik, and Michiel Noordewier. 1990. Refinement of approximate domain theories by knowledge-based neural networks. In *Proceedings of the eighth national conference on artificial intelligence*, 861–866, AAAI Press.

Turney, Peter, and Patrick Pantel. 2010. From frequency to meaning: Vector space models of semantics. *Journal of Artificial Intelligence Research* 37: 141–188.

van Lambalgen, Michiel, and Fritz Hamm. 2003. Moschovakis' notion of meaning as applied to linguistics. In *Logic colloquium '01, Lecture notes in logic*, ed. M. Baaz and J. Krajicek. Boca Raton, London, New York: CRC Press.

Vecchi, Eva, Marco Baroni, and Roberto Zamparelli. 2011. (Linear) maps of the impossible: Capturing semantic anomalies in distributional space. In *Proceedings of the workshop on distributional semantics and compositionality*, Portland, OR.

Veltman, Frank. 1996. Defaults in update semantics. *Journal of Philosophical Logic* 25 (3): 221–261.

Vilnis, Luke, and Andrew McCallum. 2015. Word representations via Gaussian embedding. In *Proceedings of ICLR*.

Vilnis, Luke, Xiang Li, Shikhar Murty, and Andrew McCallum. 2018. Probabilistic embedding of knowledge graphs with box lattice measures. In *Proceedings of ACL*, Melbourne, Australia.

Wang, Alex, Jan Hula, Patrick Xia, Raghavendra Pappagari, Thomas McCoy, Roma Patel, Najoung Kim, Ian Tenney, Yinghui Huang, Katherin Yu Shuning Jin, Berlin Chen, Benjamin Van Durme, Edouard Grave, Ellie Pavlick, and Samuel Bowman. 2019. Can you tell me how to get past Sesame Street? Sentence-level pretraining beyond language modeling. In *Proceedings of ACL*, Florence, Italy.

Wu, Ling, and Lawrence Barsalou. 2009. Perceptual simulation in conceptual combination: Evidence from property generation. *Acta Psychologica* 132 (2): 173–189.

Zamparelli, Roberto. 2000. *Layers in the determiner phrase*. New York: Garland Press.

5

Mathematical Transfers in Linguistics: A Dynamic Between *Ethos* and Formalization as a Process of Scientific Legitimization

Adrien Mathy

1 Introduction

The purpose of this paper is to lay the foundations for a model, which integrates all the aspects specific to the dynamics of conceptual transfers between two sciences, that is, from mathematics to linguistics. In order to build this model, I will be interested in the scientific production in the field of French linguistics, and more precisely, in texts produced in the sixties, which mark a turning point in French linguistics. This turning point concerns both epistemological changes and socio-professional changes in the field of linguistics, with the emergence of new language epistemologies characterized by their desire to give linguistics a rigorous

A. Mathy (✉)
Centre de Sémiotique et Rhétorique, University of Liège, Liège, Belgium
e-mail: amathy@uliege.be

R. M. Nefdt et al. (eds.), *The Philosophy and Science of Language*,
https://doi.org/10.1007/978-3-030-55438-5_5

and scientific basis, especially using mathematical tools. Therefore, this period is ideal to observe the dynamics between conceptual transfers and a priori non-scientific phenomena. In order to carry out this study, I started from observations of scientific discourse conveying conceptual transfers. I built the model on the basis of text studies, using a back and forth between model and text. In this way, I wanted to understand better the link between the use of mathematics in linguistics and complex phenomena as the legitimization of the field. I will focus this study on a publication of Antoine Culioli entitled *La formalisation en linguistique* (1968).

Antoine Culioli (1924–2018) was a central figure in the emergence of these new epistemologies. He notably participated in the restructuring of the field of French linguistics. His works on the Theory of Enunciative Operations are representative of the phenomena I wish to study, that is, the use of mathematics in linguistics. Thus, it is possible to generalize the results by proposing a model of conceptual transfers between mathematics and linguistics, and more broadly between sciences. Actually, conceptual transfers obviously raise epistemological questions. However, they also involve questions about the writing and the writing style. These questions can be strictly textual: lexical questions (what term to use?) or rhetoric question (is the use of the concept metaphorical or not?). These questions can also be semiotic, stylistic, affective, and discursive. Above all, the question of writing concerns the application of a formalization specific to the concepts transferred. As a matter of fact, concepts never come alone. When scientists operate conceptual transfers, these are never devoid of an epistemological, scriptural or sociological anchorage specific to certain practices of the former field. Thus, it has implication for the philosophy of the targeted field and for the belief system of its scientists. How to convince of the relevance of this conceptual transfer? How to integrate it into a different conceptual, emotional, stereotypical, aesthetic and practical universe? In other words, how to legitimize conceptual transfers? These implications are specific to the *ethos*. Before defining more specifically this term, let us accept it in its usual meaning of spirit or belief system.

Conversely, conceptual transfers sometimes pursue a need for legitimization. Using concepts specific to a more legitimate science is a way to legitimate one's own science. Sometimes the opposite effect can occur: either researchers in the field whose concepts have been imported deny the legitimacy of the transfers and criticize their intrinsic relevance, or researchers in the target domain are against the introduction of concepts specific to another science. For example, we can look at the controversy regarding the status of the cognitive linguistics (cf. subsection 3.2 of this paper). Scientists can be emotionally attached to their epistemological and socio-professional independence. Thereby, to transfer a concept from one science to another is not simply a transfer from episteme to episteme. With the concept comes a series of non-epistemological connotations. Because the import of this concept implies a process of legitimization, the question of the *ethos* is essential. Because it has to be as integrated in the new episteme as in a new stereotypical universe, the question of the *ethos* is even more essential. In addition, considering the *ethos* of a scientist is consistent with an observer's position of what science is. Besides, in order to better observe the science and the deployment of this *ethos*, the study of the discourse seems necessary. In the remainder of this paper, I will speak of discourse rather than text. I will use the term text to designate the empirical items in my corpus. The discourse is a production of the scientific activity which makes it possible to observe the conceptual transfers as much from the point of view of the writing (formalization, vocabulary, and so on.) as of the researcher's *ethos* and the process of legitimization.

For methodological reasons, I will present separately the construction of the model and its application to the specific case of Culioli. In the second section, I will study the scientific dynamics between conceptual transfers on the one hand and discourse, ethos, writing, formalization, and finally legitimization on the other hand (2). In the third section, I will develop the context (3). In the fourth section, I will study the *aimed ethos* of Culioli (4). In the fifth section, I will study the epistemological status of the conceptual transfers (5). In the sixth section, I will develop the concept of discourse duplication, in order to understand the dynamics between conceptual transfers and the linguistic model (6). Then, in the seventh section, I will formalize the *expected ethos* of Culioli (7). Finally, in the eighth section, I will conclude with an updated model integrating the various aspects of the dynamics according to their specific configuration proper to the text studied (8).

2 From Science to Discourse: Understanding the Dynamics

In this second section, to tackle the links between science and *ethos*, I will study the relationship between science and discourse (2.1) and discuss the contribution of discourse analysis to my work (2.2). Thereby, I will define the concept of *ethos* (2.3). Then, to study the issue of formalization, I will redefine the concept (2.4) and discuss the role of writing in mathematics and linguistics (2.5).

2.1 Science and Discourse

We have seen in the introduction that studying the *ethos* involves studying the discourse in science – and vice versa. However, even before talking about discourse, we must agree on what science is. To put it simply, when we speak of science, we conceive the object in two different ways. Either one thinks that the science *is*: that this is an entity with its own dynamic (Pestre 2006, 5); or one sees science as a construction, without singularity, without essence (Pestre 2006, 5), organized as a socio-professional field with various dynamics and social interactions (Bourdieu 1976). My paper fits in this second approach proper to the science studies. Indeed, it allows us to understand the issues of conceptual transfers beyond the idea of intrinsic logic internal to the episteme. It allows us to apprehend the individual, institutional, or socio-political dynamics. To study science in this way, we can observe various empirical elements: the discourse is one of them.

The discourse is more than just a consequence or a product of a scientific research. It is not an external concept like the functioning of a field, nor an internal concept, like the internal validity of a demonstration. It is more about the connection between the two. It is the concrete and empirical realization of an abstract epistemological fact and an encompassing sociological process – including various aspects as the history of a specific science, its epistemology, the epistemic rule of the field, its functioning, its associated imaginary and stereotypes, its material issues and so on. More importantly, the encompassing social process includes the rules of the recognition of what is true or false inside the episteme – what Foucault called *regime of truth*. In other words, the scientific discourse is an empirical object the study of which enables a further comprehension of the aforementioned dynamics. This empirical object is like the occurrence (the token) of the science as a potentiality (as a type). In other words, it is a "material device directly participating in the production of knowledge" (Lefebvre 2006)[1]. Finally, the notion of scientific discourse implies a twofold premise: first, that there is discourse in science, and secondly that scientific discourse responds to a series of properties that are sufficiently effective to distinguish scientific discourse from other discourses.

Thus, to define these properties, I choose either an internal point of view or an external point of view. Incidentally, this implies a heuristic bias: in order to define the properties of scientific discourse, it is necessary to observe a discourse defined at first glance as scientific. From then, there is circularity: the properties of scientific discourse are those of scientific discourse, as we can see it in the dictionnary of Neveu (2004). In other words, there is a postulate (of the order of beliefs) about what a scientific discourse is. Besides, this definitional difficulty becomes even more complex with the question of the type of scientific discourse. It has various forms: paper presented to a lecture, publication in a journal, column in a journal or a blog post in a scientific blog, vulgarization works, monographs, dictionaries, research notebook, and so on (Rinck 2010). These discourses can be oral or written, individual or collective, in various media. In short, there are many kinds of discourses in science. Thus, defining what is – or what is not – scientific discourse is as difficult as to define what is and what is not scientific. To establish a model, we must first look at simplified cases. I will not study the variations which are specific to the kinds of discourse or specific to the conceptions of what a science is. In this paper, I will be interested in an unambiguous case: an article published in a journal recognized by peers as scientific.

The fact remains that one can opt for an internal point of view or an external point of view. In the first approach, it is about studying the production process itself. In the second approach, it is about studying the

[1] Unless otherwise indicated, all references from French texts have been translated.

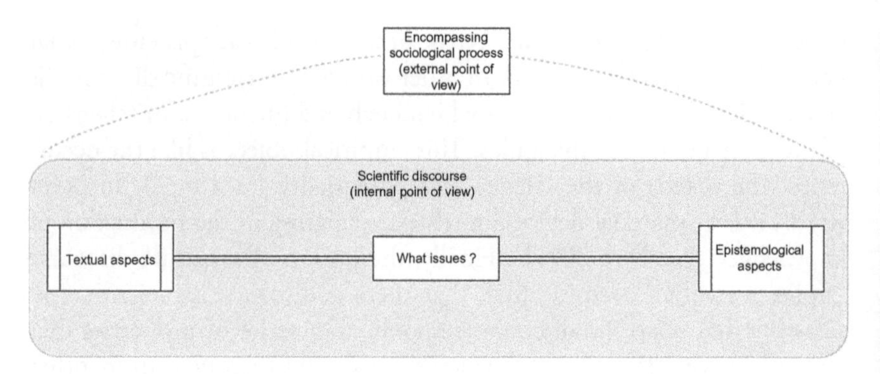

Fig. 5.1 Scientific discourse

conditions of production. The internal point of view consists of studying how the linguist created the model, how the discourse realized it, and how the medium enables and constrains its diffusion and its textual form. The external point of view consists of studying the researcher and his careers, as well as the journals where he published certain publications, or the laboratory where he was at a certain time. It implies the study of the theoretical framework of his field, its historical background, and in a larger scope, the society. While the internal point of view corresponds to the epistemological, textual and material aspects, the external point of view corresponds to the encompassing sociological process. Based on these first elements, I can establish the premise of the model. The following figure (Fig. 5.1) represents the relation between the scientific discourse from an external point of view and from an internal point of view. Within the internal point of view, the graph represents the textual aspects and the epistemological aspects (as the model, for example). Then, we must understand the issues that are at stake between these aspects.

2.2 Discourse and Discourse Analysis

The distinctions I have just made are specific to a domain called discourse analysis and a subdomain called scientific discourse analysis. Depending on the language, on a specific field or simply on the time, the terms of *analysis of discourses* or *discourse analysis* refer to relatively different things, although it is always literally to analyse discourses. When I talk about discourse analysis, my approach is not part of the work of Zellig Harris, nor that of Harold Garfinkel and his ethnomethodology, Erving Goffman, or Oliver Sacks' conversational analysis. In France, as Maingueneau explains (2016), the discourse analysis is the work of Michel Pêcheux, Michel Foucault and Jean Dubois. In this paper, I rely mainly on the work of Maingueneau (especially for the concept of *ethos*). Concerning the scientific discourse analysis: "Its purpose is to link the linguistic characteristics of texts with the practices in which researchers produced and interpreted these very discourses. In fact, this analysis impose at the same time a strong empirical anchoring and a questioning on the issues that they reveal. The empirical anchoring is the linguistic observations I can make, whereas the issues are the socio-institutional and socio-cognitive aspects of the activity of scientists", Fanny Rinck (2010, 427) explains in her state of the art. For example, the legitimization process is a socio-institutional aspect of the scientific activity while the formalization is a socio-cognitive aspect. In my model, to avoid this terminology and these epistemological implications, I prefer to define things in terms of external and internal points of view, the former corresponding to the societal aspects and the latter to the discursive aspects with the textual and epistemological aspects.

In substance, our observable is the discourse, in which we have (1), on various levels, textual components, such as lexicon, argumentative structure, tropes, narrative structure, interdiscourse, colophon, peritext, and so on; and (2) epistemological components such as the model, the concepts, the methods, and so on. For example, in the case of *La formalisation en linguistique* the discourse is the textual occurrence of the model (internal point of view), through its medium, and conditioned by a multiple layered sociological background – the encompassing sociological process (external point of view). *La formalisation en linguistique* is one instance of a model Culioli has created. A specific medium, namely a journal entitled *Les cahiers pour l'analyse*, enables the diffusion of the discourse and restrains its form. Moreover, the history of the field, the history of the institution and some happenstances condition the discourse. I just mentioned the medium, which does not appear in the Fig. 5.1. To study the effect of the medium on the textual and epistemic aspects, to study its effect on the discourse is even more complex than to study the variations specific to the kind of discourse. Moreover, in addition to being too complex, the notion of medium is too vague to be integrated (Krajewski 2015). It covers very different realities. In this case, the physical journal is a medium as a physical support, as a formal medium and as a peculiar social practice as well. Therefore, I will put aside the problem of the medium too.

2.3 Discourse Analysis and *Ethos*

I defined what I meant by discourse. Based on the works of Maingueneau, I will define the concept of *ethos*. I have previously used this word in its meaning of *spirit* or *character*. I will not use it either in its usual meaning or in the meaning of Merton and his mertonian norms – what he called the scientific ethos.[2] The concept of *ethos* is popular in the French field of discourse analysis. It is important to explain briefly its origin. Initially, Aristotle introduced the concept of *ethos* to describe the style of an orator to capture the auditor's attention. This original concept is part of the "triangle of antique rhetoric" with *pathos* for the passion, *logos* for the arguments, and the very *ethos* for the behavior.

The concept of *ethos* has evolved from rhetoric to discourse analysis (Maingueneau 2002). Because of its conceptual variation, from its Aristotelian origin to its current reinterpretation, the concept of *ethos* can be problematic (Maingueneau 2002, 117). However, in my analysis, I will use the concept in the sense Maingueneau has defined it. Moreover, the French tradition of rhetoric or discourse analysis appears to have fabricated its own concepts and words with a strong linguistic and cultural anchoring: *énonciateur, énonciataire, situation de communication, scène d'énonciation, scénographie, monde éthique*, and so on. Despite these difficulties, I will explain a pair of concepts central for the model: *aimed*

[2] Merton described the scientific ethos with the following concepts (Merton 1973): communism, universalism, disinterestedness, organized skepticism. Merton used the term *ethos* in the meaning of philosophy or dogma. It is neither the rhetorical concept, nor the discursive concept. However, these Mertonian norms can be part of the researcher's rhetorical posture, of the stereotypical conception of science. In other words, these Mertonian norms can function as an *aimed ethos* and as an *expected ethos*.

ethos [Fr. éthos visé]/*expected ethos* [Fr. éthos attendu]. Apart from this pair, I will use the concept of *ethical world* [Fr. monde éthique]. Besides, concerning the translation issues, I will use the term *enunciator* and the term *addressee* to translate the concept of *énonciateur* and *énonciataire* in the rest of my presentation.

I understand the *ethos* as something dynamic and not as a static quality of the enunciator. In other words, the *ethos* results from the participation of the enunciator but also from the presence of the addressee. In order to understand this dynamic aspect of the *ethos*, we must distinguish the *aimed ethos* and the *expected ethos*. The former is the one the enunciator wants to create. However, the *aimed ethos* is anterior to the speech. It refers to the addressee's expectations on the enunciator's *ethos*. This concept is important, because it implies that the addressee has stereotypical representations (and expectations) of the *ethos* in accordance with the *ethical world*. We do not have the same expectations for the *ethos* of a politician's speech or a scientist's speech. In addition, the concept of *ethos* is equivocal (Maingueneau 2002, 119). However, we can generalize it as "a way for the speaker to give an image of himself, likely to convince the audience by winning his trust" (Auchlin 2001, 82). It is important to understand this definition in full. We must free ourselves from the rhetoric, which has restricted the *ethos* to the orality (Maingueneau 2002, 8). However, all texts (or enunciator of a text) have a specific *ethos*. This specific *ethos* occurs in a certain context, with a certain addressee, in certain circumstances. It is the conjunction of this *aimed ethos* of the enunciator and of the *expected ethos* of the addressee.

Furthermore, this *ethos* implies for the addressees (auditor or reader) the knowledge of cultural or professional stereotypes, the *ethical world* [fr. *monde éthique*]. Maingueneau summarizes this concept by defining it as "various stereotypical situations blended with verbal and non-verbal attitudes (the ethical world of the dynamic young executives, snobs, stars, and so on)" (Maingueneau 2013, 6). The *ethical world* of the researcher will depend on this stereotypical vision of good and legitimate science. For example, the professional ethos of the mathematician will depend on a certain image of what mathematics are, what good mathematics are (Zarca 2009). The addressee and the enunciator will respectively calibrate their *expected ethos* and *aimed ethos*, according to some stereotypes that form the *ethical world*. Moreover, these very stereotypes obviously have an origin. The society and the imaginary of science that it builds through media, political discourses, and the very scientific discourses. I explained earlier (Sect. 2.1) that encompassing sociological process included imaginary and stereotypes, as well as rules to evaluate the truth and the false.[3] The imaginary, the regime of truth, the epistemological validation criteria or paradigms proper to a specific science, the socio-professional practices, and so on, define a stereotypical vision of science, or rather a stereotypical vision of good and legitimate science. This vision is none other than a form of *ethical world*. To understand the *aimed ethos* and the *expected ethos* of a researcher, one must understand the *ethical world* that nourishes this *produced ethos*. Moreover, to understand this *ethical world*, one must understand the stereotypical vision of science in force.

2.4 Formalization and Writing

We have seen in the introduction that the question of conceptual transfers implied that of writing. The question of writing is complex. In addition, it affects all aspects of the model. The writing obviously concerns the internal point of view of the discourse, mainly the textual aspects: the vernacular language, the layout, the style, the argumentative structure, and so on. Even the use of footnotes is an interesting problem. However, it also concerns the epistemological aspects: graphical representations, suitable language (mathematical writing, phonetic writing, writing in a computer language, etc.), or a particular linguistic formalization (*Transformational grammar* or *immediate constituent analysis* for example). Finally, this issue concerns external and societal aspects, such as reading comfort, stylistic rules, ease of printing, and so on (Waquet 2019). Therefore, the writing issues are present when we talk about discourse.

[3] There are various ways to describe these rules. For instance, Foucault described these rules by mobilizing the concept of a regime of truth. "Each society has its regime of truth, and by this expression Foucault means: (1) the types of discourse [society] harbours and causes to function as true. (2) The mechanisms and instances, which enable one to distinguish true from false statements. (3) The way in which each is sanctioned. (4) The techniques and procedures, which are valorised for obtaining truth. (5) The status of those who are charged to say what counts as true." Lorenzini says (2015, 2).

The writing influences the *ethos* and adapts itself to the stereotypical vision of science. Incidentally, even in the case of oral discourses the question of writing arises: (1) firstly, though the issue of the textual aspects, as the style; (2) secondly, the oral text was first written (rough draft, memo); (3) thirdly, because even an oral presentation involves writing (support for the presentation, use of a blackboard).

In this subsection, I will focus on two aspects of writing: the expected style in a scientific text and the question of formalization – which is a form of writing. I will start with the second issue. Defining this term seems to me more important as I observed distrust in formalization. This lack of confidence is all the greater because one confuses formalization with mathematical formalization. However, the problem of formalization is threefold. The first problem is the confusion between formalization and the objectivist myth. According to this myth, the facts observed are external and independent of the observer (Auchlin 2016, 116). This objectivist myth corresponds to the distinction between two scientific cultures, proposed by Bruno Latour (2001, 25). "One side thinks that science is accurate only when it has been purged of any contamination by subjectivity, politics or passion", Latour says (2001, 25). Formalization often obeys this expurgation of the subjective. Moreover, scientific writing obeys this expurgation of the subjective. The activity of the scientist represses subjectivity and affectivity (Waquet 2019). The tension between those myths, these two cultures, is strong. It appears mainly between the so-called hard sciences and so-called soft sciences, between the natural sciences and the human sciences. Obviously, this objectivist myth is part of the epistemological validation criteria of these sciences. Therefore, this myth is part of the imaginary and the stereotypical vision of these sciences. Consequently, this myth nurture the *ethical world* of its scientists. This tension is observable in the core of linguistics.

The second problem consists of an epistemological misunderstanding. It is about the fear that the formalization substitute itself to the episteme it formalizes (Martin 2001). If this fear is understandable when scientists import formal tools from another domain (in the case of conceptual transfers), it no longer makes sense when the formalization tools are specific to the episteme itself. It seems that this fear is mainly related to the fear of the imposition of mathematical, logical or computer paradigms in linguistics. Thus, the issue of the formalization, as well as the conceptual transfers, and the question of epistemological autonomy are alike. However, in linguistics, we always had formalization: making the evolution of a word is a formalization, transforming linguistic investigation into linguistic atlas is a formalization, transforming an unstructured lexicon into dictionaries is a formalization.

In other words, it is important not to make a confusion between a formalization and the logical formalization proper to the mathematics. In a previous work of mine (Mathy 2017, 320), I defined formalization as (1) the production of an utterance which structures the information – and implies a specific writing; (2) with rules shaping its own grammar; (3) in order to facilitate the analytic processing. For example, transforming a geometrical problem in algebra is a formalization; transforming textual data in numbers and vectors is a formalization. We might object that geometry is already a formalization, as well as textual data. In fact, any transition from one formal reference frame to another is a formalizing again – and therefore a formalization. I like to say that the formalization is nothing more than a tool, which can help us to uncover trends, new data or metadata. In his work, Latour describes the scientific process the same way (Latour 2001). In fact, this definition of formalization and its link with writing brings it much closer to Latour's concept of inscription (2001, 388).

The third problem of formalization arises only in the case of conceptual transfers. Conceptual transfers imply a change of formalization, since they come from another episteme, from another formal frame. Thus, the mathematical formalization and the application of a mathematical concept imply to question the status of the transfers. Is the transfer epistemological, metaphorical, or just lexical? In the case of linguistics, are the relationships between mathematical concepts and linguistic concepts part of what Lakoff calls conceptual metaphors (Lakoff and Johnson 1980, 165)? Alternatively, is it simply a terminological borrowing to designate a purely linguistic concept (Assal 1995)? What are the functions of these metaphors (Oliveira 2009; Rossi 2014)? Is it filling a lexical gap? Is it establishing conceptual analogies? Is it facilitating the understanding of an abstract concept? Is it explaining it to a secular audience? Alternatively, is it helping a learner, from an educational perspective? In short, assessing the level of formalization and the status of conceptual transfers can be difficult. Indeed, as Antoine Culioli's paper illustrates, if the conceptual origin is not explicit, the status of the concept becomes particularly vague (cf. Sect. 5).

2.5 About Writing in Mathematics and Science

I just raised the problem of formalization. The formalization is considered (wrongly) as objectivist and (wrongly) as mathematical. Furthermore, formalization sometimes implies that conceptual transfers and their status can be hard to define. However, when we talk about conceptual transfer, should not we also talk about transfer of writing practice? Does using a mathematical concept necessarily involve using mathematical writing? Does mathematical formalization as an epistemological tool implies a mathematical formalization at the level of writing? When we talk about conceptual transfers, we must define if the transfer affects only the model or the textual aspects as well, with vocabulary, or above all specific writing.

The boundary between these two phenomena is fuzzy and probably very difficult to overcome. It is nevertheless important to remember that between using mathematical concepts and using mathematical writing there is sometimes a chasm. It is quite possible to describe or use mathematical concepts without using mathematical writing – or at least without using canonical writing. If the opposite may seem more complex, it is possible to use mathematical writing to describe non-mathematical concepts. Additionally, we must wonder about the meaning of *mathematical*. In his work about Spearman, Martin defines as *mathematical* "all kinds of writing falling under today meaning of mathematics" (Martin 2003, 195). Even though Martin recognized the historicity of the concept, it seems convenient. This problem is identical to that of the definition of science and scientific discourse. *Is mathematical what mathematicians have defined as mathematical?* Indeed, what is said to be mathematical is mathematical. It is almost performative. Based on the scientific discourse and on the enunciator's *ethos*, which must legitimize his discourse as mathematical, this performative function is the core of the definition of what is mathematics, science, or even linguistics.

The issue of mathematical writing and the general issue of writing in science are alike. For that matter, we have two distinct phenomena: on the one hand, the use of mathematical writing as a more limpid way to describe something; and on the other hands the standardization of scientific writing. Regarding the standardization of scientific writing, one should not forget that the correlation between scientific validity and writing quality is not timeless and self-evident. There is a confusion between epistemic norms and writing norms. The conception of a *limpid writing* has historicity. In the end of the nineties, Jay described scientific discourse as unreadable, arid, without life, and quiet about the research process (Jay 1998). The model of scientific writing and the use of mathematical writing belong to the neutral language myth. Moreover, the myth of a neutral language is only the objectivist myth applied to writing.

Thus, some people sometimes interpret mathematical language as a *neutral language*, as the degree zero of writing [Fr. degré zéro d'écriture]. For example, Roland Barthes said that mathematics had a "functional perfection" without "accidents of forms or disposition" (Barthes 1972). This is questionable. The mathematical language knows a diastratic and diatopic variation. Mathematical language knows polysemy, even subduction. Mathematical language has poetry and sometimes prefers non-optimum writing for its beauty. The specific case of Euler's formula is well-known (Volken 2009, 122). Besides, as soon as there is a canonical writing, there is a norm, and consequently variation inside the system. We are very far from the zero degree or the alleged neutrality of mathematical language, the myth of which is still well established. In addition, it appears that the mathematical formulas participate to a formal stripping, which seeks to answer the second aspect quoted by Popper (1973, 29–30): the highlighting of the logical form (Juignet 2015). The contrast with the dominant *French theory* (or other authors in humanities, especially in its post-modern turn) is obviously striking.[4] For example,

[4] Verdès-Leroux (1998) considers the style of Bourdieu illegible. In is work, Timbal-Duclaux (1985, 13) identified four reasons of the illegibility: a noble style among scientists, the continuous abstraction among the scientist and philosophers, especially in humanities, a gloomy prose and the oratorical aspect of some texts among literati. The emergence of personal styles exists or has existed elsewhere than in the French domain. The case of some avant-garde English journals speaks for themselves (Cuny 2011).

Dominique Ducard explains how important it was for Culioli to develop a formal writing that could describe in a clear and systematic way. A formal writing that reveals what could not be observed otherwise (Ducard 2016).

On the following figure (Fig. 5.2), I represent the dynamics between mathematics and linguistics, when there are conceptual transfers. In this scientific discourse, the textual and epistemological aspects raise the question of their intelligibility. Concerning the mathematical formalization, I represented the epistemological aspects (on the right) and the mathematical writing question (on the left). Then I represented the link between these aspects: is the mathematical writing canonical? Is it visible or not in the text? As for mathematical concepts, are they complementary or substitute for linguistic concepts? Finally, there is the question of the status of this formalization. Is it metaphorical, or does it suppose mathematical phenomena behind linguistic phenomena?

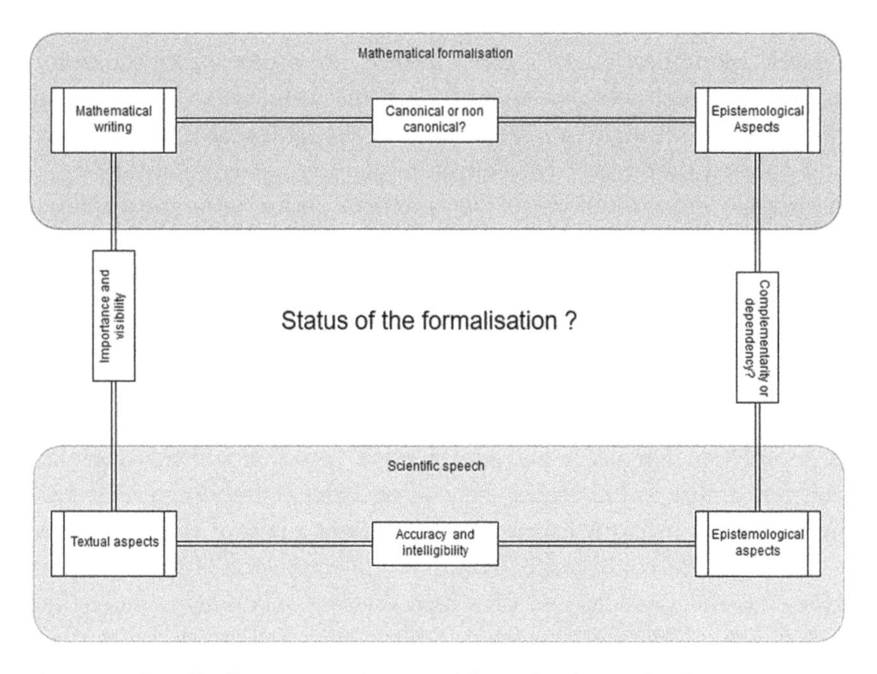

Fig. 5.2 Scientific discourse, mathematical formalization and writing

3 From Discourse to Context

In this section, I will develop the context of my case. I will first explain why linguistics is an interesting case (3.1). I will mobilize the previously developed concepts to explain how linguistics defined itself as a science in the French field (3.2). I will then develop the dynamics of this legitimization by studying the emergence of new language epistemologies giving feature to mathematical formalization (3.3). I will then descend from the linguistic field to the precise work of Antoine Culioli (3.4). I will briefly describe how this researcher and his article fit into the era of interest.

3.1 Why Is Linguistics an Interesting Case?

To apply the model to linguistics, we must understand the conditions of production of the discourse (the *encompassing sociological process*). We must firstly understand the history of the field and its emergence as a full-fledged science, as a legitimate science. This process is linked to stereotypical vision of science of a given society. The process of legitimization implies the construction of scientific legitimacy and acknowledgment by the pairs, the setting of a scientific status; the setting of an institutional and political status; and the acquisition of epistemological autonomy. A science has its own object, its own methods and its own epistemology. Regarding these aspects, linguistics is an interesting case. First, the object of linguistics does not belong only to linguistics. Various sciences can study the language from various points of view: as a biological or physiological fact, as a neurologic fact, as a sociological fact, and so on. Since the language is the conjunction of various phenomena, it is hard to claim that linguistics is a *specific approach*. It is why we observe the following phenomenon: linguistics becomes a suffix (psycholinguistics, sociolinguistics, anthropo-linguistics, neuro-linguistics, ethno-linguistics, biolinguistics, and so on). Consequently, is it not a lack of epistemological autonomy? One could argue that this is a phenomenon of specialization. However, the specialization of a field involves subdomains more than *crossdomains*. Moreover, this fear is explicit in Culioli's work. He is afraid that linguistics will lose its specificity (1999, 14). The evolution of linguistics confirmed these fears. Indeed, linguistics has dissolved "in the big bag of language sciences" (Culioli 1990, 10). This issue relates to the question of epistemological independence, which I noted in Sect. 2.4 as being one of the three problems with formalization.

3.2 The Constitution of Linguistics as a Full-Fledged Science: The French Case

Thus, what kind of science is linguistics? Depending on the type of science, the epistemological validation criteria, the imaginary along with the stereotypes will change accordingly. In addition, the functioning of the *ethos* will change too. In broad terms, on the one hand, some conceptualize linguistics as a social science. For them, the language is an empirical object we must not hypostyle. That is the approach developed by some French journals, as *Langage et société*, and by some subfields as the sociolinguistics or the ethnolinguistics. For example, in the French domain, before the emancipation of linguistics as a full-fledged science, Antoine Meillet had published "Comment les mots changent de sens" in the journals *Année sociologique*, directed by Émile Durkheim. On the other hand, some think that linguistics is a natural science: people like Harald Hammarström who asked in 1978 *Is Linguistics a natural science*? Alternatively, Jan Koster asks the same question thirty years later (2005), testifying to the durability of this questioning. Regarding Jan Koster, his views are interesting in order to understand how some scientists conceive the relation between natural science, linguistics and mathematics. Indeed, there is a third party: for them, language is a sign system describable abstractly and mathematically-like.

Actually, the fertile ground of French linguistics is philology and continental philosophy, phenomenological approach, sociology and, to a certain extent psychoanalysis and Marxism. More broadly, European linguistics is rooted in a complex tradition from the emergence of occidental grammar to the constitution of comparative philology and contrastive grammar (Swiggers 1998). In other words, initially, humanities were the theoretical background where French linguistics was rooted. However, while science and society were evolving, the epistemological validation criteria changed consequently and the linguistics approach had to look for his own specificity and methods in order to legitimate itself as a full-fledged science: that is the start of the process of scientific legitimization. Obviously, a process takes time. Various authors and researchers led it. For example, regarding the works of his pupils, Saussure wanted to define linguistics as a science in his own right. His work is like a starting point of linguistics as a science – as a *terminus ab quo*. The work of Ferdinand de Saussure is present in many linguists' works who have sought to refine the method of linguistics. In this continuity came Culioli, who wanted to create a formalization for linguistics while making a clear distinction between the work of mathematicians and linguists, in order to keep the autonomy of his science. This formalization pursued the goal of making linguistics more scientific. In other words, the aim was to better match linguistics to the criteria for epistemological validation, to better match linguistics to the dominant stereotypical vision of science.

Controversies over the legitimacy of certain approaches to linguistics are still current. In addition, they relate to its status as a science. More recently, for example, Gilbert Lazard asked if the language could be a scientific object (Lazard 2007a). Throughout his article, he poses the question of the legitimacy of approaches that claim to be linguistic. For instance, he entitled one of his publications "The cognitive linguistics does not exist" (2007b), which received a response from Catherine Fuchs (2009).[5] This normative view of a field and these kinds of controversies seem specific to a disagreement about (1) the epistemological validation rules that define the field of linguistics and (2) epistemological autonomy of the field. A lack of recognition or a lack of legitimacy explain this kind of controversy. Besides, Lazard makes a distinction between the American linguistics and an allusive non-quoted European linguistics.

The notion of cognitive linguistics is only meaningful in connection with American linguistics [...]. Out of that context, it merely means a return to the traditional conception of language and linguistics. However,

[5] In fact, in the first place, the publication of Lazard is a reaction to a book by Fuchs. Fuchs was one of the colleagues of Culioli, and her work is a sort of introduction to American linguistics and cognitive linguistics in France.

it involves the risk of downplaying the specificity of the analysis of language structures.

Therefore, the process of legitimization is an ongoing process,[6] which we cannot understood without considering the internal logics of the field and the general evolution of society. However, the legitimization process in not linear and without controversy inside the field itself. For instance, between the 1955 and 1969, we can see a change in the background. In these key years, we are in a context of intellectual excitation. Nevertheless, the Marxist and psychoanalytical movements are losing influence in the linguistics field (Bert 2014). At the same time, André Martinet came back from the USA (in 1955) and France had the first translation of Noam Chomsky in 1963 – realized by Jean-Claude Milner. In other words, we witnessed the introduction of North American linguistics approach and mathematical formalization. In this context, some researchers took advantage of the space of academic freedom to build research projects far from the monopoly of the Sorbonne. These researchers participated in the emergence of new epistemologies of language. Among them was Antoine Culioli.

3.3 Mathematical Transfers and New Epistemologies of Language

These new epistemologies of language will make several conceptual and epistemological transfers. These are conceptual transfers from one episteme to another, in this case from mathematics, computer science, logic, and analytical philosophy toward linguistics. However, as I mentioned in the introduction (Sect. 1), its transfers involve transfers of other types. First, there are transfer of practices from one epistemological validation paradigm to another and from one episteme to another. In this case from the axiomatic and inductive sciences to empirical and deductive sciences. Secondly, there are transfers of texts, ideas and individuals from one

[6] For example, the call for publication of the journal *Tetralogie* testifies the problem of autonomy and legitimacy in the field. Entitled *The deconstruction of language. Where language sciences go?*, the announcement says "Has there not been a dilution of this object that he thought [Saussure] could distinguish and made autonomous: language [*langue*] or language [*langage*] as a structural "value" (January 15, 2019).

country to another, in this case from the United States to France. Lastly, there are transfer of standards from one socio-professional context to another. Actually, socio-professional practices can travel with the concepts and the epistemological validation paradigms. In addition, these practices can be diametrically opposed to the scientific ethics of the episteme of arrival. Therefore, they can hinder the integration of these transferred concepts. We cannot neglect this last kind of transfer, since it participates directly in scientific rhetoric and ethos. Actually, there are transfers of practices from a certain way of conceiving linguistics and language toward structural linguistics. In other words, we observe here a separation between two ethical worlds specific to two ways of doing science, since the ethical world is nourished by these paradigms of validation, these socio-professional practices, and so on (see subsection 2.3).

This separation between two ways of conceiving linguistics comes on top of a geographical gap: between the United Kingdom and America from one perspective, and the European continent from another perspective. However, it comes on top of an epistemological gap as well: between analytical philosophy and continental philosophy.[7] Furthermore, a sociological framework embed these transfers. The structure of the French linguistic field translates the epistemological, conceptual and geographic divisions. The field replays internally the geographical and epistemological divisions, through the logic of sociological positioning. According to its logics, we can apprehend the epistemological and sociological tension of the proponents of structuralism and linguists who emerge around a new thought, which aim to converge analytical philosophy and structuralism, mathematics and psychoanalysis, in a new syncretic way to study language.

More precisely, it is a gap between two different stereotypical vision of science: from then on, everything that belongs to the epistemological model (methodology, concepts, heuristic), the writing (style, formalization), the ethos (expected ethos, aimed ethos and ethical world), and the process of legitimization will change accordingly. The stereotype of *good*

[7] It may seem paradoxical that French authors imported tools from the so-called analytical philosophy, even though for some analytical philosophers linguistics should not be scientific (Hubien 1968). This kind of position shows to what extent the conception of a same episteme (from a quick look) varies according to space, time, and school of thought.

science, which is historically determined (Carnino, 2015), is the foundation of this legitimization. The more a science meets the criteria of real science, the more legitimate it is. This question of legitimacy is essential to understand this gap and to understand how these new epistemologies have tried to overcome it. At the heart of this legitimacy and of the *ethical world* is the stereotypical vision of *good* science. For example, we can describe this good science as the adequacy to Popperian-like scientificity, operationality, and economic production (Mathy, *in press*).

Various ways of doing science, various epistemological validation criteria can be the realization of these norms. The epistemological validation criteria of the stereotypical *good* science (1) implies that only scientific discourses can achieve truth, and that mathematical discourse is the most suited one. (2) It promotes mathematical-like formalization. (3) It sanctions illegibility and favors analytic texts according to the criteria of information literacy. (4) It increases the value of mathematical proof-like technique and axiomatic investigation. (5) It gives a special status to hard science, natural or evidence-based science, and mathematical science. In accordance with this way, some linguists will feed on a new *ethical world*, and construct a new *ethos*.

3.4 From Linguistics to the Work of One

To study the text, I must first talk about the author: Antoine Culioli. It is important to understand how the work of Culioli enters the field of linguistics in the sixties. Sophie Fisher (*in* Ducard and Normand 2006) explains how some linguists restructured the field at the beginning of the year 1965 until 1968. Fisher defined the emergence of Culioli's linguistics with extra-scientific events that I mentioned earlier: the struggle against the so-called official linguistics beard by Martinet and the Sorbonne; innovation by outcasts as Algirdas Julien Greimas, Oswald Ducrot, Bernard Pottier, Nicolas Ruwet or Tzvetan Todorov; the creation of *the Center of Study for TAL* (traitement automatique des langues = automated language-processing) of Grenoble; and May 1968. Regarding Culioli himself, he wanted to give a specific method to linguistics. As a philologist, Culioli had a contact with the empirical aspect of language.

In his seminar called BCG (according to the initials of the surnames), Culioli and his colleagues François Bresson and Jean-Blaise Grize[8] had the custom to link philosophical aspects of language and neuropsychological or biological questions (Ducard and Normand 2006).

Moreover, Culioli is a fine connoisseur of the analytical philosophy. Finally, Culioli perpetuates the work of Saussure: "he wanted to transform the study of specific language [Fr. langue] into the study of language as a concept [Fr. langage]" (Ducard 2006, quoting Saussure 2002). Furthermore, Culioli had interest in psychoanalysis and was Louis Althusser's student. Actually, Culioli was a perfect go-between between humanities and mathematics, between French linguistics field and American linguistics field. He studied "original articles that went beyond the French base, the lessons of G. Guillaume and the lessons of Émile Benveniste, as much as the American base, and they were constantly challenged without concession, to the concepts of the new logic, that Culioli would take up minutely with his disciple, the mathematician J. P. Desclès", says Jean-Claude Chevalier about Culioli (2010, 113).

Besides, we must look at the journal in which Culioli's paper was published: *Les cahiers pour l'analyse.* Founded by young Althusserian French philosophers, the life of this journal was very short: from 1966 to 1969. Their aim was to make a fundamental contribution to philosophy by funding philosophical analysis with formal concepts. In other words, they wished to use scientific-based methodology for philosophy – it is therefore still a question of conceptual transfers and epistemological validation. Consequently, this is a part of a process of formalization, on the one hand, and of legitimization, on the other hand. Additionally, the *Cahiers* specify that they, "sought to combine structuralism and psychoanalysis with logical or mathematical formalization, generating a field of theoretical reflection". This last sentence summarizes the previous descriptions of these new epistemologies. Besides, it is interesting to note that the journal republished the texts of authors, such as Georg Cantor,

[8] François Bresson is a psychologist specializing in language development and Jean-Blaise Grize is a specialist in natural logic and epistemology.

Bertrand Russel, Kurt Gödel[9] or René Descartes, Antoine Lavoisier, Dmitri Ivanovitch Mendeleïev, and so on. The choice of the authors says a lot about the epistemological anchoring of the journal. Therefore, the fact that Culioli had chosen this journal says a lot about his intention: giving linguistics a scientific-based methodology by using mathematical formalization. This is all the more important as Culioli published very little and preferred oral presentations.

Regarding the publication itself, Culioli published it in 1968 and republished in 1999. It was an attempt to create a mathematical model for linguistics and for language – in Saussure's sense. To my knowledge, the publication has never been translated in English. We can find a republishing in the second volume of *Pour une linguistique de l'énonciation*. It is interesting to note that this publication is one of the oldest of Culioli. The majority of his other publications seem to be posterior. Indeed, among the subsequent editions of his major publications, it is the only one dating from the sixties. I have counted seven publications from the seventies, thirteen from the eighties, fifteen from the nineties; it would be possible to recount all his production to study how his thoughts have evolved in time. However, if it is not an early career for Culioli (who was 44 years old at that time), it is the starting point of a long work about the enunciation and the formalization of linguistics. Moreover, this first publication matched with the acme of the restructuration of the field of linguistics in France (Fisher in Ducard and Normand 2006, 21).

4 *Aimed Ethos*: Rigor, Mathematics and Discourse

Before further analysing some elements of Culioli's formal apparatus, I will study the introduction in the light of the concepts I have introduced beforehand. The criteria of scientificity and the standardization of

[9] Apart from these mathematicians, the journal counts among its republished authors as Louis Althusser, Georges Dumézil, Michel Foucault, Claude Lévi-Strauss for the structuralism (although some have refuted this label like Foucault), Xavier Audouard, Jacques Lacan, Jacques Nassif for the psychoanalysis, Jacques Bouveresse, Georges Canguilhem, Gaston Bachelard, for the epistemology, and other great names as Niccolò Machiavelli.

scientific writing, namely the epistemological validation criteria, imply in the researcher an attitude. I can summarize it with the term *rigor*. In short, it is about being rigorous. This applies as much to methodology and model as it does to writing. This rigor appears to contribute to the *ethos* that legitimizes the action and research of Culioli. In addition, this rigor comes along with conceptual transfers. The rigor itself raises from another science, or from another epistemological validation paradigm. I am not saying that rigor is not a quality for a good scientist. I am saying that even the rigor in science has an origin and a historicity. However, it is important to distinguish between being rigorous from saying it. In order to legitimate himself and his field, an author can create a rhetorical posture based on this expected rigor. If being rigorous is expected, saying and showing one's rigor within the discourse is a good strategy to fit in with the epistemological validation paradigm. More broadly, it is a way to fit with the stereotypical vision of science in force and the associated *ethical world*. Obviously, this rigor comes in many ways. We have seen previously how the writing is part of the *ethos*. Consequently, this *aimed ethos* fulfills itself through a precise language, little chastened – which must come as close as possible to analytic language (see subsection 2.5). Besides, the content analysis also gives a lot of information about the *aimed ethos* of Culioli. Indeed, the introduction of his publication sets the tone:

> From the beginning, it is important to set the aim of this article, in order to avoid misunderstandings and ensure the reader's approach through a composite set of epistemological and methodological thoughts, overflights or schematizations that require a good knowledge in linguistics, and finally rapid incursions into the realm of linguistic practice. (Culioli 1999, 16)

This first paragraph is interesting as it displays the aim of the publication: avoid any misunderstandings. In order to study his *aimed ethos,* we can observe the vocabulary. Here, the key words are *methodological* and *epistemological.* Culioli seems to fear that the reader misunderstands his work and does not grasp the complexity of a "composite set" [Fr. ensemble composite] of "overflights and schematization" that suppose a good knowledge of linguistics. We found this type of expression elsewhere in his scientific work. For example, he talks about "rough guidance" when he summarizes the history of linguistics (1999, 8) He is always cautious. As Chevalier says, for Culioli, a linguist is never cautious enough (2010, 113).

However, we do not have any words about the knowledge in mathematics and mathematical formalization. It is probably because Culioli writes in a philosophical journal, not in a linguistic one. The matter of the target is something important. The *aimed ethos* varies accordingly. Culioli continues this kind of analysis in the second paragraph where he expresses the aim of his publication – funding the bases of a natural language's formalization to give linguists rigorous tools to model language:

> This paper is a warning: [...]. [I want] to mark the dangers of a fascinated, multi-rooted craze that is very likely to have harmful effects. [...] The illusion that a stenographic symbolization will allow 'to see more clearly' [...] incoherence in the use of models, facilitated by the desire to be interdisciplinary, by the clutter of misunderstood mathematical concepts and by insufficient reflection on what is, the theme of linguistic science [...]. (1999, 16)

In this sequence, he gives a warning because a part of this publication is preventive. We could even say prophylactic. The warning is against the utopia of formalization and the dangers of a fascinated fury that symbolization will help to see more clearly. We observe a severe vocabulary: *fascinated, novice effects, expedite commodity, illusion, inconsistency.* At first sight, Culioli seems wary and suspicious, that he does not seem to approve the formalization. Such a stance seems paradoxical with the objective to give mathematical tools! Culioli's remark is like Benveniste's warning about the structural theories (2006, 3–45). Culioli does not only want to provide linguists with the formal tools they need, he wants these tools to be used correctly. The fascination for formalization – and mathematical formalization – is dangerous because it combines fear and admiration. For some linguists, formalization is like a diversion, and for others it is like an epistemological solutionism, as I said earlier (cf. subsection 2.4). He wants to keep the focus on relevant things rather than wasting resources in a delusional use of mathematics. We can also observe his fear of losing autonomy for interdisciplinarity (see subsection 2.4 and 3.1) and the insistence with the term *linguistic science* [Fr. science linguistique], which is rare in French. Notwithstanding, for Culioli, mathematics is the role model for linguistics:

> Now that linguistics rediscovers language, instead of constructing its object, it cleaves it into research with different intentions, which sometimes imply incompatible models: the unavoidable consequence is a reduction of language, for technical reasons that we often do not know. In particular, it is clear that irresponsible formalization [...] makes it difficult to mark the dialectical relationship between language and languages [...] The linguist's discourse ends easily in rewriting games, which, unlike mathematics, are neither rigorous nor fertile. [...] Nothing allows us to think that current mathematics is necessarily appropriate that the grammatical beings, even mathematized, with which the linguist works, have a value other than traditional. (1999, 18)

As we can see in this paragraph, for Culioli it is important to pose the theoretical problem of formalization. According to him, the linguistic formalization is nothing but a rewrite without rigor or fertility, unlike mathematics. In this passage, we can identify at least three elements. First, we can interpret this passage as a criticism of generativism. Second, Culioli seems to mix up formalization and mathematical or logical-like formalization. Even in the case where he speaks of generative grammar, it is not a mathematical formalization. Finally, mathematics is the core of his comparison. These last two works are the key: *unlike mathematics*. It is implicit that the *condition sine qua non* for a real science is mathematics. It is somewhat the same position as Milner, when he talked about linguistics as a Galilean science.[10] The relationship with Milner is all the more evident, as he was one of the animators of the journal *Cahiers pour l'analyse* and Culioli's pupil. Finally, we must study Culioli's reflection on grammatical beings. His remark is fundamental, and it is important to keep it in mind for the next section (see Sect. 6).

[10] The word *Galilean* is used by Milner, probably based on what Chomsky said (1980 quoted by Milner 1989). He used it to define a science with two traits: the mathematical formalization [mathématisation] of the empirical observation and technic as a practice of science/a science as a theory of the technic.

5 Conceptual Transfers and the Status of Mathematics

To sum up, Culioli wants to use mathematics correctly in order to create a model, which is not only a rewrite but also a heuristic. We explained that among the entourage of Culioli was the young mathematician Jean-Pierre Desclès. Together, they used the *new logic*, which Desclès himself recognizes that the border with mathematics is blurred (Schmid and Nicole 2014, 260). In the rest of his paper, he gives various examples of his mathematical formalization. What I will observe here is the status of mathematical concepts, but with the point of view of discourse analysis. I will not focus only on the epistemological aspects, but also on the rhetorical and lexical aspects of the mathematical formalization as well:

> We will reduce all unary operations of predication [...] to an application, which is nothing but very ordinary, but we will go to the end of the analysis, adding a theory of predicates. [...] a classification of operations that can be performed on the starting set and / or on the set of arrival, on the arrow symbolizing the functor. (1999, 26)

In this first extract, we see a terminology proper to mathematics. We have various terms as *mapping* [Fr. *application*], literally *starting set* and *set of arrival* [Fr. *ensemble de départ*; *ensemble d'arrivée*], the domain and the co-domain of a function, or the word *functor* [Fr. *foncteur*]. The lexicon is proper to the domain of analysis, but the word functor needs a focus because it has two meanings, two implementations. Originally, the word *functor* is a word proper to the analytical philosophy, used by Carnap in a linguistic context (Carnap 1937). Thus, mathematicians used it to express a complex idea of mapping between algebraic objects and topological space. We must ask ourselves what the semantic value of this word is. It is not a matter of meaning but a matter of epistemological integration and intertextuality. We do not know the background of the concept: Culioli does not explain it, nor do his epigones. Is it a concept proper to Culioli, a concept proper to Carnap borrowed by Culioli, or a concept proper to mathematics borrowed by Culioli? I looked in the scientific literature how the word *functor* is used. When Boisson (1999: §39) talk about the functor, he uses a univocally mathematical writing:

> In mathematics and logic, one finds [...] functors, or even morphisms. Given two functions y = f (x) and y = g (x), we can compose them as follows: y = F (x) = g (f (x)) = (g • f) (x) = gf (x), according to various notations. We see that g • f is a compound function, where f is applied first, and then g. In infixed relational notation, xRy and ySz are composing x (S • R) z.

At first sight, it is hard – especially for a non-mathematician, and linguists are not necessarily mathematicians – to understand the meaning of the formula and the link with the subject. Moreover, it seems that for Boisson, the functor corresponds to a mathematical object. Besides, this extract helps us to realize, by contrast with this ostentatious mathematical paragraph, that Culioli never used mathematical writing in this extent. In another work, from Lowrey and Toupin, the authors defined functor as the "essentially grammatical morphemes, became independent" (Lowrey and Toupin 2010, 9). Therefore, in the first case, he explains the concept with a mathematical digression, but not in the second. Is the mathematical digression a vital part of the demonstration, or is it a rhetorical part? Is it unavoidable to apply a mathematical reductionism to linguistic concepts – even though this mathematical reduction is valid? In other words, it is not because an argument is sound that it is relevant. The difficulty is not to understand the term functor. The difficulty is to understand to what extent it is a mathematical concept. I raised this question previously concerning the formalization and the status of imported concepts (see subsection 2.5).

In the next extract, we can notice many occurrences of mathematical words, as *topological representation*, *finite list of operators*, and so on:

> We can give values of a verbal system (a system being defined as a network of values) a topological representation, which makes it possible to pose some problems better [...] We can reduce the operations on the units in the starting set and in the arrival set to a finite list of operators [...]. (1999, 26)

However, what is a *topological representation*? In this sentence, is the word *topological* meaningful? Topology is an area of mathematics, which studies the space deformation. The topological approach and the topological metaphor were popular at this time in human sciences: for example, we find the word in Baudrillard (Bessis and Degryse 2003), Deleuze and Guattari (Regnauld 2012; Jedrzejewski 2017), Lacan (Granon-Lafont 1995), and so on. However, we still do not know if it is a borrowed word, a borrowed concept, an analogy, a metaphor, and so on. When Culioli talked about this use of topology, he said: "For example, there was the topology, which is the ultimate art of deformation. But we do not see why we would make language a kind of device to make fresh pasta ..." (Grésillon and Lebrave 2012, 147–155). This is not ultimately of great help. In the next extract, it is the same issue: what is the meaning and the conceptual status of *vector*, *absorbing elements*, *neutral elements*, *vector of vectors*, or *sliding vector?*

> Property vectors will represent some categories; so that we can have vector vectors [...] This vector is 'sliding' [...]. Special logic systems of the type 0, 1 (where 0 can be an absorbing element depending on the systems), * (neutral term, which means that which is neither 0 nor 1 or 0 or 1), will be constructed, ω (term that is outside of (0, 1, *). (1999, 26–27)

Let us focus on one of these terms: *sliding vectors*. In fact, this is my translation of a strange hapax of Culioli: *vecteur coulissant*. When I did some research on this word, I could only find texts written by epigones or colleagues of Culioli (as Fisher or Verón). However, the only time I found this word in a mathematical context was in a dictionary entry with the following translation: *sliding vector*. Usually this word is translated in French by the words *vecteur glissant*. It is an interesting problem to raise if we want to make the conceptual genealogy of Culioli's model and formalization. Either *vecteur coulissant* is a non-mathematical word for a non-mathematical concept; or it is a non-mathematical word for a mathematical concept (maybe the concept of *vecteur glissant*/*sliding vector*); or it is a mathematical word for the corresponding mathematical concept (maybe the very concept of *sliding vector* or another one). The status of the concept is always unknown to the reader, to the addressee. In addition to the vocabulary, it is possible to identify a certain use of mathematical writing. We can identify the use of the denomination by letter (as in an equation), the use of matrix-like expression, of tuples, and so on. For example, concerning the concept of *sliding vector* Culioli says:

> This vector is 'sliding', that each term, except for C0, can take a value of zero. We can therefore have: (Co, Ag, Th), (Co, Ag), (Co, Th), (Co). For his part, Agent is represented in another vector (Agent, Animate, Determined), also sliding. It is by chance that we have, twice, a triplet. (1999, 26)

In terms of writing, we do not have mathematical writing specific to vectors. We do not have a writing specific to mathematics. Nonetheless, what is the precise status of this concept? We have what seems to be a conceptual metaphor. The concept of vector is not applied as in mathematics. However, it makes it possible to apply the mathematical idea of vector to the linguistic domain. However, this is not explicit in the text. It is impossible for the reader to know precisely the status of the concepts. In other words, the paper is not self-sufficient.

6 Mathematics and Linguistics: The Discourse Duplication

This last paragraph raises an extremely problematic issue: what is the status of mathematics in this article? More broadly, what is the status of mathematics in Culioli's work? In order to think about this question, let us focus on the words *absorbing element* present in this paragraph. This concept as obviously a mathematical origin. More precisely, this concept comes from the set theory of which Culioli takes up many notions. To understand fully the use of this concept, we must go to a footnote:

> Thus, masculine + feminine, in French, gives masculine; in the same way, we + you + them give us. (1999, 27)

In this extract, Culioli explains the use of *absorbing* and *neutral element*. He remarks that if you have masculine words and feminine words together, the word agrees with the masculine gender. In mathematical terms, masculine is like the zero in a multiplication: an absorbing element. However, this kind of explanation tends toward a hypostasis of methodological or epistemological model. The same issue arose earlier (Culioli 1999, 18): is there a mathematical essence of a grammatical category? Has French masculine grammatical gender the mathematical-like property of being *absorbing*, as a linguistic subtype of a prototypical mathematical property? Alternatively, is it just a way to describe a linguistic phenomenon? In other words, is Culioli using the mathematics as an axiomatic-like discovery tool or as a descriptive tool? It is not the same thing at all. In the second fashion, mathematics is nothing more than a useful epistemological descriptive tool for linguistics and language. However, in the first fashion, linguistics is a subdomain of mathematics; subsequently I can deductively discover linguistic phenomena by a homothetic-like transformation of a mathematical axiomatic system. It is not the same thing as using analogy to uncover new phenomena.

According to Culioli: "The ignorance of most linguists about logic, especially mathematical logic is fantastic" (1985, 10). Grize (2006, 33) does not hesitate to compare his natural logic to Culioli's work. Moreover, the very notion of natural logic is an indicator as to the metaphorical status (or not) of mathematical concepts. When we read the framework of Culioli and his colleagues' work, the position of the logic in the analysis becomes transparent. After all, Culioli himself has defined the relation between language, logic and cognition, by making out three levels of analysis: the first level is the language, the intermediary level is the logic as metalanguage and the last level is the cognition itself (Culioli 1990, 21–24). Yet, the status of the mathematical concept is still uncertain. First, because I do not consider that mathematical and logical concepts are the same thing – even if the distinction is blurred in Culioli's work. Secondly, as we can see in the Fig. 5.3., the relation between mathematical concepts, mathematical writing, scientific writing and linguistic concepts is so complex that it is too reductive to conclude that the use of mathematical concepts is strictly logical, at face value, without any metaphorical aspect. Indeed, it is important to understand the use of mathematical concepts in relation to the epistemological constraints of the model, with the constraints specific to writing, and the entire theoretical universe specific to the *ethos*.

In other words, there are two discourses in Culioli's work. There is a mathematical discourse and there is a linguistic discourse. Both obviously work at the same time, my separation being artificial. Mathematical

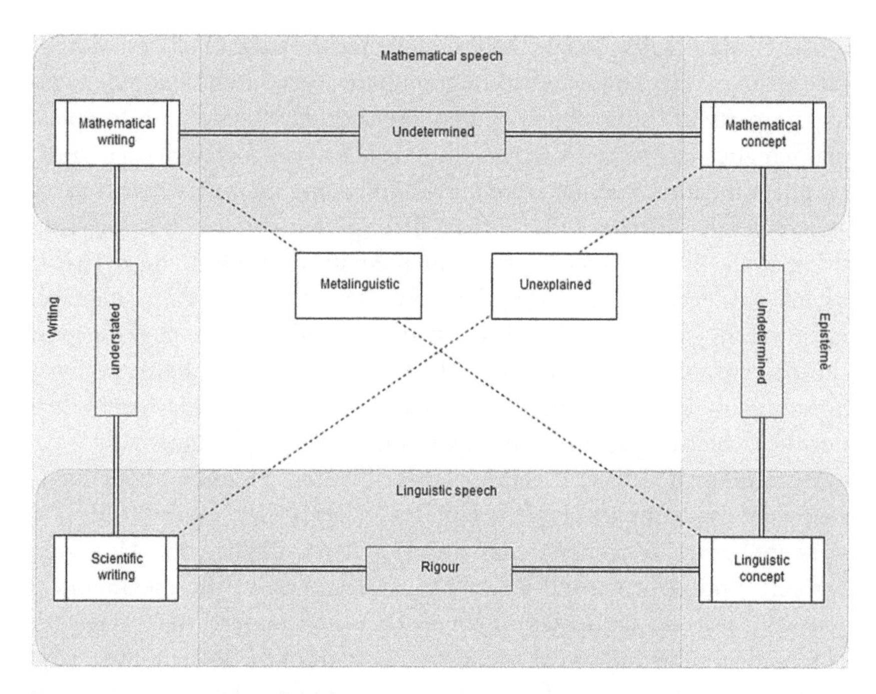

Fig. 5.3 The discourse duplication

discourse is characterized by mathematical writing and mathematical concepts, while linguistic discourse is characterized by linguistic concepts and the stylistic standard in science. The presence of mathematical writing is understated, the relation between the mathematical concept and the linguistic concept is undetermined, just like the relation between mathematical writing and mathematical concept. In addition to these relationships, diagonal relationships must be modeled. The relationship between mathematical writing and linguistic concepts is the metalanguage of Culioli, what he called formal writing [Fr. écriture formulaire/formulaïque] (1999, 71). Finally, Culioli never makes explicit the mathematical concepts. This discourse duplication creates a mathematical imaginary, which responds to the stereotypical vision of good science.

7 *Expected ethos*

Let us briefly consider the *expected ethos*. It is difficult to describe it, since it is specific to each addressee. The *expected ethos* comes from the addressee's knowledge of the stereotypes of the *ethical world*. It comes from his knowledge of epistemological validation criteria and stereotypical vision of science. It also comes from his knowledge of the *ethos* generally built by the researcher concerned. However, this *expected ethos* also comes from an imaginary world built around a science or around a researcher. Narratives, affective situations, and rumors nourish this imaginary. In short, this imaginary takes place in the socio-professional sphere. These factors nourish the imaginary within the framework of the sociability of the addressee. This imaginary belongs therefore to the encompassing sociological process. In addition, scientific discourses allow us to observe this very imaginary. Or rather, we can observe it at the margin: a *thank you* in a footer, a dedication in a monograph, an emotional expression that arises within a scientific text. We observe also this imaginary in the productions of scientists, which are not scientific. Letters, research notebooks, informal discussions, and so on. About Culioli, we read that he is *open-minded* (Ducard and Normand 2006), *rigorous*, *unique* (Ducard 2006), *original* (Achard 1992; Milner 1989; Détrie 2007), and so on. About Culioli, we read he was friend with the mathematician J.-P. Desclès. This feeds the idea that Culioli is somehow a mathematician. There is an effect of aura and authority on the *expected ethos* of Culioli.

Some type of scientific discourses nourish this imaginary more than others do. They nourish the *ethical world* and the *expected ethos* of a given researcher. The practice of homage is this very particular discourse. The tribute allows scientists to let talk about affect and emotions, often repressed and kept quiet in scientific work (Waquet 2019). For example, as part of a symposium at Cerisy on Culioli, the program explained, "all his interest and his commitment to the dialogue of disciplines give [to Culioli] an original and unique place in contemporary thought movements. For all these reasons, it seems important to me today to organize this symposium, which should interest a wide audience". This type of lecture involves a certain style, which is often panegyric. The title of the publication of Cerisy's symposium is "Antoine Culioli, un homme dans le langage: originalité, diversité, ouverture". In the overture of the symposium, Dominique Ducard said something interesting:

> Let me assume that A. Culioli is like Confucius, a thinker with a unique idea. I choose for myself the idea of bifurcation. A. Culioli recounts his interest as a child for canals, switches, and rails, which he made in his garden with a "steel forge". Interest that he shares with the mathematician and philosopher René Thom, who wrote an article, entitled "Songeries Ferroviaires", in which he makes, from childhood memories, a theoretical analysis of the railway morphology. The first element is the way, [...] whose equilibrium stabilization is subject to limits. This importance of the limits is found, [Thom] says, in linguistics, especially with the qualities likely to gradation. He mentions the limits that designate "enough" [assez] and "too much" [trop]. This reminds [me] of the analyzes that Culioli presented during the 2005 seminar [...]. (Ducard and Normand 2006, 13)

It is very peculiar to observe how the third part enunciators create, if not a myth, at least a teleological explanation. These kinds of descriptions are building the *ethos* of Culioli. As we can see in the following figure, the addressee builds the *expected ethos* of the enunciator based on all his knowing about him (the personal imaginary, which is part of the *produced ethos*). The *aimed ethos* is the mathematical imaginary, that is, the result of the discourse duplication. The formal model is use of mathematical concepts and writing to describe linguistic concepts and phenomena. The *ethos* is the use of mathematical concepts and writing as a way of writing scientifically (*aimed ethos*) and the personal imaginary (*expected ethos*), as a way of being rigorous (Fig. 5.4).

8 Conclusion

At the end of this analysis, several questions arise. First, what does the analysis of this paper explain about how conceptual transfers work in Culioli's discourse? Second, does the theoretical model presented here

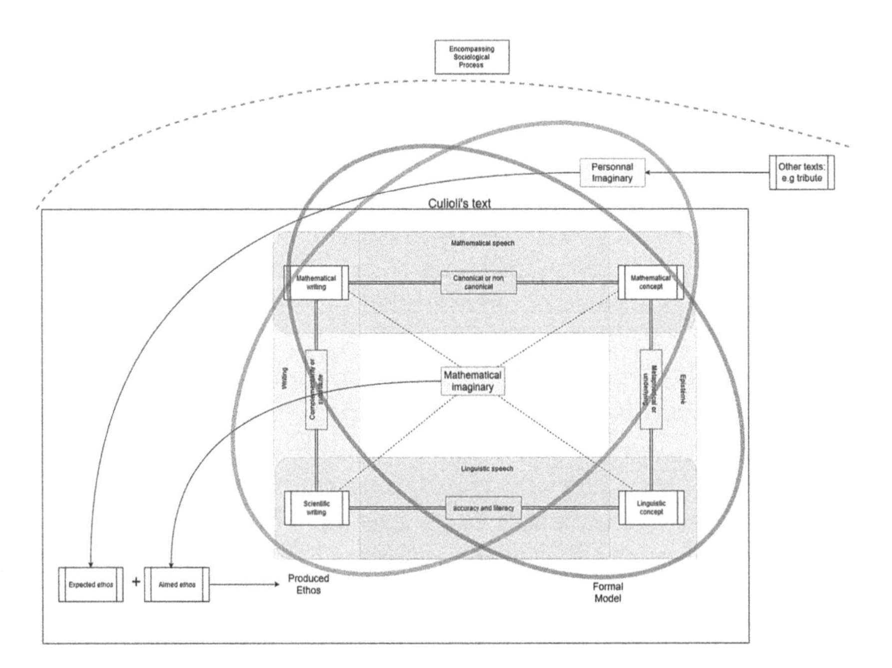

Fig. 5.4 Produced ethos

make it possible to summarize this operation? In the excerpts, we observed an attachment to the rigor of the scientist's work. This rigor is said and shown. Culioli threads carefully and uses some expression to make its rigor explicit. We observe the use of mathematical concepts, but without these being made explicit. We also observe the use of a mathematical-like writing whose status is almost that of a metalinguistic language. We are therefore in a situation where it is difficult to evaluate epistemologically the status of conceptual transfers. However, these transfers are part of Culioli's *aimed ethos*. At the crossroads of the use of mathematical writing, mathematical concepts, linguistic concepts, and Culioli's scientific writing, mathematical imaginary is formed. This mathematical imaginary is his *aimed ethos*. This does not mean that Culioli wanted to produce this *ethos*. However, these are the rhetorical effects produced by his approach.

This *aimed ethos* responds perfectly to the *expected ethos*, because this *expected ethos* comes from the expectations built, on the one hand, on an imaginary about Culioli and, on the other hand, on an *ethical world*. This imaginary about Culioli is nourished by tributes, but also by the reception of the mathematical imaginary of Culioli in other addressees. Culioli's scientific discourse, its mathematical imaginary, the imaginary about Culioli and stereotypical vision of science are part of the encompassing sociological process from which flows the *ethical world*. Thus, the three aspects of Culioli's *ethos* are the mathematical imaginary (*aimed ethos*), the Culioli's tribute and imaginary (*expected ethos*), the prevailing of a certain vision of science (stereotypical vision of science and epistemological validation criteria), which nurture the *ethical world*. The resulting *ethos* (*produced ethos*) differentiate itself from other linguists' *ethos* by the Culioli's specific mathematical writing, scientific writing and mathematical conceptualization. This *ethos* can operate because it is in harmony

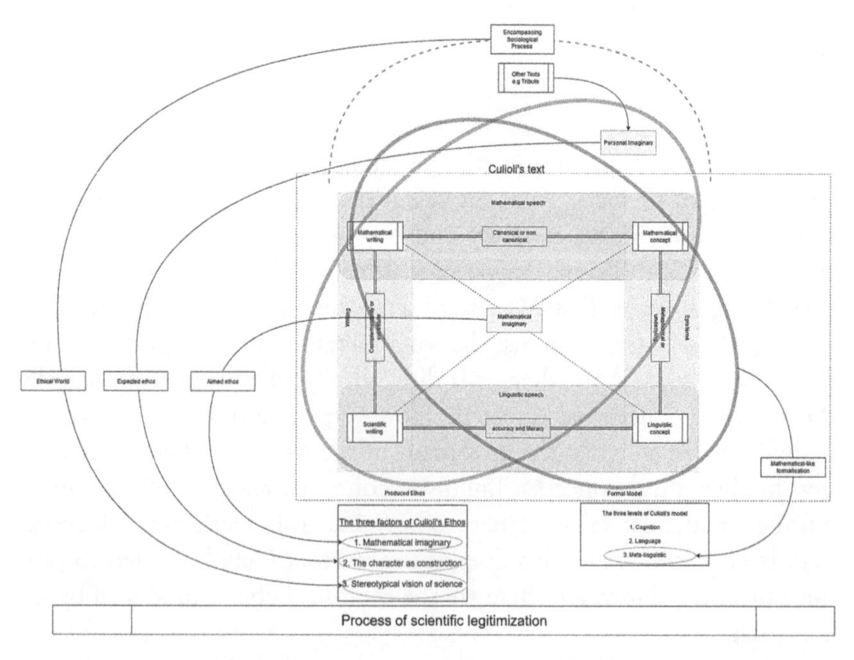

Fig. 5.5 The dynamics of conceptual transfers in Culioli's work

with the functioning of the conceptual transfers, which create Culioli's model. From a rhetorical point of view, Culioli's text is at the heart of a process of legitimization. This process works by the conjunction (1) of the use of mathematical concepts and writing to describe linguistic concepts as a formal model and (2) of the use of these mathematical concepts and writing to write scientifically as *ethos*. This conjunction responds to a certain vision of science (Fig. 5.5).

References

Achard, Pierre. 1992. Antoine Culioli, Pour une linguistique de l'énonciation. T.l: Opérations et représentations; Jean-Jacques Fraenkel et Daniel Lebeau, Les figures du sujet; Ham Adamczewski, Le français déchiffré, clé du langage et des langues. *Langage et Societe* 61: 81–85.

Assal, J.-L. 1995. La métaphorisation terminologique. *Terminology Update* XXVIII (2): 22–24.

Auchlin, Antoine. 2001. Ethos et expérience du discours: quelques remarques. In *Politesse et idéologie: Rencontres de pragmatique et de rhétorique conversationnelle*, ed. Michel Wauthion and Anne-Catherine Simon, 75–93. Leuven/Louvain: Peeters.

———. 2016. L'expérience du discours : comment et pourquoi y accrocher son attention. In Langage et savoir-faire. Des pratiques professionnelles du travail social et de la santé passées à la loupe, dir. Kim Stroumza, and Heinz Messmer, 113–145. Genève: IES Editions

———. 1972. *Le Degré zéro de l'écriture. Suivi de: Nouveaux Essais critiques*. Paris: Seuil.

Benveniste, Émile. 2006. *Problèmes de linguistique générale 1*. Paris: Gallimard.

Bert, Jean-François. 2014. La linguistique française à la lumière du marxisme. *Le Portique* 32. https://journals.openedition.org/leportique/2717

Bessis, Raphaël, and Lucas Degryse. 2003. Entretien avec Jean Baudrillard. *Le Philosophoire* 19 (1): 5–21.

Boisson, Claude. 1999. Le concept de "métalinguistique" dans la linguistique anglaise. *Anglophonia/Sigma* 3 (6): 151–198. https://journals.openedition.org/anglophonia/683

Bourdieu, Pierre. 1976. Le champ scientifique. *Actes de la Recherche en Sciences Sociales* 2 (2–3): 88–104.

Carnap, Rudolf. 1937. *The logical syntax of language*. Oxford: Harcourt.

Carnino, Guillaume. 2015. *L'invention de la science*. Paris: Seuil.

Chevalier, Jean-Claude. 2010. À un jeune linguiste: lisez Culioli ! *Modèles linguistiques* 3: 111–114.

Chomsky, Noam. 1980. *Rules and representations*. New York: Columbia University Press.

Culioli, Antoine. 1968. La formalisation en linguistique. *Cahiers pour l'analyse* 9 (7): 106–117.

———. 1985. *Notes du Séminaire de DEA 1983–1984, prises par ses étudiants*. Poitiers/Paris: Département de Recherches Linguistiques de l'Université de Paris 7.

———. 1990. *Pour une linguistique de l'énonciation 1*. Paris: Ophrys.

———. 1999. *Pour une linguistique de l'énonciation 2*. Paris: Ophrys.

Cuny, Noëlle. 2011. D'un style scientifique dans certaines revues d'avant-garde (*BLAST*, *The Signature*, *The Egoist*, 1914–1915). *Études de stylistique anglaise* 2: 23–38.

Détrie, Catherine. 2007. *Antoine Culioli, un homme dans le langage*, ed. Dominique Ducard and Claudine Normand. Paris: Ophrys, 2006, 378p. *Cahiers de praxématique* 48: 243–47.

Ducard, Dominique. 2016. La formalisation dans la théorie des opérations énonciatives: formes, formules, schémas. *Dossiers d'HEL* 9: 113–122.

Ducard, Dominique, and Normand Claudine. 2006. *Antoine Culioli, un homme dans le langage*. Paris: Ophrys.

Fuchs, Catherine. 2009. La linguistique cognitive existe-t-elle ? *Quaderns de filologia. Estudis literaris* 14: 115–133.

Granon-Lafont, Jeanne. 1995. La barre et la topologie lacanienne. *Linx* 7: 439–445.

Grésillon, Almuth, and Jean-Louis Lebrave. 2012. Antoine Culioli – "Toute théorie doit être modeste et inquiète". Entretien avec Almuth Grésillon et Jean-Louis Lebrave. *Genesis* 35: 147–155.

Grize, Jean-Blaize. 2006. *Antoine Culioli, un homme dans le langage*, ed. Dominique Ducard and Claudine Normand. Paris: Ophrys.

Hammarström, Göran. 1978. Is linguistics a natural science? *Lingua* 45: 15–31.

Hubien, Hubert. 1968. Philosophie analytique et linguistique moderne. *Dialectica* 22 (2): 96–119.

Jay, Monique. 1998. Sur l'écriture en sciences humaines. *Journal des anthropologues. Association française des anthropologues* 75: 109–128.

Jedrzejewski, Franck. 2017. Deleuze et la Géométrie Riemannienne: Une Topologie des Multiplicités. In *From Riemann to differential geometry and*

relativity, ed. Lizhen Ji, Athanase Papadopoulos, and Sumio Yamada, 311–358. Cham: Springer.
Juignet, Patrick. 2015. Philosophie, science et société [en ligne]. https://philosciences.com/index.php?option=com_content&view=article&id=112:karl-popper-et-les-criteres-de-la-scientificite&catid=15
Koster, Jan. 2005. Is linguistics a natural science? http://www.let.rug.nl/koster/papers/KOSTERsquib.pdf
Krajewski, Pascal. 2015. Qu'appelle-t-on un medium ? *Appareil*. https://doi.org/10.4000/appareil.2152.
Lakoff, George, and Mark Johnson. 1980. *Metaphors we live by*. Chicago: University of Chicago Press.
Latour, Bruno. 2001. *L'espoir de Pandore: pour une version réaliste de l'activité scientifique*. Paris: La Découverte.
Lazard, Gilles. 2007a. Le langage peut-il être objet de science ? *Comptes rendus des séances de l'Académie des Inscriptions et Belles-Lettres* 15: 347–363.
———. 2007b. La linguistique cognitive n'existe pas. *Bulletin de la Société de Linguistique de Paris* 107: 3–16.
Lefebvre, Muriel. 2006. Les écrits scientifiques en action: Pluralité des écritures et enjeux mobilisés. *Sciences de la société* 67: 3–16.
Lorenzini, Daniele. 2015. What is a "regime of truth"? *Le Foucaldien* 1 (1). https://doi.org/10.16995/lefou.2.
Lowrey, Brian, and Fabienne Toupin. 2010. L'invariant à l'épreuve de la diachronie. *Corela. Cognition, représentation, langage* 8 (2). https://doi.org/10.4000/corela.1853.
Maingueneau, Dominique. 2002. L'ethos, de la rhétorique à l'analyse du discours, Version raccourcie et légèrement modifiée de "Problèmes d'ethos". *Pratiques* 113–114. http://dominique.maingueneau.pagesperso-orange.fr/pdf/Ethos.pdf
———. 2013. L'èthos: un articulateur. *COnTEXTES* 13. https://journals.openedition.org/contextes/5772
———. 2016. Énonciation et analyse du discours. *Corela. Cognition, Représentation, Langage 19* (*HS*). http://journals.openedition.org/corela/4446
Martin, Robert. 2001. *Sémantique et Automate. L'apport du dictionnaire informatisé*. Paris: Presses Universitaires de France.
Martin, Olivier. 2003. Les mathématiques dans l'écriture en sciences humaines. Evolutions textuelles, transformations conceptuelles et épistémologiques. In *Figures du texte scientifique*, ed. Jean-Michel Berthelot, 193–223. Paris: Presses Universitaires de France.

Mathy, Adrien. 2017. Formaliser le trope. De la subjectivité linguistique à la subjectivité épilinguistique. *Signata. Annales des sémiotiques/Annals of Semiotics* 8: 313–340.

———. *In press*. What do the Digital Humanities and the Language Sciences say about the evolution of SHS? A comparative approach. In *Langues et linguistique.*

Merton, K.Robert. 1973. The normative structure of science. In *The sociology of science*, ed. Norman W. Storer, 267–278. Chicago: University of Chicago Press.

Milner, Jean-Claude. 1989. *Introduction à une science du langage*. Paris: Seuil.

Neveu, Franck. 2004. *Dictionnaire des sciences du langage*. Paris: Armand Colin.

Oliveira, Isabelle. 2009. *Nature et fonctions de la métaphore en science. L'exemple de la cardiologie*. Paris: L'Harmattan.

Pestre, Dominique. 2006. *Introduction aux Science Studies*. Paris: La Découverte.

Popper, Karl. 1973. La logique de la découverte scientifique. Paris : Payot.

Regnauld, Hervé. 2012. Les concepts de Félix Guattari et Gilles Deleuze et l'espace des géographes. *Chimères* 76 (1): 195–204.

Rinck, Fanny. 2010. L'analyse linguistique des enjeux de connaissance dans le discours scientifique. *Revue d'anthropologie des connaissances* 4 (3): 427–450.

Rossi, Micaela. 2014. Métaphores terminologiques: fonctions et statut dans les langues de spécialité. In *4e Congrès Mondial de Linguistique Française. SHS web of conferences*, 8. https://doi.org/10.1051/shsconf/20140801268.

de Saussure, Ferdinand. 2002. *Écrits de linguistique générale*. Paris: Gallimard.

Schmid, Anne-Françoise, and Mathieu Nicole. 2014. *Modélisation et interdisciplinarité: Six disciplines en quête d'épistémologie*. Versailles: Quae.

Swiggers, Pierre. 1998. *Histoire de la pensée linguistique*. Paris: Presses Universitaires de France.

Timbal-Duclaux, Louis. 1985. Textes "inlisable" et lisible. *Communication & Langages* 66: 13–31.

Verdès-Leroux, Jeannine. 1998. 'Il n'est pas scientifique. *LExpress.fr*, Octobre 1. https://www.lexpress.fr/culture/livre/il-n-est-pas-scientifique_802517.html

Volken, Henri. 2009. Orientation émotionnelle mathématique: la raison esthétique. *Revue européenne des sciences sociales* XLVII: 121–134.

Waquet, Françoise. 2019. *Une histoire émotionnelle du savoir. XVIIe-XXIe siècle.* Paris: CNRS Éditions.

Zarca, Bernard. 2009. L'ethos professionnel des mathématiciens. *Revue Française de Sociologie* 50 (2): 351–384.

Part II

Linguistics and the Natural Sciences

6

Scientific Realism and Linguistics: Two Stories of Scientific Progress

Johannes Woschitz

1 Introduction

Discussions about scientific progress have been predominantly focused on physics and chemistry, without making significant inroads into other scientific disciplines. It seems that this has started to change recently, as discussions of scientific progress now tend to go beyond the realm of particles (see e.g. Alexandrova 2016; Kincaid 2008). Linguistics has become part of the same trend. Nefdt (2016a, b), for instance, has pondered continuity in generativism. He argues that even though we cannot infer ontological continuity from the history of generativism (minimalism has arguably little to do with the generativism of the 1960s), we can find continuity in its scientific modeling practice of *idealization*. In the context of sociolinguistics, I have laid out how, in the past 50 years, the explanans of large-scale sound change has shifted from language internal

J. Woschitz (✉)
School of Philosophy, Psychology, and Language Sciences, University of Edinburgh, Edinburgh, UK
e-mail: johannes.woschitz@gmx.net

R. M. Nefdt et al. (eds.), *The Philosophy and Science of Language*,
https://doi.org/10.1007/978-3-030-55438-5_6

factors to socially meaningful language behavior that is logically prior to what Neogrammarians theorized as rigid 'sound laws' (Woschitz 2019).

The goal of this paper is to take these descriptions one step further and ask what they can tell us about progress in their respective fields. While Nefdt (2016a, b, 2019) lays out modeling continuity in generativism quite convincingly, he is hesitant in passing judgment on the ontological breaches he discusses. If we want to ponder scientific progress, however, we need something to measure these ontological breaches against; something by virtue of which we are then able to say: "Theory A is better than Theory B, and *hence* the ontological breach", without ignoring the fact that what counts as better, truer or more real is, in part, theory-dependent (Kuhn 1970 [1962]).

In ways which will be laid out in the following two sections, scientific realism has the potential to offer such a measure, and it provides a theoretical toolkit that lets one break down seemingly diverse driving forces behind theory change in contemporary linguistics to broader underlying philosophical criteria. Two of these criteria shall be discussed in detail: *explanatory power/value*, where I shall draw on the philosophy of Hempel (1965); and *the avoidance of value-laden presumptions*, where I shall draw on Nagel (1961). When assessed against these two criteria, I argue that a theory's 'relational objectivity' (Nagel 1961) becomes evident.

2 A Short History of Scientific Realism

In a nutshell, scientific realists believe that the aim of science is to provide us with theories which, literally construed, can be accepted as true or approximately true (see e.g. Boyd 2010). 'True' theories tell us something about how the world is structured, and how physical, chemical or other phenomena are subject to lawful behavior that can be described with the help of scientific scrutiny. 'True' theories bring about a state of affairs in which students can, in good conscience, read a chemistry textbook and take the existence of molecules at face value. Our best theories point to the fact that our concept of molecules corresponds to mind-independent facts of the world, and the same holds for atoms in physics and chromosomes in biology.

From this, it is only a small step to account for why some theories are preferred over others in the school syllabus: the preferred ones are taken to correspond better to mind-independent facts than others. Newtonian mechanics, for instance, describes the falling of a stone with recourse to gravity, the physical force between any two objects with mass. Aristotle, on the other hand, held that a stone falls to the ground because it moves toward its natural place (Kuhn 1970 [1962], p. 104). From today's point of view, this is an approach that does not reach the standard of scientific scrutiny because we no longer believe in entities having such intrinsic properties by which they act (see Chomsky 2006, p. 7, chapter 1).[1]

Similar examples from other fields are abundant. Before the Chemical Revolution in the seventeenth and eighteenth centuries, for instance, people believed that so-called phlogiston was released during combustion. However, many inconsistencies arose with such a view. Metals, for instance, gain mass during combustion. This means that, under certain circumstances, phlogiston needed to have variable mass; sometimes negative, sometimes positive, dependent on the material in question (see e.g. Partington and Mckie 1937). Lavoisier's account of combustion as a chemical reaction with oxygen dispenses with such irregularities. In his theory, combustion requires a gas with mass (oxygen). Burning metals combine with oxygen, which explains the mass increase. Therefore, when one sends one's child to school, they will learn about oxidation theory, not phlogiston.

The crux of the matter is that, in the late eighteenth century, the general belief was that phlogiston was *real* – corresponded to mind-independent facts – much as today's chemists believe in the actual reality of oxidation. Throughout ancient and medieval times, both in the West and the East, all material things, in line with Aristotle's view, were believed to be a combination of the four elements water, earth, air and fire (see Joseph 2018, pp. 49–50). What convinced people to give these theories up? What gives us the epistemic authority to deem these theories less

[1] James McElvenny has pointed out to me that Newton's treatment of gravity is in its own way mysterious because it posits gravity as an irreducible force that acts at a distance. That is a strong metaphysical statement that Newton himself has been criticized for in his day. Einstein's general theory of relativity does away with this metaphysical baggage because it describes gravity not as a force but as a consequence of the curvature of spacetime.

trustworthy than our state-of-the-art theories? To address this issue, scientific realists have begun scrutinising the characteristics in virtue of which some theories are better than others at explaining a given phenomenon. This is important if one wants to avoid conceiving of scientific realism as merely grounded in hindsight. While it is comparatively easy to point one's finger at errors and misconceptions in the past, when faced with contemporary theories, things get more complicated.

Psillos (2000, p. 706) has identified three assumptions most schools of scientific realism share in their assessment of theories: the *metaphysical thesis*, which holds that '[t]he world has a definite and mind-independent structure'; the *semantic thesis*, which holds that '[s]cientific theories should be taken at face value'; and the *epistemic thesis*, which holds that entities posited by a theory really inhabit the world. Most discussions, it seems, are centered upon either metaphysical or semantic claims and often, these two are intertwined. If I, on a hike, pick up a stone and say, 'this is a stone', I (a) imply that this stone constitutes a mind-independent fact (metaphysical), and (b), that the word 'stone' in English accurately refers to this fact (semantic). Or to put it in a more sophisticated way, in so doing I claim that a term we use denotes truthfully an entity that exists mind-independently. A possible caveat is that, intuitively, this only pertains to what analytic philosophers call 'natural kinds'. Can the same logic apply when one talks about abstract terms such as *love*, *society* or *language*? Given that it is in the spirit of this paper not to jump the gun, I do not want to take social scientific explananda out of the equation too hastily. The Durkheimian term 'social facts' implies the mind-independence of what it denotes, the key word being *facts*; or, to give another example, no linguists have expressed doubts that there is a mind-independent shift in pronunciation of millions of people in the United States (see e.g. Labov 2007), even though it is a social product and has nothing to do with 'natural laws' in the strict physical sense.

If we focus on the *semantic thesis* for the moment: as stated above, it holds that our scientific theories, including the terms within our theories, should be taken at face value; or, in philosophical jargon, that they refer to a mind-independent state of affairs either truthfully or not. What is the source of the link between the word and the referent; a jargon term and the entity existing mind-independently? Typically, this link is established via a truth-value, which, under the traditional semantic view, is fixed in a sentence (see Putnam 1981, pp. 32–33): *X is a stone* is true if and only if x belongs to the set of stones. Such a view is problematic, however, because 'truth-conditions of whole sentences underdetermine reference' (Putnam 1981, p. 35, emphasis omitted). This was shown in Putnam's (1981) famous permutation argument: we can switch the referents in the building blocks of a sentence while the truth value of the sentence remains the same, such that the truth values of the sentences *the cat is on the mat* and *the cat* is on the mat** are equivalent, even though *cat** picks out *cherries* and **mat* picks out *trees* (while *cat* and *mat* pick out their usual referents as expected). Truth-conditions of whole sentences, then, do not tell us anything about how reference works. But this is important, since one wants to take a word to truthfully (semantically) denote an actual mind-independent state of affairs – just as one wants whole statements of a scientific theory to do.

Philosophers have long engaged in heated debates over this issue. Putnam's permutation argument was a fatal argument against semantic descriptivism (on which see Michaelson and Reimer 2019, section 2.1). Descriptivism is a view attributed to Frege and Russell which holds that speakers have, in their minds, a variety of descriptions such as 'the man who is currently president of the USA' and 'the man who was born in June 1946'. When speakers hear the name Donald Trump, they associate it with these descriptions and are therefore able to pick out the correct referent. Meaning, in this view, resides in the head, while 'referential success hinges on speakers attaching to each name in their repertoire some descriptive content *F* which uniquely singles out a specific object in the world' (Michaelson and Reimer 2019, section 2.1, italics in original). An alternative proposed in the Kripke/Putnam causal view of reference (e.g. Putnam 1975) holds that meaning lies instead in the referent and is chained to it via causal baptism. That is, reference is fixed externally, and meaning depends on a correspondence to a class of objects which shares likeness with this referent. For instance, at some point, someone decided to call H2O water[2] (even if that someone was not necessarily aware of its

[2] Or perhaps something like *h_2ék^weh_2 in the case of Proto-Indo-European, but these details tend to be overlooked by people who want to make a philosophical point.

chemical formula), and from then on, people have used the term water to denote a class of objects which shares likeness with the referent (in this case, H_2O).

Others, such as Quine (1968), have contested this view by arguing that reference cannot be fixed externally that easily. Put simply, what guarantees that when someone points at the ocean and says "Water!" that a non-English speaking witness does not think the person means the movement of the waves or the color of the ocean? We always need a background 'coordinate system' (Quine 1968, p. 200) to regress into, something the interlocutors share and with recourse to which they are able to disambiguate "Water!". In a first instance, Quine argues, it is the language they share. In a second instance, it is the ontological system we acquire with language.[3] Since, so the argument goes, different ontologies come with different languages, the best we can do is translate a term unknown to us with reference to our own ontology. Or put differently, '[a] question of the form "What is an *F*?" can be answered only by recourse to a further term: "An *F* is a *G*". The answer makes only relative sense: sense relative to an uncritical acceptance of "*G*"' (Quine 1968, p. 204, italics in original). Such a view relativizes ontology because reference becomes dependent on something that is not set in stone but acquired in our upbringing. Interlocutors are treated as translators between different world-views; and ontology, Quine argues by quoting Kant, becomes part of transcendental metaphysics.

Quine's postulate of a background coordinate system has often been rejected as too radical – it would heavily threaten the *metaphysical thesis* stated above – but it has had considerable impact on competing theories. Putnam's (1982) internal realism, for instance, incorporates it by holding that there might be a mind-independent state of affairs (here he remains faithful to the *metaphysical thesis*), but that it is the scientific method that puts our thought in correspondence with it, not language, as Quine would have it (see Putnam 1982, pp. 145, 162–163). With this change, a hard-line approach to truth, in which a term can truthfully denote a

[3] Here, Quine seems to follow ideas commonly associated with the Sapir-Whorf-hypothesis, which, strictly speaking, is not one coherent theory but a collection of ideas that posit that our language at least inclines us to perceive a thing in a certain way (see Joseph 2002, chapter 4).

mind-independent entity or not, gives way to a softened approach in which 'rational acceptability' (Putnam 1981, p. 49) decides on the value of a theory. Quine's critique has evidently rubbed off on Putnam, along with many others, because correspondence between a jargon term and a mind-independent state of affairs is no longer something that is solely determined externally, but something that scientists co-construct in their theorizing.

From this point of view, 'progress' could be conceived of as a process in which scientific theories incrementally get better at world-representation. But more often than not, theory changes go hand in hand with radical changes in world-representation, and the continuity one would expect under such a view is practically non-existent. In his famous *The Structure of Scientific Revolutions*, Kuhn (1970 [1962]) argued that we cannot infer ontological continuity (and therefore 'progress' in a streamlined sense) from the history of physics or chemistry. To return to the *water* example above: according to Putnam, the term *water* has always picked out H_2O, even when people had yet not found out about the chemical formula empirically. But in the 1750s, Kuhn (1990, pp. 310–312) convincingly argues, the referent of *water* picked out only liquid water. Only as late as in the 1780s did chemists begin to distinguish between states of aggregates (Kuhn 1990, pp. 310–312). When people in the 1750s spoke of *water*, they denoted a subset of what *water* denotes today. What is more, before the Chemical Revolution, chemists distinguished between species more or less by 'what are now called the states of aggregation' (Kuhn 1990, p. 311). Water was subsumed under other liquids on account of having the same aggregate, which is no longer the case. So, can we really say that *water* has always denoted H_2O? Kuhn's answer is no. While he acknowledges that Putnam's logic might still pertain to natural kinds (*water* in the 1750s still "kind of" picked out what it picks out nowadays), when analysing the historical development of terms like *planet*, *star*, *force* or *weight*, one can no longer speak of minor adjustments (Kuhn 1957, 1990, p. 313). Aristotle's definition of *motion*, for instance, is characterized by two endpoints (Kuhn 1990, p. 299): if change of quality happens between two contraries, such as *black* and *white*, Aristotle would speak of *motion* (Rosen 2012, p. 65). If change happens between endpoints that are 'neither contraries nor intermediates but rather contradictories', such as *white and not white*, Aristotle would not classify that as *motion* (Rosen 2012, p. 65). Are these criteria still relevant when Newton talks about *motion*? This time, Kuhn's answer would be a definite no.

Consequently, much as chemists nowadays think about *water* within different paradigms from those of pre-Chemical Revolution times, physics underwent a paradigm shift between Aristotle's *Physics* and Newton, and again when Einstein jettisoned some key assumptions of Newton in his theory of relativity (and yet again with the emergence of Quantum Theory). These competing paradigms (our background languages in Quine's terminology), Kuhn (1970 [1962], 1990) argues, are incommensurable, because they are not translatable into one another. In Kuhn's view, Quine's translator is therefore really a language learner (Kuhn 1990, p. 300). When physics students nowadays study Aristotle, they have to acquire a whole lexicon of the categories Aristotle thought in. It is not enough for them to trace the putative development of *motion*, because he and Newton conceived of *motion* in such different terms. Even if two competing paradigms use the same term, there would never be an exact 1:1 correspondence, because with new lexicons, new referents are introduced (Kuhn 1970 [1962], pp. 200–201). And, per Saussure's principle of contrast, when new referents (or more precisely, signifieds) are introduced to the lexicon, the whole coordinate system changes (Joseph 2012, pp. 597–600).

3 Of Superseding and Superseded Theories

This short history of scientific realism shows us two important things for the following discussion of scientific progress. One is that many philosophers of science want to hold that correct theories correspond to mind-independent facts, as per the *metaphysical thesis* (also subsumed under 'the correspondence theory of truth', see Marian 2016). However, there is no God's-eye view, and we can only assess scientific claims about the world against the backdrop of our own scientific theorizing. This is complicated by the fact that, as Kuhn argued, there is no ontological continuity in scientific theorizing. Is this a nail in the coffin for belief in scientific progress?

Kuhn *did* believe in scientific progress; his criticism just happened to complicate the picture. What he rejected is progress in the form of 'evolution-toward-what-we-wish-to-know' (Kuhn 1970 [1962], p. 171), as if scientists incrementally worked toward ideal world-representation. But he did hold that scientific revolutions happen for a reason: a novel paradigm can for instance lead to better accuracy of prediction, an increased number of problems solved or better compatibility with other specialities when compared to the superseded paradigm (Kuhn 1970 [1962], p. 206). From this perspective, nothing prohibits one from stating that Galileo's analysis of the pendulum movement improved on Aristotle's, or that Einstein's operationalisation of gravity allowed for a better account of light deflection than Newton's (see Kuhn 1970 [1962], pp. 206–207). Kuhn's sense of progress is then an 'evolution-from-what-we-do-know' (Kuhn 1970 [1962], p. 171). Some paradigms are better equipped to solve puzzles than others, but the puzzles are ultimately constrained by the paradigm one is working in. From this it follows that one can and should measure the success of theories against each other, not against an idealized notion of truth.

A new paradigm, for Kuhn (1970 [1962], pp. 23–24), does not gain ground because it is necessarily successful, but because it *promises* success. The *actual* scientific success can often take years, decades, sometimes centuries to manifest. But oftentimes, scientists need to choose among two or more competing theories rather spontaneously. It is hard for them to estimate the success potential of an emerging paradigm, so their decision often requires a leap of faith. This allowed scientists to commit to the wave theory of light, even though the corpuscular theory was initially more successful in resolving, for instance, polarization effects (Kuhn 1970 [1962], p. 154). This allowed them to commit to the heliocentric model, even though the geocentric view, meticulously discussed in Kuhn (1957) as the 'two-sphere universe', was very successful in predicting trajectories of stars and the sun.

It is scientific crises that get people to convert from an old paradigm to a new one nonetheless (Kuhn 1970 [1962], p. 77). A new paradigm could, for instance, promise to explain phenomena that, in the current paradigm, cannot be accounted for accurately enough (or, in more extreme cases, are treated axiomatically). In the example just mentioned, two-sphere theories – increasing theoretical complexity notwithstanding – never really achieved measurement accuracy of planetary movement. This led to the so-called Copernican revolution (Kuhn 1957), whose heliocentric model outperforms the geocentricism of the two-sphere model on the grounds of its *explanatory power/value*: if we treat the earth as a planet that itself moves, the movement of other planets becomes less opaque. Such explanatory power is moreover often linked to the new paradigm outperforming the old by virtue of being less dependent on *value-laden presumptions*, examples of which can be found below.

Both notions are philosophically rather vague. Every description, even if taxonomic in nature, creates knowledge. The Bible, arguably, explains *everything*, from the creation of the universe to the moral understanding of humanity. But explanatory value and value-laden presumptions are still utilized and endorsed by philosophers and scientists alike; the following two chapters are cases in point. When it comes to explanatory value, Chomsky argued in the late 1950s that Neo-Bloomfieldian distributionalism had mere *descriptive* adequacy (see Chomsky 1965, pp. 26–27; also Matthews 1993, pp. 31–32), while his notion of a mental grammar shed light on the intricacies these descriptions are born from (and thus has *explanatory* adequacy). So-called third-wave variationists claim that what Labovian variationists do is merely *correlational* (they establish patterns like: working-class males are more likely to say *I'm walkin' down the street* instead of *I'm walking down the street*), while their own approach traces these large-scale patterns back to socially meaningful persona projection. In other words, correlating language variation and change with social macro-categories such as social class or gender does not *explain* the underlying dynamics that make them important. In this regard, the generative and third-wave variationist criticisms are similar: predecessors are credited for their meticulous description of the phenomenon in question, but now it is time to explain how the described regularities come about.

Hempel (1965) argues that scientific progress is essentially a teleological process from descriptive to explanatory quality:

> Broadly speaking, the vocabulary of science has two basic functions: first, to permit an adequate *description* of the things and events that are the objects of scientific investigation; second, to permit the establishment of general laws or theories by means of which particular events may be *explained* and *predicted* and thus *scientifically understood*; for to understand a phenomenon scientifically is to show that it occurs in accordance with general laws or theoretical principles. In fact, granting some oversimplification, the development of a scientific discipline may often be said to proceed from an initial "natural history" stage, which primarily seeks to describe the phenomena under study and to establish simple empirical generalizations concerning them, to subsequent more and more "theoretical" stages, in which increasing emphasis is placed upon the attainment of comprehensive theoretical accounts of the empirical subject matter under investigation. (Hempel 1965, pp. 139–140, emphasis in original, footnote omitted)

For the reasons discussed in the preceding chapter, such a view seems idealistic and falls short of addressing either the problems circling Psillos' semantic thesis (p. 146 above), or the Kuhnian notion of theory-dependence in general. In the majority of cases, the 'general laws or theoretical principles' alluded to in the quote are projections from within theories themselves, rather than what Hempel calls phenomena. But the essence of the quote seems reasonable nonetheless: if an observer on earth predicts the sunrise from the crowing of a rooster, they have established a descriptive correlation, but that does not explain *why*, from their perspective, the sun seems to move (Craver 2006, pp. 357–358). It is a mere starting point for in-depth reasoning, and one could come up with a range of hypotheses to try to account for the factors that could underlie the perceived movement of the sun. For instance, one could postulate that (1) the sun tailors its movement to the activity of roosters, or (2) that what we perceive as a sunrise is the result of earth's eastward rotation on its axis, and that roosters tailor their crow to the presence or absence of sun exposure.

Few readers will take the first hypothesis seriously, and to account for their intuition, Nagel's (1961) notion of 'value assumptions' is useful.[4]

[4] I am using this term interchangeably with *value-laden presumptions* (motivated by Alexandrova's 2016 use of the term).

Nagel's (1961, p. 582) account of scientific explanation is less idealistic than Hempel's, because it is not reliant on objective, theory-external truth. '[T]hough absolutely objective knowledge of human affairs is unattainable', he argues, 'a "relational" form of objectivity called "relationism" can nevertheless be achieved'. A theory with relational objectivity should be able to state all its claims in the form of conditionals: if we operationalize X as Y, Z holds. What can then be criticized is the operationalisation of X as Y, as other operationalisations might be better or more accurate. To return to the examples above, hypothesis (1) hinges on roosters having extraterrestrial powers, while (2) does not require the assumption that what we perceive as the sun's movement is influenced by such small-scale factors. We have abundant reason to believe that the former operationalisation is incorrect, and hence is a value-laden presumption (perhaps motivated by a desire for this distant, mysterious, powerful object that is the sun to be under the control of creatures over whom we ourselves have control) that should be dropped.[5]

Notice that the notion of value-laden presumptions is necessarily intertwined with explanatory value. What validates or invalidates a value assumption is, in the spirit of scientific realism, in part dependent on a mind-independent reality, in this case the modern view that the sun does not move at all and is instead orbited by planets. For the following discussion of Chomsky and Labov, I will keep the notions of explanatory value and value-laden presumptions separate nonetheless, for heuristic purposes. Both scholars, it seems, have attributed less and less explanatory power to certain entities over the years, until they have been identified by some as value-laden presumptions that need to be challenged or dropped entirely. This led to ontological breaches/discontinuities typically associated with scientific revolutions. Both disciplines, in their own ways,

[5] Another, rather bizarre example showing that predictive success does not necessarily warrant accurate world-representation comes from econophysics, a branch of economics that applies models from physics to solve problems in economics: One can model an economy's income distribution by treating economic agents as if they were two gas molecules colliding (see Bradley and Thébault 2019, pp. 84–86). This model presupposes that all 'zero intelligence' agents carry all their money with them in cash. If they meet another agent, they both drop their entire belongings and then pick up a random amount. This process is then repeated. While no one seriously believes that this is how a *homo œconomicus* actually behaves, the result is accurate enough to depict an economy's income distribution.

tell the story of theories being superseded by competing approaches because they rely on debatable value assumptions that bring their validity limits (Rohrlich and Hardin 1983) to the fore.

4 Chomsky: From a Formal Description of Language to Biolinguistics

A word of caution: Chomsky's approach is rational, hence non-empirical (though he would deny that, see e.g. Chomsky 1993 [1981], p. 35, 1995, pp. 171–172; 2006, p. 168), and therefore essentially not falsifiable (Itkonen 1978; Katz 1981). Granted, Chomsky has claimed over the years that the structures posited by the generativist paradigm can be falsified by data from a previously understudied language, or indeed from a well-known language. The replacement of earlier phrase-structure grammars by the Principles and Parameters framework (discussed below) is a prominent example of such a falsification-driven development. But the posited structures are always inferred and never directly observable, and therefore potentially at odds with the epistemic thesis outlined in Sect. 2 – we can never be sure if they actually describe how the brain works, or if they are merely part of a formalized description. This makes generativism rather hard to assess against the above-mentioned criteria of explanatory value and value-laden presumptions. I will try nonetheless, given that Chomsky himself, as alluded to above, grounded his criticism of structuralism and behaviorism in the fact the he assigned them mere *descriptive* adequacy. Generative Grammar, on the other hand, claims *explanatory* adequacy, because it ought to – theoretically – account for underlying principles of natural languages (see, for instance, Chomsky 1993 [1981], p. 87). The distinction between descriptive and explanatory adequacy is laid out in Chomsky's *Aspects of the Theory of Syntax* (1965).

Chomsky's œuvre seems to be a good example of what Kuhn (1970 [1962], p. 184) referred to when he said that there is a continuum between heuristic and ontological models. More often than not, changes in the ontological commitments of generativism have been papered over by downgrading past accounts to heuristic approximations – as if all theoretical detours in the past 60 years logically lead to the present view of things. *Syntactic Structures* (2002 [1957]) started out as a formal description of language, exemplified largely by English, in part as a reaction to Hockett's statistical information theory. Radick (2016) shows that Hockett was also the principal target of Chomsky's (1959) aggressive review of Skinner. At this stage, we read of no competence/performance dichotomy, no Language Acquisition Device, no Universal Grammar, let alone I-language. Only in *Aspects of the Theory of Syntax* (1965), and later in *Lectures on Government and Binding* (1993 [1981]) and in the *Minimalist Program* (1995) does he speak about language as biological endowment in its fully-fledged form. This is a first, relatively coarse-grained ontological reorientation: a formal description of language became a biological property of humans, and thus ontologically *real* (see Chomsky 2006, chapter 7; Nefdt 2019, pp. 178–179).

Other, more fine-grained ontological reorientations can be found in his treatment of so-called representational systems of grammar. In the so-called 'Standard Theory' developed in *Aspects*, Chomsky posits that every speaker possesses a mental linguistic capacity called "grammar" that allows them to generate language. This capacity "base-generates" a deep structure with the help of the syntactical component of grammar, and then derives from it a surface structure by various transformations.[6] This is typically exemplified by the following sentences: *John is easy to please* and *John is eager to please* have essentially identical surface structures, as both main clauses take infinitival complements. Their deep structures, however, are different, as *eager* takes a deep structure subject denoting the person who experiences the eagerness, whereas *easy* takes a clausal deep subject but undergoes subject-raising. In the course of transformations that map these deep structures to their surface structures, it just so happens that they have the same surface structure. Similar examples include passivization, emphatic affirmatives, negation or interrogatives. The latter two, Chomsky (2002 [1957], pp. 64–65) derives from the same

[6] It should be noted that Chomsky has never put forward the idea that for every sentence in every language, a deep structure is wholly fixed by UG. Rather, according to Chomsky, some sort of innate formal schematism is universal and present in the mind of every new-born human, but a distinction needs to be made between universal and parochial schematisms. I would like to express my gratitude to Geoff Pullum for pointing this out to me.

underlying structure, which he again takes as proof of the distinct reality of the two representational systems.

Over the years, the transformations between the representational systems have been refined in numerous ways. While in *Syntactic Structures*, Chomsky describes grammatical functions such as passivization and negation directly, in *Lectures on Government and Binding* he argues that grammatical functionality is subordinate to principles such as case and binding theory, which themselves can be reduced to the theory of government (Chomsky 1993 [1981], pp. 7, 14, 44, 121). Transformations, or movement in general, become more abstract in nature, and, for Chomsky, more explanatory because they describe abstract regularities that seem to underlie each and every language (and are therefore part of Universal Grammar). At the same time, more importance is attributed to semantics. In the Standard Theory, for instance, syntax is treated as relatively autonomous, with semantics playing a subordinate role of interpreting these syntactic structures (Matthews 1993, pp. 39–41). Many scholars deemed this a fallacious assumption, and after the so-called 'linguistics wars' in the 1960s and 70s (Harris 1993), Chomsky developed as an answer the Extended Standard Theory, in which semantic interpretation is determined by multiple levels of representation, not just one, via the so-called projection principle and the assignment of theta-roles. Syntactic behavior, for instance the ability or inability of a given verb to take a clausal complement (*I promised him [to leave]* is fine, but **I gave him [to leave]* is not), in this view, is lexically specified along with other morphophonological properties (Chomsky 1993 [1981], p. 5). Deep and surface structures are then essentially treated as 'projections of lexical properties' (Chomsky 1993 [1981], p. 39) – a considerable change of mind.

In a further step, in the Minimalist Program (Chomsky 1995), deep and surface structure are dropped, and with them the notion of government, which had been so carefully fleshed out in the *Lectures on Government and Binding*. The derivations – present since 1950s generativism – are instead entirely 'driven by morphological properties to which syntactic variation of languages is restricted' (Chomsky 1995, p. 194).

This concludes our brief philosophical journey: first, language-specific transformations were abandoned in favor of more abstract principles; then these abstract principles were dropped in favor of morphological properties of the lexicon, and deep and surface structure with them. Taken at face value, it is hard to infer any ontological continuity from this development, and we could thus speak about them as examples of Kuhnian scientific revolutions. But many generativists spot continuity elsewhere (here, Nefdt 2016a, b, gives a good overview). The main driving force behind this development, arguably, is Chomsky's conviction that humans are born with a language faculty that enables them, and only them (not animals), to create language (Chomsky 2006, chapter 3). The argument outlined in *Language and Mind*, and similarly in numerous other publications by Chomsky, is as follows: certain intricacies in language use are never available for introspection by the native speaker. However, every speaker of English, independent of intelligence, is able to assign complex stress patterns to sentences they produce, without being aware of the regularities. They are not learned, therefore they must be innate. In the 1960s and 70s, this led Chomsky and other generativists to posit that people are born with an innate schematism that consists of, for instance, the principle of cyclic application (which underlies phonological stress-assignment) and the so-called A-over-A principle (which prohibits the extraction **London, I enjoyed my trip to*, but allows *My trip to London, I enjoyed*[7]) (Chomsky 2006, pp. 39–40, 46). As principles like these proliferated, the sentiment grew stronger within linguistics that they would render language acquisition a trivial phenomenon (see Nefdt 2019, p. 181). When they were later dropped in favor of lexically specified morphological properties, the complexity behind language acquisition grew so large that it would become hard to account for the speed and ease of language acquisition (Joseph 2002, pp. 62–65).

Chomsky's answer to this impasse was that if there indeed exists a language faculty in the brain, its properties must be abstract and general enough so as to balance the complexity of natural languages with the ease

[7] I have been told that most native English speakers find the former sentence perfectly fine, which shows that grammaticality judgments made in syntactic theory are often peculiar to linguists themselves.

of language acquisition. When one describes how stress is assigned to English sentences, for instance, the scope of one's claims is likely to be limited to English alone, and one therefore runs the risk of describing an idiosyncrasy rather than a property shared by all natural languages. If, on the other hand, one wishes to uncover Universal Grammar as something that 'determine[s] the class of possible languages' (Chomsky 2006, p. 155), the characteristics must become necessarily more general, so the logic goes. This of course papers over the fact that the principle of cyclic application and the A-over-A principle were in fact believed to be part of that very UG, and it would be inadequate to not mention the fact that the ontological commitments UG eventually boils down to have changed considerably in the last 60 years.[8] But in a way, these reorientations are justified by the fact – and Chomsky acknowledges this from the start – that different theories of grammar can be descriptively adequate. In theory, they can all account for the same corpus of data, and they can all predict ungrammaticality. The principle of cyclic application outlined in Chomsky and Halle (1968) still holds – is still descriptively adequate – for stress patterns in English, but it would be too specific a principle to hold for natural languages in general, hence it lacks explanatory adequacy and cannot be part of Universal Grammar. There must, therefore, be a principle underlying it, which *is* part of UG. This captures the essence of the Principles and Parameters framework developed in the *Lectures on Government and Binding* (1993 [1981]). Principles and Parameters holds that natural languages share certain abstract principles, and it is these abstract principles that need to be captured by UG. In this vein, the principle of cyclic application can be re-theorized as the principle that transformations apply in embedded domains before applying in superordinate domains – in phonological and syntactic environments alike. A prominent example of a parameter is the existence of so-called empty categories, one of which is *PRO*, a phonetically unrealized feature sharing syntactic properties of pronouns which allows us to resolve pronoun antecedents in sentences like *I expected to leave* as *I_j expected PRO_j to leave.*

[8]Whether a cyclic principle is part of UG or not has been a controversial topic within generative literature (see Pullum 1992). One could make a case that it survived in Minimalism in the notion of 'phase', but only when acknowledging that it is not a phenomenon restricted to phonetic form (PF). I would like to express my gratitude to Geoff Pullum for pointing this out to me.

How Chomsky justifies the existence of such empty categories can be inferred from the following quotes. Notice that he still adheres to the idea of a deep and surface structure at this stage.

> Note that the distribution of the empty categories, the differences among them and their similarities to and differences from overt elements are determined through the interaction of quite simple principles that belong to several different subtheories, in accordance with the modular approach to grammar that we are pursuing throughout. (Chomsky 1993 [1981], p. 72)

> The range of phenomena – by no means exhausted in the preceding discussion – is fairly complex, but it follows, assuming the projection principle, from quite simple and for the most part independently-motivated principles of the interacting theories of bounding, government, binding, Case and control, a fact of some significance, I believe. (Chomsky 1993 [1981], p. 85)

What is remarkable is that he actually seems to follow Kuhn (1970 [1962], p. 171) here in his *modus operandi*. When it comes to improving theories, his concept of 'progress' seems to be one of 'evolution-from-what-we-do-know', rather than one of 'evolution-toward-what-we-wish-to-know'. The empty categories he posits follow from 'quite simple' and 'independently-motivated' principles. If we operationalize UG as consisting of these independent principles, it may no longer seem so outlandish to add empty categories which build them into our theoretical apparatus.

It is here that Chomsky's epistemological tension comes to the fore. His ontological commitment to grammar being a biological endowment separate from other cognitive traits does not seem to fit the *modus operandi* underlying his theorizing. His theorizing broadly falls under a coherence theoretical approach of truth (Young 2016), in which ontological commitments are not inferred from a mind-independent reality but from how well they fit with the applied theoretical apparatus. Such approaches to truth typically do not state an epistemological "endpoint" because they are at odds with the metaphysical thesis outlined in Sect. 2. For Chomsky, however, the epistemological endpoint is set: Universal Grammar is a biolinguistic capacity, separate from other cognitive faculties. From this perspective, it becomes quite illuminating why the study of Chomsky's œuvre tells us a story of abstraction and, in the end, one of widening scope. As generativists became aware that many properties they had previously attributed to the autonomy of language(s) might actually be part of more general, abstract cognitive capacities (Chomsky 2006, p. 183), the principles underlying each and every language (UG) have become broader and broader. In the end, it might all boil down to Merge, the capacity of natural languages to combine constituents to form larger structures. 'The simplest account of the "Great Leap Forward" in the evolution of humans would be that the brain was rewired, perhaps by some slight mutation, to provide the operation Merge', Chomsky (2006, p. 184) says. What is left of elaborated notions of I-languages as developed in the Standard Theory and Extended Standard Theory is a mere linkage of sound and meaning, which takes us back to a Saussurean doctrine, and ultimately to the behaviorists who Chomsky was so eager to resist in the first place (see Joseph 2002, p. 154).

5 Labov: From Linguistic Autonomy to Sociolinguistic Performativity

In many regards, William Labov can be regarded as a successor to the Neogrammarians, a late nineteenth century German school of linguists including, among others, Karl Brugmann (1849–1919), Hermann Osthoff (1847–1909) and Hermann Paul (1846–1921) (see Morpurgo Davies 1998, chapter 9). He himself praised them as the 'heroes of the story' (Labov 2006 [1966], p. 9) for their approach to the study of phonetic change, i.e. the study of changes in pronunciation. The Neogrammarians held that 'every sound change, inasmuch it occurs mechanically, takes place according to laws that admit no exception' (Osthoff and Brugmann 1878, p. XIII). If the pronunciation of /a/ changes, it will do so in every word in which /a/ occurs, without exception. 'Mechanical' needs to be understood as the opposite of the 'organic' approach which they attributed to the work of August Schleicher (1821–1868), who conceived of language as a biological entity that evolves independently of its speakers (Morpurgo Davies 1998, chapter 9). The Neogrammarians instead proposed that the emphasis should shift to the speakers who actually produce and change language on a daily basis; and that the explanation of language change needs to be sought in the change of the psychological 'sense of movement', or *Bewegungsgefühl* in Paul's (1920 [1880]) terminology. Only if the *Bewegungsgefühl* of e.g. the pronunciation of a vowel changes in many individuals at the same time can one speak of a change in the 'language custom' (*Sprachusus*). Readers familiar with the work of Labov will immediately recognize parallels between his and their theorizing. In fact, Paul (1920 [1880]) even mentioned repeatedly that changes in the sense of movement can go entirely unnoticed, an important determinant in Labov's (1972, 2006 [1966]) early work to distinguish between *change from above* and *change from below*.

These similarities notwithstanding, the modern paradigm of language variation and change (the Labovian school, nowadays often called 'first wave variationism') was essentially born out of a criticism of the above-quoted Hermann Paul, whose *Principien der Sprachgeschichte* 'at some stage counted as the Bible of the contemporary linguist' (Morpurgo Davies 1998, p. 235). Weinreich, Labov, and Herzog (1968) argue that Paul overemphasized the role of the individual:

> Jede sprachliche Schöpfung ist stets nur das Werk eines Individuums. Es können mehrere das gleiche schaffen, und das ist sehr häufig der Fall. Aber der Akt des Schaffens ist darum kein anderer und das Produkt kein anderes. Niemals schaffen mehrere Individuen etwas zusammen, mit vereinigten Kräften, mit verteilten Rollen. (Paul 1920 [1880], p. 18)

> Every linguistic creation is always the work of an individual. Many can create the same, and this happens very often. But the act of creation, for this reason, is not different and the product not different either. Never do many individuals create something together with united forces or with allotted roles. (translation by the author)

Weinreich et al. (1968, pp. 107–108) argue, for instance, that, in this approach, Paul trivializes sociological issues concerning the relationship between the individual and the society they live in – particularly in relation to how a 'language custom'/*Sprachusus* comes about; and they further accuse him of woefully undertheorizing the mechanisms underlying idiolectal change. It should be noted at this point that, to the linguistic historian, this critique seems rather anachronistic. Paul's focus on individuals as the driving forces behind language change is essentially a counterproposal to H. Steinthal's (1823–1899), and later Wilhelm Wundt's (1832–1920) *Völkerpsychologie* (Klautke 2013, pp. 30–32; Morpurgo Davies 1998, pp. 247–250) – much as the whole Neogrammarian project was essentially an attempt to rid linguistics of romantic baggage and render it a modern, empirical science. But Weinreich et al. (1968, p. 125) saw in this an opportunity to propose a powerful alternative to the then-blossoming Chomskyan approach to grammar. A linguistic system, so the argument goes, always systematically correlates with non-linguistic factors such as social class, age or gender; and we must come to the conclusion that Chomsky's (1965, p. 3) 'ideal speaker-listener [...] in a completely homogeneous speech-community' is not only ideal but impossible (see Weinreich et al. 1968, pp. 101, 125). In every speech community one can establish social stratifications along lines of social class, gender, or age; where older people speak differently from the young, men differently from women, or the working class from the upper class. It is these social categories, contra Paul, that allow us to identify the driving forces behind language change. If the *Bewegungsgefühl* of a single individual changes, it is merely a token of a social type.[9]

With meticulous attention to detail, Labov studied the social stratification of language empirically. For instance, he soon established that when the collective pronunciation of a certain vowel changes without being noticed by the speech community, it is likely that upper-working class or lower-middle class women are leading the change (Labov 2001,

[9] In some regards, this logic seems to echo Wundt's criticism of Paul between the 1880s and the 1910s (see Klautke 2013, pp. 30–32, 64–65, 71). Wundt held that a language is the product of a *Volk*, not something that is created individually (Klautke 2013, chapter 2). A *Volk* is always more than the sum of its parts and a phenomenon Herbartian individual psychology (which Paul followed) cannot account for – hence the need for *Völkerpsychologie* (Klautke 2013, pp. 64–65). This view is preserved in the sociology of Durkheim (1938 [1895]), who similarly held that 'social facts' cannot be explained solely by the behavior of individual members of a society – they exist *sui generis*. Durkheim's sociological method underlies Labov's work down to the present day (see Figueroa 1994, chapter 4).

chapter 8). If a certain pronunciation is used by upper-class speakers, the lower middle class will likely hypercorrect its own pronunciation to match theirs, with women once again in the lead (Labov 1972, chapter 5). The former example is a case of 'change from below', the latter of 'change from above'. Soon, these empirical generalizations evolved into fully-fledged principles of linguistic change, to which Labov devoted an entire trilogy (1994, 2001, 2010).

Similarly to the case of Chomsky described in the preceding chapter, several important Labovian concepts have undergone considerable change to arrive at their current state – sometimes on account of inaccurate interpretations of Labov's work by his readers, at other times because Labov himself dramatically shifted the focus of his thinking (a detailed discussion can be found in Woschitz 2019, sections 2 and 4). A case in point is his treatment of change from below. Ask a linguist who is only peripherally concerned with sociolinguistics to define it, and they are likely to say that it refers to changes initiated by the lower class. This, Labov (2006 [1966], p. 203) admits in retrospect, has been an egregious misreading all along, and that he should have used 'change from within' and 'without' the linguistic system instead (in the spirit of Labov 2007). But in earlier work (Labov 1972, 2006 [1966]), whether a change in the linguistic system has its origin within or outside its own linguistic system played only a subordinate role. What mattered more as a distinguishing feature was the change's availability to social commentary and social awareness more generally: if speakers are not aware of their own changing pronunciation, Labov (1972, p. 123) correspondingly spoke of change from 'below the level of conscious awareness'; if speakers are aware of their own changing pronunciation, Labov spoke of change from above the level of conscious awareness.

Recent critiques by the so-called third-wave (e.g. Eckert 2016, pp. 77–79; Zimman 2017, p. 366, footnote 1) have pointed out that the conscious/unconscious distinction does not adequately capture socially meaningful aspects of behavior. Body language is a case in point: if I have a hunched posture, this could stem from a lack of self-esteem which I unconsciously project to the outside; and even if I have it because of orthopedic reasons, other people could still pick up on it to form their opinion about me. Why should linguistic behavior be any different? After all, the way I talk can involve my own desire to project an identity to the outside (for instance, that I am a Western, urban male); and even if not, others can still pick up on the way I talk to see if I fit a social stereotype. It should then not be too hard to introduce similar kinds of optional performativity into the above-mentioned distinction between change from above and below.

The stakes, however, are surprisingly high, not because Labov denies the importance of social meaningfulness, but because in the orthodox framework, the distinction between language and society, in other words between internal (Labov 1994) and external (Labov 2001) factors, largely hinges on the distinction between conscious and unconscious awareness (see Woschitz 2019). If not for language-internal organizing principles, so the logic goes, how can a complex reorganization of the phonological space in a chain shift come about if speakers are not even aware of it? For example, it seems to be a pattern in Indo-European languages that when raising vowels reach the point where they can go no higher (e.g. when a raising /æ/ has come to be pronounced as a tense [i]), they can diphthongize to continue their movement (Labov 1994, pp. 120, 122, 248–249). The prime example of this is, of course, the English Great Vowel Shift, but we also find present-day varieties of English that have [miən] where others have [mæn] or [men] for *man*. Another tendency is that vowels positioned at the peripheral track of the vowel space become less open (Labov 1994, p. 262): on average, we are much more likely to witness raising of long peripheral /æ/ toward [i], as in the above example, than a lowering and backing to long [ɑ], as in Southern British English *bath*.

Principles such as these do not seem to be attributable to social meaningfulness or performativity in the sense described above. They just 'happen' to a speech community undergoing multigenerational language change (Labov 1994, pp. 264–265), by placing restrictions on which change in pronunciation is likely to happen, and in which direction it is likely to go (Labov 1994, p. 115). Labov (2001, p. xv) accordingly treats the social structure as the 'material substratum' of language change, the latter being ontologically prior to what speakers do when they open their mouth to speak. Notice the parallel to Chomsky's justification of the innateness of certain linguistic principles discussed in the preceding chapter: the principle of cyclic application is too abstract for native speakers of English to be aware of, so the argument goes; thus, it needs to have an ontological existence separate from other, inductively learned parts of language.

It should be noted that the term 'material substratum' is rather ambiguous, and it allows for diametrically opposed readings. It could address Locke's substance/substratum distinction (which is a matter of controversy itself; see Robinson 2018, section 2.5). 'Properties – or, in Locke's terms qualities – must belong to something – cannot "subsist… without something to support them"' (Robinson 2018, section 2.5). In this logic, if one abstracts all properties away from an orange, such as its weight, its color or its shape; there is still something left that functions as the carrier of these properties – the substratum. In the context of above, the 'material substratum of a language change' would then be what is left when one abstracts all linguistic regularities such as the described restrictions on vowel-raising or diphthongisation away from speakers engaging in communication. 'Social structure', the substratum, functions as the carrier of language-internal processes, much like the substratum of an orange is the carrier of roughly 130 g, a round shape and orange color.

In a more colloquial reading, on the other hand, 'material substratum' can also denote 'something important from which something else develops, but that is not immediately obvious' (Cambridge Dictionary, retrieved February 22, 2019). This could be read as the opposite of the Lockean conception because it does not attribute to the substratum the role of a carrier or that of a mere recipient of properties. Instead, the substratum constitutes the observed properties. In the case of the above-described phonological restrictions, this would mean that the diphthongisation of [i] or the raising of peripheral [æ] is somehow the product of a speech community engaging in communication, rather than an ontologically prior restriction of language use. This is Labov's intended reading in the quote above.

Still, that he would use 'material substratum' to make his point is interesting because the epistemological tension between the two readings is also somewhat reflected in the contrast between his later and earlier work, and, to some extent, in his whole œuvre. In sum, what seems to have changed in Labov's theorizing is how much in language change we can attribute to language-internal organizing principles, and how much we can attribute to the social structure (the 'material substratum') in which it unfolds. In his earlier phase, Labov (1994, p. 3) argued that 'the two sets of factors – internal and external – are effectively independent of each other', and he attributed quite a lot to the former. All a linguistic community had to do is take up a change in vowel production (that came about because of linguistic pressures such as a phonemic merger, or 'in response to social motivations which are relatively obscure', Labov 1972, p. 123) and carry it along a pre-set path to its conclusion. In Labov (2001), this continuation is explained by the so-called 'nonconformity principle'. Children, in this view, interpret more advanced tokens of a vowel as informal/nonstandard speech, which they then reinterpret as not conforming to sociolinguistic norms (Labov 2001, p. 513). Therefore, juveniles from a social stratum more likely to resist these kinds of norms make comparatively greater use of such variants (Labov 2001, p. 516); they are the 'leaders' of the change.

These 'abstract polarities' (Labov 2001, p. 515) of standard/non-standard speech and conforming/non-conforming behavior then allow us to account, at least in part, for how large-scale regularities in e.g. vowel raising come about:

> [I]t may be observed that the very generality of these abstract polarities makes it possible to conceive of how linguistic change may be driven in the same directions across thousands of miles in many separate communities. [...] The general principles of chain shifting outlined in volume 1 give us one clue to the uniformity of these great regional developments. If this volume has added anything to our understanding of linguistic change, there must also be an explanation for the uniformity of socially motivated projection throughout these vast areas. The nonconformity hypothesis is submitted as a first step in that direction. (Labov 2001, p. 515)

In Labov's later phase, much more is attributed to less abstract socially meaningful/performative language behavior, with the nonconformity principle being downgraded to one possible explanation among many:

> In one form or another, [driving forces which may be responsible for the continuation, acceleration or completion of change] involve the associa-

tion of social attributes with the more advanced forms of a change in progress: local identity, membership in communities of practice, social class, age or gender. (Labov 2010, p. 368)

In this passage, and in fact in the whole volume, the social structure is no longer treated as something in which language change is merely embedded. We witness instead an orientation away from a Lockean 'material substratum' to the colloquial reading mentioned above. The social structure is now in part constitutive of that change because advanced tokens carry performative meaning, often more locally relevant than an analogous equation of advanced tokens with nonconformity to sociolinguistic norms would suggest. And as third-wave variationists and socio-phoneticians continue to discover social performativity in small-scale features involved in sound change (see Eckert 2010), less and less in language change can be captured under linguistic structure 'unfolding itself' in a social structure. Or, conversely, linguistic structure becomes more and more restricted in its application domain.

6 A Similar Kind of Progress?

From 1965 on, Chomsky persuaded a critical mass of scholars to seek explanations for linguistic phenomena, including language acquisition, in a Universal Grammar located in the brain. As history unfolded, however, the representational systems needed to capture UG in a form that would account for all the syntactic structures of known languages became ever more complex. He adjusted his framework numerous times to rise to the challenge, but this has had important implications for the ontological commitment of his theory. Since the 1970s, fewer and fewer specific characteristics have been attributed to a linguistic genetic endowment, to a point where other scholars, such as Kirby, Smith, and Brighton (2004, p. 587), have started to attribute many properties previously attributed to UG to a 'prior learning bias' which seeks to explain intricacies of language entirely by general cognitive mechanisms.

I have argued that the theoretical reorientations in generativist theory can be understood as driven by the conviction that language is something biological and "natural"; that, at the end of the day, there is at least *something* in our brain that is distinctively linguistic and allows human beings alone to speak. But from a philosophical perspective, this is a strong assumption that is at odds with the applied methodology. If the preceding discussion of Kuhn and scientific realism has taught us anything, it is that mind-independent entities are projected in part out of our own theorizing. But there is nothing in the *modus operandi* of Chomsky from which we can infer that the described regularities are distinctively linguistic, let alone a genetic endowment. Abstract regularities do exist, but we can never tell for sure if they belong to a distinct biolinguistic endowment or to other cognitive or social capacities. That the treatment of UG has become more and more general is then essentially an epistemological tension playing out in the pursuit by generativists to safeguard the autonomy of linguistics. If one axiomatically believes in the existence of a biolinguistic entity, one must live with the fact that such an entity, if it exists at all, is so simple that others would call it trivial.

'Progress', to return to the scope of the paper, can then be found in the developing awareness over the last 60 years that less and less in language behavior can be explained by UG, and that it should be treated as a last resort rather than an explanation for everything. To attribute too much explanatory importance to it, as was done in the 1960s and 70s, is nowadays considered a value-laden presumption in the sense outlined in Sect. 3; an ontological commitment that academics have started giving up in favor of more general cognitive principles on the one hand, and social and cultural aspects on the other hand. Chomsky seems to be at ease with this:

> The quest for principled explanation faces daunting tasks. We can formulate the goals with reasonable clarity. We cannot, of course, know in advance how well they can be attained – that is, to what extent the states of the language faculty are attributable to general principles, possibly even holding for organisms generally. (Chomsky 2006, p. 185)

He seems to be satisfied with the fact that his work of the previous 50 years, in which he tried so eagerly to bring linguistics to the level of a natural science, is now considered the harbinger of a modern cognitive science in which his linguistics has been swallowed up. Where some see in this the ultimate self-abolition of generativism, Chomsky sees biolinguistics brought to its logical conclusion. Both, in a way, are right in their assessment.

A similar kind of epistemological tension resolving itself seems to underlie the historical development of Labovian sociolinguistics. On the one hand, 'first-wave variationists' celebrate Neogrammarians for their analysis of sound change. On the other hand, however, early Labov has failed to live up to their conviction that each and every sound change is a psychological phenomenon. Large-scale regularities notwithstanding, the Neogrammarians never attributed driving forces of sound change to abstract principles like "if the vowel is tense, raise it". If anything, they resisted such claims on account of their empiricism: if something is not regular sound change, it is the result of analogy. We can never go further than this if we want to remain faithful to the data. All other things that we extrapolate from the data are entirely speculative.

It just so happened that as descriptions of large-scale phonological changes proliferated, linguists began to abstract from them principles that were then taken to be ontologically prior to how speakers communicate in everyday life. Crucially, however, all proposed explanations of how internal factors come about – for instance, that, in many Indo-European languages, tense vowels seem to raise and diphthongize if need be – have proved inadequate. Laziness or carelessness, for instance, cannot account for the fact that some English varieties diphthongize tense [i] (see Labov 2001, p. 25). Diphthongisation is generally treated as a change across two levels of phonological subsystems, which makes the pronunciation *more* complicated for speakers. Labov's loophole is to follow Meillet in his assumption that 'the sporadic character of language change can only be explained by correlations with the social structure of the speech community in which it takes place' (Labov 2001, p. xv). What a speech community does is to break loose a change in a linguistic system which then adjusts itself to the disequilibrium (Labov 2010, pp. 369–370). This allows Labov to have his cake and eat it too: on the one hand, he can maintain the ontological status of internal factors, on the other hand, he is able to foreground the importance of the speech community in bringing changes about.

As, over the years, increasingly more importance has been attributed to social meaningfulness and performativity associated with changes in progress, Labov's notion of internal factors has been constrained in the process. Remnants of internal factors can, for instance, be found in so-called near-mergers, which Eckert and Labov (2017, p. 486) describe as 'the clearest case of the divergence between social meaning and linguistic structure'. Near-mergers are when speakers cannot perceive a phonemic contrast but still produce it, or vice versa (Labov et al. 1972, 1991). They are 'motivated by more abstract principles of change' because members in a speech community never direct their attention to 'the more abstract levels of phonological organization' (Eckert and Labov 2017, pp. 467, 491). However, as mentioned in the previous chapters, sociolinguists have started to criticize theorizing agency or performativity along the lines of conscious awareness (see Eckert 2016, pp. 77–79). Lack of awareness does not preclude social performativity, of which Nycz's (2018) study is a good example. She studied seven Canadians who moved to New York and had been living there for at least ten years before data collection. Speakers of Canadian English, unlike speakers of New York City English, typically present with the low-back merger, so that they cannot distinguish their vowel-pronunciation of *lot* and *thought*. Sophie, one of her interviewees, however, shows incipient unmerging without being aware of it (Nycz 2018, p. 189): she shows a clear distinction between *lot* and *thought*, albeit not as clear as normally found in New York City English. She had been living in New York for 27 years and, incidentally, had been married to a New Yorker for almost two decades. For interviewees with Canadian spouses, unmerging is still present, but less distinct (Nycz 2018, pp. 189, 198). The fact that Sophie's unmerging is more advanced is undoubtedly meaningful or performative. She could, for instance, express intimacy by aligning her low-back vowel production with her spouse's, regardless of whether she is consciously aware of it or not. Findings like these motivate Eckert (2019, p. 1) to argue quite radically that she 'would not be satisfied with the claim that a sound change was meaningless unless every effort had been made to prove otherwise'. We can see that the tables have turned: social meaning and linguistic structure are so hard to separate that what was once theorized as a 'material substratum' merely carrying forward language-internal processes is no longer accepted. This view is instead downgraded to a heuristic approximation of a phenomenon that, in reality, follows different rules that are driven by social meaningfulness.

In conclusion, we can see interesting philosophical parallels between the Labovian and the Chomskyan paradigms: internal factors and UG were attributed much explanatory power in the late twentieth century, but recently they have been treated as stand-ins for other processes at work (socially meaningful language behavior and cognitive principles respectively). The explanatory value attributed to them has decreased as other things began 'doing the explaining'. Many scholars have started treating them as value-laden presumptions. As I have discussed in Sects. 2 and 3, this is progress in the scientific realist and Kuhnian sense. Scientific progress in generativism is reflected in the fact that it did away with UG as an explanation for linguistic intricacies, and instead finds its new home in cognitive science. To some this appears to represent failure and abdication (or usurpation); but in fact, it has freed linguistics from the constraints of having to seek explanations for a variety of linguistic phenomena in one alleged entity in the brain, which no one to date has been able to locate on an fMRI scan. Scientific progress in the study of phonological change is reflected in the fact that the almost axiomatic assumption of internal factors unfolding themselves in a social structure is being overthrown.

Both generativism and sociolinguistics have increased their relational objectivity in the process, as epistemologically 'expensive' concepts have been questioned and re-theorized. That this has led, for many, to the self-abolition of generativism in the one case, and a less radical but still significant ontological reorientation in variationist sociolinguistics in the other, is a byproduct of disciplines resolving epistemological tensions.

7 Conclusion

In the spirit of a scientific realism that is faithful to Kuhn, we cannot measure the epistemic worth of a theory solely on the grounds of how well it corresponds to a mind-independent reality. We also need to consider how phenomena are in part projections of our own theorizing. It is these projections that can be improved on; and whether theories have addressed specific epistemological tensions or not can then become a yardstick of scientific progress. Chomskyan syntax and Labovian study of phonological change were found to be cases where theoretical constructs (Universal Grammar or internal factors) have been attributed less explanatory value over the years. The described historical trajectories – from formal descriptions of language to biolinguistics, and from linguistic autonomy to sociolinguistic performativity respectively – have been told as stories in which scholars have sorted out epistemological tensions by identifying and addressing value-laden presumptions. In Chomsky's work, the axiomatic belief in a biolinguistic entity led to the trivialisation of his former claims; in Labov's work, social performativity challenged the questionable primacy of internal factors.

Acknowledgments This work was supported by the Arts and Humanities Research Council, grant number AH/L503915. I would like to express my thanks to John E. Joseph, James McElvenny and Ryan Nefdt for their help and comments on earlier drafts of this paper.

References

Alexandrova, Anna. 2016. Can the science of well-being be objective? *British Journal for the Philosophy of Science* 6(2): 1–25.

Boyd, Richard. 2010. Scientific realism. *The Stanford encyclopedia of philosophy.* Retrieved from https://plato.stanford.edu/archives/sum2010/entries/scientific-realism/

Bradley, Seamus, and Karim Thébault. 2019. Models on the move: Migration and imperialism. *Studies in History and Philosophy of Science Part A* 77: 81–92.

Chomsky, Noam. 1959. Verbal behavior by B. F. Skinner (Review). *Language* 35 (1): 26–58.

———. 1965. *Aspects of the theory of syntax*. Cambridge, MA: The M.I.T. Press.

———. 1993 [1981]. *Lectures on government and binding*. Berlin/New York: Mouton de Gruyter.

———. 1995. *The minimalist program*. Cambridge, MA: The MIT Press.

———. 2002 [1957]. *Syntactic structures*. Berlin/New York: Mouton de Gruyter.

———. 2006. *Language and mind*. 3rd ed. Cambridge: Cambridge University Press.
Chomsky, Noam, and Morris Halle. 1968. *The sound pattern of English*. New York/Evanston/London: Harper & Row.
Craver, Carl. 2006. When mechanistic models explain. *Synthese* 153 (3): 355–376.
Durkheim, Émile. 1938 [1895]. *Rules of sociological method*. Glencoe: The Free Press.
Eckert, Penelope. 2010. Affect, sound symbolism, and variation. *University of Pennsylvania Working Papers in Linguistics* 15 (2): 9.
———. 2016. Variation, meaning and social change. In *Sociolinguistics: Theoretical debates*, ed. N. Coupland, 68–85. Cambridge: Cambridge University Press.
———. 2019. The individual in the semiotic landscape. *Glossa: A Journal of General Linguistics* 4 (1): 1–15.
Eckert, Penelope, and William Labov. 2017. Phonetics, phonology and social meaning. *Journal of SocioLinguistics* 21 (4): 467–496.
Figueroa, Esther. 1994. *Sociolinguistic metatheory*. Oxford/New York: Pergamon/ Elsevier Science.
Harris, Randy. 1993. *The linguistics wars*. New York/Oxford: Oxford University Press.
Hempel, Carl. 1965. Fundamentals of taxonomy. In *Aspects of scientific explanation and other essays in the philosophy of science*, ed. C.G. Hempel, 137–154. New York/London: The Free Press.
Itkonen, Esa. 1978. *Grammatical theory and metascience: A critical investigation into the methodological and philosophical foundations of 'autonomous' linguistics*. Amsterdam: John Benjamins.
Joseph, John. 2002. *From Whitney to Chomsky: Essays in the history of American linguistics*. Amsterdam/Philadelphia: John Benjamins.
———. 2012. *Saussure*. Oxford: Oxford University Press.
———. 2018. *Language, mind and body: A conceptual history*. Cambridge: Cambridge University Press.
Katz, Jerrold. 1981. *Language and other abstract objects*. Totowa: Rowman and Littlefield.
Kincaid, Harrold. 2008. Structural realism and the social sciences. *Philosophy of Science* 75 (5): 720–731.

Kirby, Simon, Kenny Smith, and Henry Brighton. 2004. From UG to universals: Linguistic adaptation through iterated learning. *Studies in Language* 28 (3): 587–607.

Klautke, Egbert. 2013. *The mind of the nation: Völkerpsychologie in Germany, 1851–1955*. New York: Berghahn.

Kuhn, Thomas. 1957. *The Copernican revolution*. Cambridge, MA: Harvard University Press.

———. 1970 [1962]. *The structure of scientific revolutions*. Chicago/London: The University of Chicago Press.

———. 1990. Dubbing and redubbing: The vulnerability of rigid designation. In *Minnesota studies in the philosophy of science*, ed. C.W. Savage, J. Conant, and J. Haugeland. Minneapolis: University of Minnesota Press.

Labov, William. 1972. *Sociolinguistic patterns*. Philadelphia: University of Pennsylvania Press.

———. 1994. *Principles of linguistic change, volume 1: Internal factors*. Cambridge, MA: Blackwell Publishers.

———. 2001. *Principles of linguistic change, volume 2: Social factors*. Malden: Blackwell Publishers.

———. 2006 [1966]. *The social stratification of English in New York City*. Cambridge: Cambridge University Press.

———. 2007. Transmission and diffusion. *Language* 83 (2): 344–387.

———. 2010. *Principles of linguistic change, volume 3: Cognitive and cultural factors*. Chichester: Wiley-Blackwell.

Labov, William, Malcah Yaeger, and Richard Steiner. 1972. *A quantitative study of sound change in progress*. Philadelphia: US Regional Survey.

Labov, William, Mark Karen, and Corey Miller. 1991. Near-mergers and the suspension of phonemic contrast. *Language Variation and Change* 3 (1): 33–74.

Marian, David. 2016. The correspondence theory of truth. *The Stanford encyclopedia of philosophy*. Retrieved from https://plato.stanford.edu/entries/truth-correspondence/

Matthews, Peter. 1993. *Grammatical theory in the United States from Bloomfield to Chomsky*. Cambridge: Cambridge University Press.

Michaelson, Eliot, and Marga Reimer. 2019. Reference. *The Stanford encyclopedia of philosophy*. Retrieved from https://plato.stanford.edu/archives/spr2019/entries/reference/

Morpurgo Davies, Anna. 1998. *History of linguistics, volume IV: Nineteenth-century linguistics*. London: Longman.

Nagel, Ernest. 1961. The value-oriented bias of social inquiry. In *The structure of science: Problems in the logic of scientific explanation*, ed. E. Nagel, 571–584. London: Routledge & Kegan Paul.

Nefdt, Ryan. 2016a. Linguistic modelling and the scientific enterprise. *Language Sciences* 54: 43–57.

———. 2016b. Scientific modelling in generative grammar and the dynamic turn in syntax. *Linguistics and Philosophy* 39: 357–394.

———. 2019. Linguistics as a science of structure. In *Form and formalism in linguistics*, ed. J. McElvenny, 175–195. Berlin: Language Sciences Press.

Nycz, Jennifer. 2018. Stylistic variation among mobile speakers: Using old and new regional variables to construct complex place identity. *Language Variation and Change* 30 (2): 175–202.

Osthoff, Hermann, and Karl Brugmann. 1878. *Morphologische Untersuchungen auf dem Gebiete der indogermanischen Sprachen*. Leipzig: Verlag von S. Hirzel.

Partington, James, and Douglas Mckie. 1937. Historical studies on the phlogiston theory. – I. The levity of phlogiston. *Annals of Science* 2 (4): 361–404.

Paul, Hermann. 1920 [1880]. *Prinzipien der Sprachgeschichte*. Halle: Verlag von Max Niemeyer.

Psillos, Stathis. 2000. The present state of the scientific realism debate. *The British Journal for the Philosophy of Science* 51: 705–728.

Pullum, Geoffrey. 1992. The origins of the cyclic principle. In *CLS 28: Papers from the 28th regional meeting of the Chicago linguistic society, 1992; volume 2: The parasession on the cycle in linguistic theory*, ed. J. Marshall Denton, G.P. Chan, and C.P. Canakis, 209–236. Chicago: Chicago Linguistic Society.

Putnam, Hilary. 1975. The meaning of 'meaning'. In *Philosophical papers. Volume 2: Mind, language and reality*, ed. H. Putnam, 215–271. Cambridge: Cambridge University Press.

———. 1981. *Reason, truth and history*. Cambridge: Cambridge University Press.

———. 1982. Why there isn't a ready-made world. *Synthese* 51 (2): 141–167.

Quine, Willard. 1968. Ontological relativity. *The Journal of Philosophy* 65 (7): 185–212.

Radick, Gregory. 2016. The unmaking of a modern synthesis: Noam Chomsky, Charles Hockett, and the politics of behaviorism, 1955–1965. *Isis* 107 (1): 49–73.

Robinson, Howard. 2018. Substance. *The Stanford encyclopedia of philosophy*. Retrieved from https://plato.stanford.edu/archives/win2018/entries/substance/

Rohrlich, Fritz, and Larry Hardin. 1983. Established theories. *Philosophy of Science* 50: 603–617.

Rosen, Jacob. 2012. Motion and change in Aristotle's *Physics* 5. 1. *Phronesis* 57: 63–99.

Weinreich, Uriel, William Labov, and Martin Herzog. 1968. Empirical foundations for a theory of language change. In *Directions for historical linguistics*, ed. W.P. Lehmann and Y. Malkiel, 95–195. Austin: University of Texas Press.

Woschitz, Johannes. 2019. Language in and out of society: Converging critiques of the Labovian paradigm. *Language & Communication* 64: 53–67.

Young, James. 2016. The coherence theory of truth. *The Stanford encyclopedia of philosophy*. Retrieved from https://plato.stanford.edu/entries/truth-coherence/

Zimman, Lal. 2017. Gender as stylistic bricolage: Transmasculine voices and the relationship between fundamental frequency and /s/. *Language in Society* 46 (3): 339–370.

7

Linguistic Change and Biological Evolution

Unni Leino, Kaj Syrjänen, and Outi Vesakoski

1 Background

The similarity between the differentiation of languages and biological species is not a new idea but rather something that was observed already at the initial stages of evolutionary biology:

U. Leino (✉) • K. Syrjänen
Tampere University, Tampere, Finland

BEDLAN (Biological Evolution and Diversification of Languages), Turku, Finland
e-mail: unni.leino@tuni.fi; kaj.syrjanen@tuni.fi

O. Vesakoski
BEDLAN (Biological Evolution and Diversification of Languages), Turku, Finland

University of Turku, Turku, Finland
e-mail: outi.vesakoski@utu.fi

R. M. Nefdt et al. (eds.), *The Philosophy and Science of Language*,
https://doi.org/10.1007/978-3-030-55438-5_7

> *The formation of different languages and of distinct species, and the proofs that both have been developed through a gradual process, are curiously the same. [...] We find in distinct languages striking homologies due to community of descent, and analogies due to a similar process of formation.* (Darwin 1871, I:59–60)

This notion of similarity can also be seen in historical terminology of language research, as the field of diachronic linguistics was called 'evolutionary linguistics' in the nineteenth century; diachronic linguistics became popularized later alongside Saussure's dichotomy between synchrony and diachrony (Aronoff 2017). Indeed, early historical linguists considered this similarity as yet another aspect of how historical linguistics should be considered a science and use methods that are scientifically rigorous:

> *Dass bei den Sprachforschern die naturwissenschaftliche Methode mehr und mehr Eingang finde, ist ebenfalls einer meiner lebhaftesten Wünsche. Vielleicht vermögen die folgenden Zeilen einen oder den andern angehenden Sprachforscher dazu in Betreff der Methode bei tüchtigen Botanikern und Zoologen in die Schule zu gehen.* (Schleicher 1863, 5–6)
>
> ----
>
> *It is also one of my most spirited hopes that the methodology of natural sciences finds increasingly its way into linguistics. Perhaps the next lines can persuade a budding linguist to learn methods from great botanists and zoologists.*

It should also be noted that the phylogenetic tree itself was not invented by Darwin, even though he was instrumental in spreading the concept into biology. Similar trees were used to represent other types of historical lineages in linguistics, in the form of family trees, as well as manuscript lineage trees even before the publication of *The Origin of Species* (Atkinson and Gray 2005; O'Hara 1996). Even earlier, Leibniz (1710) had used linguistic and onomastic evidence in his study on the history and origins of different peoples. Of course, this was not strictly speaking an analogy between linguistic and biological evolution, as the evolution of species had not yet been proposed. Still, on the level of human populations Leibniz noted how the origins and divergence of different peoples was coupled to that of the languages, and in this his work can be seen as a precursor to modern evolutionary theories (e.g. Cavalli-Sforza 2000).

In the early twentieth century, the importance of historical linguistics decreased markedly while the emphasis shifted toward structuralist approaches, focusing on describing present-day languages. As a consequence, work on language change as a type of evolution was mostly discontinued, and similarities between language change and evolution generally downplayed. In the shift from 'evolutionary' to 'diachronic' linguistics—as in other aspects of the early development of general linguistics—the influence of Saussure (1916) appears to have been notable. The rise of generative grammar further shifted focus away from linguistic evolution, and in fact Chomsky (2017) states clearly his view that languages 'do not evolve, in the technical sense of the term'. In his view, a full-fledged concept of linguistic evolution would appear to be incompatible with a universal grammar and language capacity as an autonomous system in human cognition. On the other hand, the 'evolution' that historical linguistics studies is different from what Chomsky argues against—historical linguists are not talking about the evolution of the cognitive mechanisms involved in human language but rather the evolution-like changes and divergence seen in languages, in a similar way that evolution-like change is observed in the history of other cultural objects.

In the last decades of the century, sociolinguistics started to look at the mechanics of language change and its relationship with variation. The way this relationship is seen (e.g. Labov 1994) is not all that different from the way biologists see the relationship between variation within a population and the divergence of populations into species. Both are also open toward computational methods and a data-driven approach to research.

Since the late twentieth century, evolutionary approaches to the study of language have become more visible. While the concept of linguistic evolution is not new and the recent developments are best seen as a matter of a renaissance, the trend raises not only methodological but also deeper theoretical questions. General meta-theories have been proposed, most notably by Dawkins (1976 onward) and Hull (1988) to cover evolutionary change in biology and the domain of human culture. These are by their very nature very general, which can make them difficult to understand. They appear to be in line with what is known of language change, but the concepts used in the general meta-theory, biology, and linguistics have subtle – and sometimes not so subtle – differences.

From the point of view of linguists trying to make sense of the development and divergence of languages, the key question is whether traditional, biology-based views of evolution are sufficiently compatible with the phenomena in linguistic 'evolution', and thus whether these linguistic processes can be called evolution in a meaningful fashion and studied from this perspective. In order to do so one must be confident of two things: first, that the tools used in evolutionary biology are adaptable for research of linguistic variation and change, and second, that the underlying concepts have sufficient similarities that this kind of approach is honest scholarship. To do the first, one needs a team with expertise in computational methods and linguistics; to do the second, a team with expertise in linguistics and biology.

2 Computational Approaches to Language Change

As computers were developed, their applicability to linguistic analysis was seen almost immediately. As Weaver (1949) proposed, the principal goal was to develop tools for automatic translation, and among the methods suggested were linguistic and semantic analysis, utilizing both language universals and contextual information. Later on, this developed into the more full-fledged modeling and analysis of language structure that became the primary meaning for 'computational linguistics'.

In studying language variation, applying computational methods to dialectology led to the development of dialectometry in the 1970's (Goebl 1982), even as traditional dialectology itself was giving way to sociolinguistics. The starting point of dialectometry was the realization that a dialect atlas was the compiler's interpretation of a massive amount of data, and that this data could be analysed more thoroughly. More recently, research has started to expand, for instance to generalizing the relationship of dialectal and geographic distance over languages (Wieling and

Nerbonne 2011) or looking at explanatory factors other than simple geographic distance (Honkola et al. 2018).

Meanwhile, sociolinguistics has moved the focus of studying linguistic variation from geographically-defined dialects to include such variational axes as social class or gender, and further to seeing contextual and stylistic variation as different aspects of the same phenomenon as the variation seen between dialects (e.g. Eckert 2012). Some sociolinguistic corpora (such as Helpuhe 2014) are useful for studying ongoing language change, partly using computational data analysis similar to that used in population biology (e.g. Kuparinen et al. 2019).

Around the end of twentieth century, computational linguistics also started to get interested in language divergence in ways that resemble earlier research in language evolution. A major feature in the current renaissance is the use of computational methods (e.g. McMahon and McMahon 2005). While some of the earlier quantitative linguistic methods, such as tree-building tools, bore striking resemblance with their contemporary biological counterparts (see e.g. Embleton 1986; Felsenstein 2004; Atkinson and Gray 2005), the tools became much more popular and developed more rapidly in biology than in the linguistic domain. In this endeavor arguably linguistics is catching up to decades of theoretical and methodological advances within biology.

As was apparent already in the mid-nineteenth century, the phylogenetic tree is a conceptual abstraction that appears to work in cases where we are summarizing historical processes characterized by branching events, which include, among others, both linguistics and biology. Such trees have been in use continuously since their introduction, and they have proved useful in not only these fields but also in a wide range of other contexts. Nevertheless, the question remains whether a tree showing the relationships between biological species is similar enough to one showing relationships between languages that essentially the same tools can be used for studying both.

It's important to keep in mind that what we are looking at is not just one single, clear-cut claim that the 'evolution' apparent in the divergence of languages is similar to that apparent in the divergence of species. The similarity can be seen from different perspectives and its existence is not a yes/no question. At one end, one can go deep into meta-theoretical studies of evolution (Dawkins 1976; Hull 1988); at the other end, one can take a more pragmatic approach and see whether methods used in one field can be adapted to work in the other.

While the pragmatic approach sounds naive it is supported by cognitive theories of how this kind of analogy is at the center of human abstract thought (Fauconnier and Turner 2003; Lakoff and Johnson 1980). Analogies between two different domains are never perfect or complete, but they can still be useful, and in the case of applying an evolutionary view to language divergence, there is already a long history of some of the tools being used in both biology and linguistics. Because of this, it doesn't seem too far-fetched to look into the rest of the toolbox.

3 Ontological Versus Methodological Similarity

When one talks about an evolutionary view of language one can mean a wide range of different things. For instance Croft (2006) lists three main approaches. First, one can take an approach similar to evolutionary psychology that attempts to study the human capacity for language. However interesting such research may be, Chomsky (2017) is correct in considering this literal approach to language evolution to be outside the scope of linguistics.

Second, one can look at the correlations and interactions between the diversification of language, culture, and human populations. In this approach biological evolution and language change are seen as partially intertwined historical processes that also resemble one another in how they operate. This approach has been actively pursued by scholars in one form or another since Leibniz (1710), and as data sets are becoming available in such fields as linguistics, genetics, and archaeology, research along these lines is increasing in popularity (see Pakendorf 2014 for review of studies focusing on parallel diversification of genetic and linguistic populations).

Third, language change itself can be studied as a process that has similarities with biological evolution—and as a consequence, research methods and tools can be adapted from one of these fields to the other. Croft (2006) points out that using simple analogies is limited and limiting, and that a genuinely evolutionary theory of language change is only possible if it is derived from a systematic evolutionary framework or meta-theory. His choice is to turn toward Hull (Hull 1988), and this choice appears sound.

Croft ends up requiring a generalized theory of evolution, of which both the biological and the linguistic are specific cases. His reasoning (Croft 2008, 220) is that such a framework is needed to determine which parallels can be legitimately drawn and what is needed for the analogy to make sense. We feel that his position is at the same time too lax and too strict.

On the one hand, a generalized theory will by necessity be quite general. If one wants to use a method developed for biological evolution to analyse language change, it is usually not enough to just point out that both these changes are evolutionary in the sense Croft or Hull use the term, but rather one needs to go much deeper into analysing how the assumptions of the method relate to the realities of the phenomenon it is used to study.

On the other hand, using a method first developed for biology for linguistic analysis does not require a deeper general relationship between the two fields. It is sufficient that the concepts used in the computational tool can be properly mapped to the concepts in the linguistic theory it is used to support; that the tool was originally developed for something else is at this point irrelevant.

Biological and linguistic evolution are similar enough that tools developed for one can be adapted for the other. This does not mean that the two phenomena are similar in all respects—for instance, the replication of genes behaves rather differently from that of linguistic features. All in all, as Andersen (2006) shows, the mechanics of language change are very different from those of biological evolution. However, it is not necessary that the underlying mechanisms are completely analogous; in order to use a computational method, it is enough that the assumptions of the specific method that is being applied hold in the new target field.

As an example, Syrjänen et al. (2016) studied variation within Finnish using the population genetic clustering tool STRUCTURE to analyse how dialectal variation could be clustered into dialects. STRUCTURE—similarly to many of the population genetic computational methods—assumes that the study object could follow the mathematical theorem of Hardy-Weinberg equilibrium, which states that allele frequencies of the sample remain constant over generations and no geographical variation exists. This assumption is a null model assuming no mutation, no natural selection and infinite population size (Pritchard et al. 2000).

Adopting such a model to dialect data requires a clear understanding of what the Hardy-Weinberg equilibrium means in terms of linguistic variation. In terms of population biology, it describes an ideal state where the allele frequencies do not change over generations; however, it is not immediately obvious how this translates into linguistic variation. Moreover, this assumption does not hold for typical biological populations but the method still yields results that are useful within limitations. As it turns out, this is what happens with linguistic data as well: the lack of Hardy-Weinberg equilibrium does not mean the method cannot be used, although it reduces the strength of the results. However, this level of detail cannot be derived from a reasonable generalized theory, but instead one needs to look rather closely into the specific fields and study the behavior of the data under the restrictions of the model.

Not all computational models adoptable from biology to linguistics have deep-level evolutionary assumptions built within them—many of them aim to model a phenomenon as simply as possible, using reasonably simple mathematical assumptions. An example of a fairly simple method that does not rely on strong evolutionary underpinnings is the TIGER algorithm, a metric originally designed for fine-tuning phylogenetic analyses (Cummins and McInerney 2011), somewhat similarly to e.g. how phylogenetic models can assume either a constant rate of change for the data, or a relaxed rate of change (see Maurits et al. 2019). The TIGER algorithm itself is free of deep-level evolutionary assumptions, as it is based on measuring how consistently the characters represent the data as subsets. In linguistics TIGER has been tested for studying cognate character data, which is typically used with phylogenetic tools to study the evolutionary patterns of language families. Syrjänen et al. (submitted ms.) analysed basic vocabulary cognate data with the TIGER algorithm. Intriguingly, unlike in biological studies which use TIGER rates to optimize tree-building, they focused on exploring the nature of the metric itself and noted that the TIGER rates can also be used as a measure of historical heterogeneity, or treelikeness, for a data set—i.e. which linguistic features in the data carry more treelike patterns and which less treelike. This shows yet another interesting factor regarding the adoption of evolutionary methods: they do not need to be applied in exactly the same way as they are used in biology, but rather adapted in a way that makes sense in the linguistic domain.

The application of evolutionary approaches may become challenging in situations where a conflict exists between how cultural history and biological evolution operate. Horizontal transmission—i.e. the transmission of material between generations in a non-treelike pattern—is one such case where incompatibility is often seen between biology and culture, and horizontal transmission is indeed extensive in cultural evolution, including linguistic evolution (e.g. Greenhill et al. 2009; Wichmann et al. 2011). This not only undermines the results of tree-based analysis techniques, such as Bayesian phylogenetic inference, but also calls for methods and even frameworks that would be better suited for describing and studying phenomena such as language and culture. Identifying a phenomenon is the first step in ensuring it is taken into account, and the aforementioned TIGER rate, for instance, provides a new tool for separating vertically non-treelike and treelike transmission of linguistic material from one another.

In general, discussion in the linguistic domain about the evolutionary parallels between biology and linguistics has tended to lack state-of-the-art knowledge of evolutionary biology. For example, also within the biological domain the tree model of evolution is regarded as insufficient: modern genetic analyses reveal that lateral transfer of genetic material between species through hybridization and introgression is in fact more common than traditionally assumed. Extensive horizontal transfer is present not only in viruses, bacteria and fungi (Boto 2009), but also in eukaryotes (e.g. Neelapu et al. 2019; Taylor and Larson 2019). Hybridization of species applies also to humans as ancient DNA analyses have revealed that the evolution of *Homo sapiens* includes admixture with Neanderthals and Denisovans (Sankararaman et al. 2016). Indeed, one current challenge in evolutionary biology is to build techniques for addressing horizontal transfer or material alongside vertical evolution (Liu et al. 2019).

With this in mind, what is needed in linguistics is further development of the methods borrowed from biology rather than a complete abandonment of an entire range of evolutionary tools. Rather than abandoning the endeavor altogether because of this, a more productive approach is to develop the techniques to address these issues. For the case of vertical vs. horizontal transfer of linguistic material, various solutions have been proposed to address horizontal transmission, including the use of metrics of treelikeness (Gray et al. 2010; Wichmann et al. 2011; Syrjänen et al. submitted ms.), techniques for visualizing non-treelike patterns, (see e.g. Bryant and Moulton 2003; McMahon and McMahon 2005; Nelson-Sathi et al. 2011), as well as techniques for exploring non-treelike patterns with the help of multiple tree-based analyses (Verkerk 2019).

4 Conclusions

The parallels between biological evolution and language diversification have been known since the mid-nineteenth century. Over time, these parallels have been used to develop both fields, and they have also been left unused for long stretches of time. In recent years, their importance has again become more pronounced. This is partly due to improvement in computational capacity, the availability of large data sets in digital humanities, and general renewed interest in interdisciplinary research.

General evolutionary meta-theories give support for finding and using similarities between biology and linguistics. Such general theories are, however, too general for the task of adapting specific tools from one field to another, so that such methodological work requires analysing both fields to see whether the specific case has requirements that cannot be analogically projected from the originating field to the other one. Here, it is useful to look not only into evolutionary theories but also to other, less clearly evolutionary sub-fields where similar parallels exist. Between biology and linguistics, the network of functional similarities extends from evolutionary biology and historical linguistics to population biology and dialectology/sociolinguistics.

Combining linguistic and biological evolution does not necessarily mean forcing the entirety of linguistic change into an evolutionary frame, but it can mean simply exchanging and adapting computational approaches and modeling solutions from natural sciences to humanities, building interdisciplinary co-operation between linguists and computational scientists. Seen this way, the goal is to build computational methods to specifically study linguistic change from the ever-growing selection of data sets for historical linguistics, dialectology, corpus linguistics, sociolinguistics and so on.

As an overall summary, there is clearly enough common ground between biology and linguistics that analytical tools and to some extent even research methodologies can be adapted from one field to the other. However, whether or not this common ground is due to the fundamental similarity of linguistic and biological evolution is irrelevant: in practice, the real requirement is to understand the method and its relationship to the target field. Even if the fundamental similarity exists one needs a thorough methodological understanding in the context of the field it is applied to, and with the support of such understanding one can confidently use the method even in the absence of a more fundamental similarity between the two fields.

References

Andersen, Henning. 2006. Synchrony, diachrony, and evolution. In *Competing Models of Linguistic Change: Evolution and Beyond*, ed. Ole Nedergaard Thomsen, 59–90. Amsterdam, John Benjamins.

Aronoff, Mark. 2017. Darwinism tested by the science of language. In *On looking into words (and beyond)*, ed. Claire Bowern, Laurence Horn, and Raffaella Zanuttini, 443–456. Berlin: Language Science Press.

Atkinson, Quentin D., and Russell D. Gray. 2005. Curious parallels and curious connections – Phylogenetic thinking in biology and historical linguistics. *Systematic Biology* 54 (4): 513–526. https://doi.org/10.1080/10635150590950317.

Boto, Luis. 2009. Horizontal gene transfer in evolution: Facts and schallenges. *Proceedings of the Royal Society B: Biological Sciences* 277: 819–827. https://doi.org/10.1098/rspb.2009.1679.

Bryant, David, and Vincent Moulton. 2003. Neighbor-net: An agglomerative method for the construction of phylogenetic networks. *Molecular Biology and Evolution* 21 (2): 255–265. https://doi.org/10.1093/molbev/msh018.

Cavalli-Sforza, Luigi Luca. 2000. *Genes, peoples, and languages*. New York: North Point Press.

Chomsky, Noam. 2017. The language capacity: Architecture and evolution. *Psychonomic Bulletin & Review* 24: 200–203. https://doi.org/10.3758/s13423-016-1078-6.

Croft, William. 2006. The relevance of an evolutionary model to historical linguistics. In *Competing models of linguistic change: Evolution and beyond*, Amsterdam studies in the theory and history of linguistic science. Series IV – Current issues in linguistic theory 279, ed. Ole Nedergaard Thomsen. Amsterdam/Philadelphia: John Benjamins.

———. 2008. Evolutionary linguistics. *Annual Review of Anthropology* 37: 219–234. https://doi.org/10.1146/annurev.anthro.37.081407.085156.

Cummins, Carla A., and James O. McInerney. 2011. A method for inferring the rate of evolution of homologous characters that can potentially improve phylogenetic inference, resolve deep divergence and correct systematic biases. *Systematic Biology* 60 (6): 833–844. https://doi.org/10.1093/sysbio/syr064.

Darwin, Charles. 1871. *The descent of man and selection in relation to sex*. Vol. I. London: John Murray. http://darwin-online.org.uk/converted/pdf/1871_Descent_F937.1.pdf.

Dawkins, Richard. 1976. *The selfish gene*. Oxford: Oxford University Press.

Eckert, Penelope. 2012. Three waves of variation study: The emergence of meaning in the study of sociolinguistic variation. *Annual Review of Anthropology* 41: 87–100. https://doi.org/10.1146/annurev-anthro-092611-145828.

Embleton, Sheila M. 1986. *Statistics in historical linguistics*. Bochum: Brockmeyer.

Fauconnier, Gilles, and Mark Turner. 2003. *The way we think: Conceptual blending and the mind's hidden complexities*. New York: Basic Books.

Felsenstein, Joseph. 2004. *Inferring phylogenies*. Sunderland: Sinauer Associates Inc.

Goebl, Hans. 1982. *Dialektometrie: Prinzipien Und Methoden Des Einsatzes Der Numerischen Taxonomie Im Bereich Der Dialektgeographie*. Wien: Österreichischen Akademie der Wissenschaften.

Gray, Russell D., David Bryant, and Simon J. Greenhill. 2010. On the shape and fabric of human history. *Philosophical Transactions of the Royal Society B: Biological Sciences* 365: 3923–3933. https://doi.org/10.1098/rstb.2010.0162.

Greenhill, Simon J., Thomas E. Currie, and Russell D. Gray. 2009. Does horizontal transmission invalidate cultural phylogenies? *Proceedings of the Royal Society B: Biological Sciences* 276: 2299–2306. https://doi.org/10.1098/rspb.2008.1944.

Helpuhe. 2014. The longitudinal corpus of Finnish spoken in Helsinki (1970, 1990, 2010). http://urn.fi/urn:nbn:fi:lb-2014073041.

Honkola, Terhi, Kalle Ruokolainen, Kaj Syrjänen, Unni-Päivä Leino, Ilpo Tammi, Niklas Wahlberg, and Outi Vesakoski. 2018. Evolution within a language: Environmental differences contribute to divergence of dialect groups. *BMC Evolutionary Biology* 18 (1): Article 132. https://doi.org/10.1186/s12862-018-1238-6.

Hull, David L. 1988. *Science as a process: An evolutionary account of the social and conceptual development of science*. Chicago: University of Chicago Press.

Kuparinen, Olli, Liisa Mustanoja, Jaakko Peltonen, Jenni Santaharju, and Unni Leino. 2019. Muutosmallit Helsingin Puhekielessä. *Sananjalka* 61: 30–56. https://doi.org/10.30673/sja.80056.

Labov, William. 1994. *Principles of linguistic change. Vol 1. Internal Factors*. Oxford: Blackwell.

Lakoff, George, and Mark Johnson. 1980. *Metaphors we live by*. Chicago: University of Chicago Press.

Leibniz, Gottfried Wilhelm, Freiherr von. 1710. Brevis Designatio Meditationum de Originibus Gentium, Ductis Potissimum Ex Indicio Linguarum. In *Miscellanea Berolinensia Ad Incrementum Scientiarum, Ex Scriptis Societati Regiae Scientiarum Exhibitis Edita* I: 1–16. Berolinum: Johan. Christ. Papenius. urn:nbn:de:kobv:b4360–1004858.

Liu, Liang, Christian Anderson, Dennis Pearl, and Scott V. Edwards. 2019. Modern phylogenomics: Building phylogenetic trees using the multispecies coalescent model. In *Evolutionary genomics*, Methods in molecular biology 1910, ed. Maria Anisimova. New York: Humana Press.

Maurits, Luke, Mervi de Heer, Terhi Honkola, Michael Dunn, and Outi Vesakoski. 2019. Best practices in justifying calibrations for dating language families. *Journal of Language Evolution* 5 (1): 17–38. https://doi.org/10.1093/jole/lzz009.

McMahon, April, and Robert McMahon. 2005. *Language classification by numbers*, Oxford Linguistics. Oxford: Oxford University Press.

Neelapu, Nageswara Rao Reddy, Malay Ranjan Mishra, Titash Dutta, and Surekha Challa. 2019. Role of horizontal Gene transfer in evolution of the plant genome. In *Horizontal Gene Transfer: Breaking borders between living kingdoms*, ed. Tomás Gonzalez Villa and Miguel Viñas, 291–314. Cham: Springer.

Nelson-Sathi, Shijulal, Johann-Mattis List, Hans Geisler, Heiner Fangerau, Russell D. Gray, William Martin, and Tal Dagan. 2011. Networks uncover hidden lexical borrowing in indo-European language evolution. *Proceedings of the Royal Society B: Biological Sciences* 278: 1794–1803. https://doi.org/10.1098/rspb.2010.1917.

O'Hara, Robert. 1996. Trees of history in systematics and philology. *Memorie Della Società Italiana Di Scienze Naturali e Del Museo Civico Di Storia Naturale Di Milano* 27: 81–88.

Pakendorf, Brigitte. 2014. Coevolution of languages and genes. *Current Opinion in Genetics & Development* 29: 39–44. https://doi.org/10.1016/j.gde.2014.07.006.

Pritchard, Jonathan K., Matthew Stephens, and Peter Donnelly. 2000. Inference of population structure using multilocus genotype data. *Genetics* 155: 945–959.

Sankararaman, Siriam, Swapan Mallick, Nick Patterson, and David Reich. 2016. The combined landscape of Denisovan and Neanderthal ancestry in present-day humans. *Current Biology* 26 (9) https://doi.org/10.1016/j.cub.2016.03.037.

Saussure, Ferdinand de. 1916. *Cours de Linguistique Générale*, ed. Charles Bally, Albert Sechehaye, and Albert Riedlinger. Lausanne/Paris: Payot.

Schleicher, August. 1863. *Die Darwinsche Theorie Und Die Sprachwissenschaft*. Weimar: Hermann Böhlau.

Syrjänen, Kaj, Terhi Honkola, Jyri Lehtinen, Antti Leino, and Outi Vesakoski. 2016. Applying population genetic approaches within languages: Finnish dialects as linguistic populations. *Language Dynamics and Change* 6: 235–283. https://doi.org/10.1163/22105832-00602002.

Syrjänen, Kaj, Luke Maurits, Unni Leino, Terhi Honkola, Jadranka Rota, and Outi Vesakoski. Submitted ms. 'Crouching TIGER, hidden structure: Exploring the nature of linguistic data using TIGER rates'. *Journal of Language Evolution*.

Taylor, Scott A., and Erica L. Larson. 2019. Insights from genomes into the evolutionary importance and prevalence of hybridization in nature. *Nature Ecology & Evolution* 3: 170–177. https://doi.org/10.1038/s41559-018-0777-y.

Verkerk, Annemarie. 2019. Detecting non-tree-like signal using multiple tree topologies. *Journal of Historical Linguistics* 9 (1): 9–69. https://doi.org/10.1075/jhl.17009.ver.

Weaver, Warren. 1949. Translation. *Memorandum*. New York: The Rockefeller Foundation. http://www.mt-archive.info/Weaver-1949.pdf.

Wichmann, Søren, Eric W. Holman, Taraka Rama, and Robert Walker. 2011. Correlates of reticulation in linguistic phylogenies. *Language Dynamics and Change* 1 (2): 205–240. https://doi.org/10.1163/221058212X648072.

Wieling, Martijn, and John Nerbonne. 2011. Measuring linguistic variation commensurably. *Dialectologia* (Special issue II): 141–162.

8

Three Models for Linguistics: Newtonian Mechanics, Darwinism, Axiomatics

Esa Itkonen

1 Preliminary Remarks

A scientific discipline can develop autonomously, but it can also imitate the model set by some other discipline which is (thought to be) more well-established or otherwise more prestigious. It has been shown, for instance, that the linguistic traditions in India and in Arabia originally adopted some elements, respectively, from the science of ritual and from jurisprudence (cf. Itkonen 1991: 78, 128). More recently, the Western tradition of linguistics has been influenced in different ways by at least the three models that figure in the title of this paper. A few qualifications are now in order.

First: It seems fair to say that no noticeable mark has been left on linguistics either by the Theory of Relativity or by quantum physics, which are surely among the most prestigious scientific theories today.

E. Itkonen (✉)
University of Turku, Turku, Finland
e-mail: eitkonen@utu.fi

R. M. Nefdt et al. (eds.), *The Philosophy and Science of Language*,
https://doi.org/10.1007/978-3-030-55438-5_8

Second: Whenever A is shaped on the model set by B, there must be an **analogy** between A and B. In the name of conceptual clarity, these two questions must be kept apart, even if the answers to them overlap to some extent: 'what is **language** analogous to?' vs. 'what is **linguistics** analogous to?' (cf. Itkonen 2005: 186–190). It is the latter question which will be addressed in this paper.

Third: It is not only the case that linguistics has been, rightly or wrongly, influenced by extraneous models. It is also the case that linguistics has in turn influenced other disciplines. This is why it has become customary, within the larger framework of **semiotics**, to speak of the 'grammar' of practically any aspect of culture.

2 Newtonian Mechanics

The core of Newtonian (= 'classical') mechanics is constituted by the three 'laws of motion' (cf. Nagel 1961: 158–159). The first one, as formulated by Newton himself, is as follows: (i) "Every body perseveres in its state of rest, or of uniform motion in a straight line, unless it is compelled to change that state by forces impelled thereon." The gist of the other two laws, expressed more succinctly, is as follows: (ii) "The alteration of motion is ever proportional to the motive force impressed." (iii) "To every action there is always opposed an equal reaction." In order to be effective, these laws demand the corresponding type of mathematics, i.e. differential calculus. So formulated, the laws are of **deterministic** nature: they have no gaps or exceptions; rather, they are valid 'always and everywhere'.

The promise that Newtonian mechanics has held out for the human sciences, including linguistics, has been described by Berlin (1980/1960) as follows: "If only we could find a series of **natural laws** connecting at one end the biological and physiological states and processes of human beings with, at the other, the equally observable patterns of their conduct – their social activities in the wider sense – and so establish a coherent system of regularities, deducible from a comparatively small number of general laws (as Newton, it is held, had so triumphantly done in physics), we should have in our hands a science of human behavior" (p. 105; emphasis added).

Distinct attitudes have been adopted vis-à-vis the Newtonian model. Those who have put it to a test have either correctly concluded that it is not valid (Sect. 2.1) or incorrectly concluded that it is valid (Sect. 2.2). Others have (incorrectly) assumed that it is valid *a priori* (Sect. 2.3).

2.1 The Chimera of Deterministic Laws

Within **historiography**, the Newtonian model turned out to be an impossible dream: "All seemed ready, particularly in the nineteenth century, for the formulation of this new, powerful, and illuminating discipline, ... The stage was set, but virtually nothing materialised. No general laws were formulated ... Neither psychologists nor sociologists ... had been able to create the new mechanism: the 'nomothetic' sciences ... remained stillborn" (Berlin 1980/1960: 110).

The development of **diachronic linguistics** in the twentieth century fully confirms Berlin's diagnosis insofar as no deterministic 'laws of linguistic change' have ever been discovered: "All the laws that have been or will be proposed have a common defect: what they assert is only possible, not necessary. ... The laws of general-historical phonetics or morphology do not suffice to explain a single fact ... we are not able to **predict** a single future event" (Meillet 1921: 15–16; emphasis added). Samuels (1972) explains why this is so: "[E]very linguistic change involves at least some degree of **choice**, ..." (p. 3; emphasis added); and he also draws the inevitable conclusion: "**predictive** power in historical linguistics ... is an impossibility" (1987: 239; emphasis added). According to Lass (1997), there has been no progress in the "post-Neogrammarian historical linguistics", precisely because "we still have no convincing [= deterministic] explanations for change" (pp. 386–387).

In fact, Lass (1980) had already expatiated on the malaise that he shares with Meillet. In his review of this book, Itkonen (1981) strikes a more optimistic note. From the qualitative point of view, matter-in-motion is hugely different from the data of (diachronic) linguistics. This is, very simply, why practitioners of diachronic linguistics must rely on **non**-deterministic 'functional' explanations (cf. also Itkonen 1984, 2011a). Samuels (1987) reaches the same conclusion in his assessment of the "status of the functional approach".

Let us add, for completeness, that the search for deterministic laws has been abandoned also in **sociology**: "I shall dogmatically assert the need for an account of **agent causality** ..., according to which causality does not presuppose 'laws' of invariant connection (if anything, the reverse is the case) ..." (Giddens 1976: 84; original emphasis).

2.2 Pseudo-Determinism

Greenberg-type **implicational universals** are the cornerstone of modern typological linguistics, for instance: (i) If a language has initial consonant clusters, then it has medial consonant clusters (but not vice versa). More formally: for all L, if L has CC-, then L has -CC- = if A, then B.

Now, (i) certainly looks analogous to prototypical deterministic laws of physics, for instance: (ii) For all x, if x is a heated piece of metal, then x expands = if C, then D.

But there are important **dis**analogies as well. The reason why (ii) is (rightly) felt to be **explanatory** is that (ii), in addition to being deterministic, expresses a relation of **causality**: being heated (= C) is the cause while expanding (= D) is the effect. By contrast, (i), or 'if A, then B', does **not** express a relation of causality. Rather, (i) expresses the fact that either -CC- and CC- occur together or -CC-, being more frequent than CC-, occurs also in languages where CC- does not occur. Hence, (i) expresses a relation of **presupposition**: the presence of CC- presupposes the presence of -CC- (but not vice versa).

But **why** is -CC- more frequent than CC-? Because it is **easier** to pronounce. This is the **explanatory** notion. But notice that it is not explicitly mentioned in (i) at all. Hence, (i) does not qualify as explanatory; nor does it qualify as a (deterministic) **law**, because 'law' is generally meant to be a (causal-) explanatory notion.

What was just said about implicational universals like (i), is true of relations of presupposition in general. Let us consider one more example: (iii) If one has money to buy a diamond necklace (= A), then (it can be inferred that) one has money to buy food (= B), but not vice versa: if A, then B.

Again, (iii) does not express a relation of causality. In the background of (iii), to be sure, there must be an entire causal network of human abilities and motivations, summarized as follows: 'If secondary needs have been satisfied, then (it can be inferred that) primary needs have been satisfied as well.' It is this network which qualifies as explanatory, but it is not even hinted at in (iii).

An additional reason why the relation of presupposition cannot be explanatory is that it is general enough to be of purely **conceptual** nature: 'If John spent the last week in London, then (it can be inferred that) he spent the last Tuesday in London', 'If John has 5 children, then (it can be inferred that) he has at least 3 children', etc.

All Greenberg-type universals state **correlations** between A and B. Many of them are presuppositional, but some are not. Significantly, the latter tend to be **non**-deterministic (= statistical) as well; this is true e.g. of generalizations concerning word (or constituent) order. Detecting correlations is only the first step; next, they need to be (causally) explained. Unexplained correlations do not qualify as laws; for more discussion, see Itkonen (1998, 2013a, 2019b).

2.3 Determinism by Fiat

If one decides that the data one is dealing with must be of **physical** nature, and if one further decides to discard the existence of **statistical** variation in one's data, it follows that the theory one is striving after is of **deterministic** nature. This characterization summarizes the common core of Bloomfield's, Harris', and Chomsky's conceptions of linguistics:

> All scientifically meaningful statements must be translatable into **physical** terms – that is, into statements about movements which can be observed and described in coordinates of space and time. (Bloomfield 1936: 90; emphasis added)

> In defining elements for each language, the linguist relates them to the **physiological** activities or **sound waves** of speech, ... (Harris 1951: 16; emphasis added)

> [S]uch notions as 'phoneme in L', 'phrase in L', 'transformation in L' are defined for an arbitrary language L in terms of **physical** and distributional properties of utterances of L and formal properties of grammars of L. (Chomsky 1957: 54; emphasis added)

At first, there seems to be absolutely no possibility of a physical theory of linguistic behavior, as demanded by Bloomfield, Harris, and Chomsky. But what seems (and in fact, is) impossible, can be imagined to be achieved by means of a spurious analogy:

> Any scientific theory is based on a finite number of observations, and it seeks to relate the observed phenomena and to **predict** new phenomena by constructing general **laws** in terms of hypothetical constructs such as (in **physics**, for example) 'mass' and 'electron'. Similarly, a grammar of **English** is based on a finite corpus of utterances (observations), and it will contain certain **grammatical rules (laws)** stated in terms of the particular phonemes, phrases, etc. of English (hypothetical constructs). (Chomsky 1957: 49; emphasis added)

An earlier passage contains the same methodologically crucial equation: "grammatical rules or the laws of the theory" (Chomsky 1975/1955: 77). It is important to clearly understand what is being claimed here. A set of grammatical rules is given e.g. in Chomsky (1957: 26–27), starting with *Sentence* → *NP* + *VP*. As we just saw, they are assigned the same methodological status as the laws of physics; and this is also true, more particularly, of such 'hypothetical constructs' as 'electron' and *NP*. Thus, anybody is free to construct any number of 'physical theories of linguistic behavior' in a matter of seconds. This is nothing sort of ludicrous.

During its history of 60-plus years, the physicalist credo of generative linguistics has remained the same: "[Elements of I-language] are real elements of particular minds/brains, aspects of the **physical** world, …" (Chomsky 1986: 26; emphasis added). "If [the strong minimalist thesis] were true, language would be something like a snowflake, taking the form it does by virtue of natural [= physical] **law**, in which case UG would be very limited" (Chomsky 2011: 26; emphasis added). Physicalism, moreover, is illustrated by means of examples that remain in

2011 on the same level of (non-)sophistication as in 1957: "External Merge takes two objects, say *eat* and *apples*, and forms the new object corresponding to *eat apples*" (op. cit., p. 28). "If computation is efficient, then when X and Y are merged neither will change so that the outcome can be taken to be simply the set {X,Y}. That is … a natural principle of efficient computation, perhaps a special case of **laws** of nature" (*ibidem*; p. 26; emphasis added).

While the laws of nature can in principle be either deterministic or statistical in character, those postulated by Chomsky must be deterministic, given his well-advertised disdain for statistics: "We see, however, …that a structural analysis cannot be understood as a schematic summary developed by sharpening the blurred edges in a full statistical picture" (Chomsky 1957: 16). Of course, this claim has been strongly contested. The following seems a fair assessment: "Assuming that T[ransformational] G[rammar] is an empirical science, it is the **only** empirical science which excludes statistical considerations on *a priori* grounds" (Itkonen 1978: 231). That is, according to Chomsky, pure philosophical reflection enables us to **see** that statistics is not needed in linguistics. This position has been reaffirmed: "The existence of a discipline called 'sociolinguistics' remains for me an obscure matter" (Chomsky 1979: 56). "That is the worst of all; presenting, say, statistical analyses on subjects that are without interest" (p. 59).

Finally, a couple of obvious **dis**analogies between linguistics and Newtonian mechanics need to be mentioned. First: As Hermann Paul (1975/1880: 30) already pointed out, linguistic data is primarily gained by means of **self**-observation, which is never the case in physics: a planet or atom does not observe itself. Second, the research object of linguistics has an inherently **normative** character, which, again, is never the case in physics. A certain amount of conceptual incoherence results from the fact that some representatives of generative linguistics are not totally unaware of these disanalogies (cf. Itkonen 1996, 2019a; Carr 2019).

Whitney (1979/1875) already saw very clearly both the justification and the ultimate fate of the 'Newtonian dream' in linguistics: "What [the linguist] does need to insist upon is that the character of his department of study be not misrepresented … – by declaring it, for example, a physical or natural science, in these days when the physical sciences are filling men's minds with wonder at their achievements, and almost presuming to claim the title of science as belonging to themselves alone. … There is no way of claiming a physical character for the study of such [linguistic] phenomena except by a thorough misapprehension of their nature, a perversion of their analogies with the facts of physical science" (pp. 310–311).

3 Darwinism

'Evolution of language' can mean two different things: either phylogenetic development out of 'pre-language' or linguistic change (= the subject matter of diachronic linguistics). 'Darwinism in linguistics' admits of the same interpretations, and can therefore be judged either a success or a failure depending on which interpretation is chosen. Needless to say, Darwinism represents the only viable approach to the phylogenesis of language, as argued e.g. by Hurford et al. (eds.) (1998) and Mufwene (2013). It is the second interpretation of 'Darwinism in linguistics' that will be addressed in this section.

In the light of our discussion in Sect. 2, the following assessment is fully justified: "Biology is a much more realistic metaphor for linguistics than is physics" (Givón 1984: 24). On the one hand, biological evolution is driven by two principal forces, namely random **mutation** and natural **selection**. On the other hand, linguistic change too can be conceptualized as driven by two principal forces, namely (individual) **innovation** and (social) **acceptance**. It is easy to understand those representatives of Darwinist linguistics who claim that this is, ultimately, one and the same process (cf. Haspelmath 1999a, b; Kirby 1999; Croft 2000).

There is certainly a *prima facie* analogy between mutation-*cum*-selection and innovation-*cum*-acceptance. But, again, there are **dis**analogies as well. The most important of them is easily summed up: "No evolutionary change of any kind came about through the application of intelligence and knowledge to the **solution of a problem**. This was at the heart of Darwin's idea" (Cohen 1986: 125; emphasis added). The importance of this remark resides in the fact that explanations given in diachronic and/or typological linguistics are nowadays customarily conceptualized as, precisely, solutions to a problem, with the consequence that Darwinism is automatically **disqualified** in this field of study:

"It is possible to view the various types of coding of the same functional domain as alternative **solutions** to the same communicative task" (Givón 1984: 145; original emphasis). "Grammaticalization can be interpreted as a process that has **problem solving** as its main goal, whereby one object is expressed in terms of another" (Heine et al. 1991: 29; emphasis added). In Itkonen (2013a), for instance, the problem-solving strategy has been applied to the following topics: the *das* > *dass* grammaticalization; overt case-marking in intransitive vs. transitive subjects; expression of plurality in nouns; *N1-and-N2* constructions; converb constructions; the distribution of personal endings in the Hua aorist paradigm. (cf. also Itkonen 2004).

The problem-solving explanations exemplify what is loosely called the 'functional' approach. More precisely, we have to do here with **rational explanation**. In outline: Because the agent has a goal G and believes that an action A is the means to achieve G, s/he chooses A from among possible actions and sets out to do A; as indicated by the use of *because*, this is a type of **causal** explanation, i.e. the agent's goals and beliefs are (assumed to be) causally effective entities; in the problem-solving scenario, to solve a problem is G, and A is the actual solution. Much more expansive accounts are given in Itkonen (1983: Ch. 3, 2013a). Let us add that presuppositional universals (discussed here in Sect. 2.2) turn out to exemplify the following figure of thought: if complex means-to-ends can be chosen, simple means-to-ends can be, and are, chosen as well (but not vice versa).

The same **dis**analogies are valid here as in connection with Newtonian mechanics: there is no room for (conscious) self-observation or for normativity (cf. Itkonen 2008, 2019b; also the other contributions to Mäkilähde et al. (eds.) 2019). There is no 'intelligent design' in biological evolution (let alone in mechanics), but it pervades the subject matter of diachronic and/or typological linguistics. To repeat: "The link between biological and linguistic changes is **metaphorical**" (Itkonen 1984: 209; original emphasis). A succinct refutation of Darwinist linguistics is given in Itkonen (1999).

4 Axiomatics

Euclid's (c. 300 BC) systematization of plane geometry contains five (general) axioms and five (more specific) postulates (cf. Sklar 1974: 13–16). For more than 2000 years, it provided the paradigm example of what a scientific study ought to be. The fate of the fifth postulate has been particularly interesting. It is equivalent to the statement that through a point outside a given line one and only one (parallel) line can be drawn which does not intersect the given line. Unlike the other axioms and postulates, this was not experienced as intuitively self-evident. In the beginning of the nineteenth century it was indeed demonstrated that both the geometry with a 'many-parallel postulate' and the geometry with a 'no-parallel postulate' are logically consistent. In the beginning of the twentieth century, moreover, the 'curvature of space' necessitated by the Relativity Theory proved that, on the universal scale, non-Euclidean geometry is not only logically possible but also actually true.

Euclid's fifth axiom states that a whole is greater than any of its parts. But can a whole be greater than the sum of its parts, and if so, when? This question has been addressed and answered in Itkonen (2016).

Euclid's geometry has been a constant source of inspiration. The best known, and also the most ambitious, attempt to imitate it has been Spinoza's (1632–1677) *Ethics*, which – in spite of its title – actually tries to give an exhaustive account of the entire universe. In this book, which has 5 parts, a total of 259 'propositions' (= theorems) are proved by means of 17 axioms (plus 2 'postulates') and 25 definitions (plus 48 'definitions of emotions'). For instance, the proposition III of Part I (= 'Things which have nothing in common cannot be one the cause of the other') is deduced jointly from the axioms V and IV (= 'Things which have nothing in common cannot be understood one by means of the other' and 'The knowledge of an effect depends on and involves the knowledge of a cause'). To properly understand these statements, it needs to be known that 'the cause of X' means here any Y which **explains** why X is the way it is; and to Spinoza, more particularly, 'if Y, then X' expresses a relation of **entailment**, not a causal relation (as this term is understood today): "Euclid's geometry provides the standard example of genuine explanation, ..." (Hampshire 1962/1951: 35). It is easy to understand that, when dealing with such a hugely complicated topic, the order of exposition cannot entirely coincide with the logical order: "These first definitions and propositions can be properly understood only in the light of the propositions which follow them in the order of exposition; they form a system of mutually supporting propositions, ..." (p. 30).

In linguistics too, there have been more or less systematic attempts at axiomatization. One of the best known is Bloomfield's (1957/1926) set of 13 postulates (= "assumptions or axioms") and 50 definitions (not including those which deal with historical linguistics). For instance, Definition 26 = "A maximum X is an X which is not part of a larger X." Assumption 10 = "Each position in a construction can be filled only by certain forms." It is the merit of the postulational method that "it forces us to state explicitly whatever we assume, to define our terms, and to decide what things may exist independently and what things are interdependent" (p. 26).

This is how Bloomfield evaluates his own achievement: "The assumptions and definitions so far made will probably make it easy to define the grammatical phenomena of any language, both morphologic (affixation, reduplication, composition) and syntactic (cross-reference, concord, government, word-order), ..." (p. 30). The central importance of the following remarks will become evident at the end of this section: "Also, the postulational method ... cuts off from psychological dispute. ... From our point of view, [Delbrück] was ... right in saying that it is indifferent what system of psychology a linguist believes in" (p. 26).

A different notion of axiomatics is introduced by Harris (1951): "The work of analysis leads right up to the statements which enable anyone to **synthesize** or **predict** utterances in the language. These statements form a deductive system with axiomatically defined initial elements and with theorems concerning relations among them. The final theorems would indicate the structure of the utterances of the language... The system can be presented most baldly in an **ordered set** of statements defining the elements at each **successive** level or stating the sequences which occur at that level" (pp. 372–373; emphasis added). On the one hand, the term *statement* is ambiguous: it may or may not presuppose Euclid-type axiomatics; on the other hand, the notion of 'synthesizing utterances' clearly anticipates the idea of **generativity**.

At the time, this idea was in the air. When recommending that the 'Item-and-Process' model be formalized, Hockett (1954) demands that "one must be able to **generate** any number of utterances in the language, above and beyond those observed in advance by the analyst" (p. 398; emphasis added). While Harris' (1951) idea of "an ordered set of successive levels" leading up to utterances is an exact formulation of the Item-and-Process model, it is at the same time an informal description of a generative grammar. And just like Hockett (1954), Harris (1951) had explicitly rejected any form of narrow corpus-thinking: "When the linguist offers his results as a system representing the language as a whole, he is predicting that the elements set up for his corpus will satisfy all other bits of talking in that language" (p. 16; also p. 372).

Euclid's and Spinoza's axioms are (assumed to be) **true**. Theorems are **deduced** (or proved) from axioms. Deduction is a truth-preserving type of inference: it **must** be the case that if the premises (here: axioms) are true, then the conclusion (here: each particular theorem) is true.

In formal logic, the axioms are not only true, but **necessarily** true, i.e. **valid**. For simplicity, we shall concentrate here on the basic type of logic, namely (two-valued) propositional logic. It was axiomatized, in 1910, by Russell & Whitehead by means of five axioms and two (validity-preserving) rules of inference, i.e. Modus Ponens (= 'if you have *A* and *if A, then B*, you are allowed to infer *B*') and Rule of Substitution (= 'in a valid formula, you are allowed to substitute any formula for any atomary formula'). By 1929, the number of the axioms had been reduced to three. Here, progress is measured by the gain in **simplicity**, and the transition from five to three is genuine progress as long as the structural complexity of the axioms does not increase. But if it does, problems arise. It is possible to replace three 'ordinary' axioms by one hugely complicated axiom; but is this progress? Ultimately, this turns out to be an insoluble problem (cf. Itkonen 2003: Ch. VI, 2011b: 144–149).

The notion of axiomatics was extended from 'theory' to 'system', i.e. from deducing valid formulae (as in formal logic) to generating meaningless strings of symbols (as in grammatical description): "A grammar is a device for generating sentences ... [i.e.] a sequence of statements of the form *X-i* → *Y-i* interpreted as the instruction 'rewrite *X-i* as *Y-i*' where *X-i* and *Y-i* are strings. … Call each statement of [this] form a **conversion** … A derivation is roughly **analogous** to a **proof**, with *Sentence* playing the role of the single **axiom**, and the conversions corresponding roughly to **rules of inference**" (Chomsky 1975/1955: 67; only the first emphasis in the original).

The word *analogous* is crucial here: axiomatics provided the (analogous) **model** which was imitated by generative grammar. This (historical) interpretation has been confirmed by Katz (1981): "The idea of generative grammar emerged from an analogy with categorial systems in logic. The idea was to treat grammaticality as theoremhood in logistic [= axiomatic] systems and to treat grammatical structure like proof structure in derivations" (p. 36). Analogical extensions (or generalizations) of this type have been rather customary in formal logic, for instance, from two-valued to many-valued logic, or from modal to deontic logic (cf. Itkonen 2005: 19–20). More generally, analogy is seen to be the driving force behind scientific discovery and/or invention (op. cit., pp. 176–198).

The role of simplicity is the same in grammatical description as it was just seen to be in axiomatic logic: "That solution is considered the correct one which complies in the highest degree with the simplicity principle" (Hjemslev 1961/1943: 18). "We may define the phonemes and morphemes of a language as the tentative phonemes and morphemes which … jointly lead to the simplest grammar" (Chomsky 1957: 57).

Clearly, generative linguistics, as we know it, would not have come into being without the existence of the axiomatic model. Therefore, it is only the more interesting to note that a slightly different type of generative linguistics **has** actually come into being, a long time ago and under quite different circumstances: "Modern linguistics acknowledges [Pāṇini's (c. 400 BC) grammar] as the most complete generative grammar of any language yet written" (Kiparsky 1993: 2912). This *prima facie* astonishing claim has been defended by several writers, including Itkonen (1991: Ch. 2). In the present context the following sub-sub-section (= 2.3.4) is especially relevant: 'On the 'axiomatic' character of Pāṇini's grammar' (pp. 38–44). This axiomatic character is illustrated quite concretely by the derivation of the verb form *náyanti* ('they lead') (pp. 63–67). The derivation starts with the corresponding abstract meaning and ends with the concrete (= allophonic) form; its gradual 'meaning-to-form' movement contains 16 distinct stages; and it implicitly refers to 65 rules out of those c. 4000 rules that constitute the total of Pāṇini's grammar. Let us add that this monumental work was originally a spoken description of spoken language (= Classical Sanskrit), i.e. it was composed before the advent of literacy.

It may sound odd that even a work of such complexity strives after simplicity, but this is indeed the case: what is meant, is simplicity within extremely complex self-imposed constraints. Anybody else would need about 10,000 rules to do what Pāṇini does with 4000 rules. In the Indian tradition, parsimony was the generally accepted descriptive ideal: "Grammarians rejoice over the saving of even the length of half a short vowel as much as over the birth of a son" (cf. Itkonen 1991: 21).

Up to this point in our historical overview, the influence of axiomatics on linguistics was exclusively beneficial. The situation changed when Chomsky (1965) introduced "the psychological interpretation of grammars", or "the interpretation of a generative grammar as a system of knowledge in the mind of the [ideal] speaker" (Freidin 2013: 453). The problem was that this "psychological interpretation" was super-imposed upon the grammar in a purely mechanical way. The actual grammatical formalization remained exactly the same as before, and when it changed in the sequel, it did so for purely grammar-internal reasons, not because of any psychological considerations. This maneuver was justified (or so it seemed) by contriving a new kind of (pseudo)psychological entity, i.e. by "treat[ing] linguistic knowledge as a separate entity of the faculty of language, distinct from linguistic behavior – ... thereby formulating a *competence/performance* distinction" (*ibidem*).

The end result might be called an 'axiomatic fallacy': "The 'ideal speaker' possesses no properties over and above those belonging to an axiomatic system: in fact, **the two are identical**" (Itkonen 1976: 214; original emphasis). Nothing changed apart from terminology when, in Chomsky (1986), 'competence' was replaced by 'I-language', understood as a property of "particular minds/brains, aspects of the physical world" (cf. Sect. 2.3): "It is safe to assume that *I-language* is equivalent to *competence*" (ten Hacken 2019: 554). Replacing axiomatic derivations by 'computations' was another cosmetic change. The rise of cognitive linguistics in the 1980s was a natural reaction against the artificial and aprioristic type of psychology advocated by generativism.

5 Conclusion

Two desiderata stand out in the history of linguistics: axiomatic systematization, on the one hand, and causal explanation, on the other (cf. Itkonen 2013b: 774). Newtonian mechanics is a bad model to achieve causal explanation of linguistic behavior. Darwinism is better but still inadequate. Axiomatics is a good model for linguistics (in the sense of grammatical description), but a bad model for psycholinguistics: the (untrained) human mind just does not operate by means of few axioms and long derivations.

References

Allan, Keith (ed.) 2013. *The Oxford handbook of the history of linguistics*. Oxford: Oxford University Press.

Berlin, Isaiah. 1980/1960. The concept of scientific history. *Concepts and categories: Philosophical essays*, 103–142. Oxford: Oxford University Press.

Bloomfield, Leonard. 1936. Language or ideas? *Language* 12 (1): 89–95.

———. 1957. A set of postulates for the science of language. In, ed. M. Joos, 26–31. Originally published in 1926, *Language* 2/2: 153–164.

Carr, Philip. 2019. Syntactic knowledge and intersubjectivity. In, ed. A. Kertész et al., 423–440.

Chomsky, Noam. 1957. *Syntactic structures*. The Hague: Mouton.

———. 1965. *Aspects of the theory of syntax*. Cambridge, MA: MIT Press.

———. 1975/1955. *The logical structure of linguistic theory*. New York: The Plenum Press.

———. 1979. *Language and responsibility*. New York: Pantheon Books.

———. 1986. *Knowledge of language: Its nature, origin, and use*. New York: Praeger.

———. 2011. Opening remarks. In *Of minds and language: A dialogue with Noam Chomsky in the Basque country*, ed. Massimo Piattelli-Palmarini, Pello Salaburu, and Juan Uriagereka, 13–32. Oxford: Oxford University Press.

Cohen, Jonathan L. 1986. *The dialogue of reason: An analysis of analytical philosophy*. Oxford: Oxford University Press.
Croft, William. 2000. *Explaining language change: An evolutionary approach*. London: Longman.
de Spinoza, Benedict. 1974/1960/1677. *The ethics*. Trans. R.H.M. Elwes. *Rationalists*, 179–406. Garden City: Anchor Books, Doubleday & Company.
Freidin, Robert. 2013. Noam Chomsky's contribution to linguistics: A sketch. In, ed. K. Allan, 439–468.
Giddens, Anthony. 1976. *New rules of sociological method*. London: Hutchinson.
Givón, T. 1984. *Syntax: A functional-typological introduction, vol. I*. Amsterdam: Benjamins.
Hampshire, Stuart. 1962/1951. *Spinoza*. Harmondsworth: Penguin Books.
Harris, Zellig S. 1951. *Structural linguistics*. Chicago: University of Chicago Press.
Haspelmath, Martin. 1999a. Optimality and diachronic adaptation. *Zeitschrift für Sprachwissenschaft* 18 (2): 180–205.
———. 1999b. Some issues concerning optimality and diachronic adaptation. *Zeitschrift für Sprachwissenschaft* 18 (2): 251–268.
Heine, Bernd, Ulrike Claudi, and Friederike Hünnemeyer. 1991. *Grammaticalization: A conceptual framework*. Chicago: University of Chicago Press.
Hjemslev, Louis. 1961/1943. *Prolegomena to a theory of language*. Trans. Francis J. Whitfield. Madison: University of Wisconsin Press.
Hockett, Charles F. 1957. Two models of grammatical description. In, ed. M. Joos, 386–399. Originally published in 1954, *Word* 10: 210–231.
Hurford, James R., Michael Studdert-Kennedy, and Chris Knight, eds. 1998. *Approaches to the evolution of language*. Cambridge: Cambridge University Press.
Itkonen, Esa. 1976. The use and misuse of the principle of axiomatics in linguistics. *Lingua* 38 (2): 185–220.
———. 1978. *Grammatical theory and metascience: A critical investigation into the methodological and philosophical foundations of 'autonomous' linguistics*. Amsterdam: Benjamins.
———. 1981. Review article on Lass 1980. *Language* 57 (3): 688–697.
———. 1983. *Causality in linguistic theory: A critical investigation into the philosophical and methodological foundations of 'non-autonomous' linguistics*. London: Croom Helm.
———. 1984. On the 'rationalist' conception of linguistic change. *Diachronica* 1 (2): 203–216.

———. 1991. *Universal history of linguistics: India, China, Arabia, Europe*. Amsterdam: Benjamins.
———. 1996. Concerning the generative paradigm. *Journal of Pragmatics* 25: 471–501.
———. 1998. Concerning the status of implicational universals. *Sprachtypologie und Universalienforschung* 51 (2): 157–163. Reprinted in Itkonen 2011b, 5–13.
———. 1999. Functionalism yes, biologism no: A reply to 'Optimality and diachronic adaptation' by M. Haspelmath. *Zeitschrift für Sprachwissenschaft* 18 (2): 219–221. Reprinted in Itkonen 2011b, 14–16.
———. 2003. *Methods of formalization beside and inside both autonomous and non-autonomous linguistics*, 6. University of Turku: Publications in General Linguistics.
———. 2004. Typological explanation and iconicity. *Logos and Language* 5 (1): 21–33. Reprinted in Itkonen 2011b, 42–61.
———. 2005. *Analogy as structure and process: Approaches in linguistics, cognitive psychology, and philosophy of science*. Amsterdam: Benjamins.
———. 2008. The central role of normativity in language and linguistics. In, ed. A. Zlatev et al., 279–305.
———. 2011a. On Coseriu's legacy. *Energeia* 3: 1–29. Reprinted in Itkonen 2011b, 191–227.
———. 2011b. Prolegomena to any discussion of simplicity vs. complexity in linguistics. In, ed. E. Itkonen 2011b, 140–152.
———. 2013a. Functional explanation and its uses. In *Functional approaches to language*, ed. Shannon T. Bischoff and Carmen Jany, 31–69. Berlin: De Gruyter Mouton.
———. 2013b. Philosophy of linguistics. In, ed. K. Allan, 747–774.
———. 2016. A whole is greater than the sum of its parts': True, false, or meaningless? *Public Journal of Semiotics* 7 (2): 20–51.
———. 2019a. Hermeneutics and generative linguistics. In, ed. A. Kertész et al., 441–467.
———. 2019b. Concerning the scope of normativity. In, ed. A. Mäkilähde et al., 29–67.
Joos, Martin (ed.) (1957). *Readings in linguistics: The development of descriptive linguistics in America, 1925–56*. Chicago: The University of Chicago Press.
Katz, Jerrold J. 1981. *Language and other abstract objects*. Oxford: Blackwell.
Kertész, András, Moravcsik, Edith & Rákosi, Csilla (eds.) 2019. *Current approaches to syntax: A comparative handbook*. Berlin: De Gruyter Mouton.

Kiparsky, Paul. 1993. Paninian linguistics. In *The encyclopedia of language and linguistics, Vol. 1(6)*, ed. R.E. Asher, 2918–2923. Oxford: Pergamon Press.
Kirby, Simon. 1999. *Function, selection, and innateness: The emergence of language universals*. Oxford: Oxford University Press.
Lass, Roger. 1980. *On explaining language change*. Cambridge: Cambridge University Press.
———. 1997. *Historical linguistics and language change*. Cambridge: Cambridge University Press.
Mäkilähde, Aleksi, Ville Leppänen, and Esa Itkonen, eds. 2019. *Normativity in language and linguistics*. Amsterdam: Benjamins.
Meillet, Antoine. 1921. *Linguistique historique and linguistique générale*. Paris: H. Champion.
Mufwene, Salikoko S. 2013. The origins and the evolution of language. In, ed. K. Allan, 13–52.
Nagel, Ernest. 1961. *The structure of science: Problems in the logic of scientific explanation*. New York: Harcourt, Brace & World.
Paul, Hermann. 1975/1880. *Prinzipien der Sprachgeschichte*. Tübingen: Niemeyer.
Samuels, M.L. 1972. *Linguistic evolution (with special reference to English)*. Cambridge: Cambridge University Press.
———. 1987. The status of the functional approach. In *Explanation and linguistic change*, ed. Willem Koopman, Frederike van der Leek, Olga Fischer, and Roger Eaton, 238–258. Amsterdam: Benjamins.
Sklar, Lawrence. 1974. *Space, time, and space-time*. Berkeley: University of California Press.
ten Hacken, Pius. 2019. The research programme of Chomskyan linguistics. In, ed. A. Kertész et al., 549–572.
Whitney, William D. 1979/1875. *The life and growth of language*. New York: Dover Publications.
Zlatev, Jordan, Racine, Timothy, Sinha, Chris & Itkonen, Esa (eds.) 2008. *The shared mind: Perspectives on intersubjectivity*. Amsterdam: Benjamins.

Part III

Linguistics and the Cognitive Sciences

9

The Role of Language in the Cognitive Sciences

Ryan M. Nefdt

1 Introduction

Language has had a central place in the emerging field of cognitive science since the latter's inception in the late 1950s. The centrality of language has owed its position largely to the influence and success of generative linguistics during that period. However, in part due to challenges to the generative paradigm from within linguistics and in part due to the integrationist agenda within the cognitive sciences, language is becoming more peripheral in the so-called Second Generation Cognitive Science (Sinha 2010) of the early 2000s with cognitive psychology taking center stage. In this chapter, I caution against this current trajectory and argue that language is still well-placed to act as a conduit to accessing general cognition despite certain problems with generative linguistics. In Sect. 2, I briefly discuss the history of mentalistic linguistics and in Sect. 3 do the same for cognitive science. In Sect. 4, I analyze different cognitive

R. M. Nefdt (✉)
Philosophy Department, University of Cape Town, Cape Town, South Africa
e-mail: ryan.nefdt@uct.ac.za

R. M. Nefdt et al. (eds.), *The Philosophy and Science of Language*,
https://doi.org/10.1007/978-3-030-55438-5_9

architectures for the field and suggest an intersectional model. Lastly in Sect. 5, I make a case for why language should not be so easily discarded as central to the cognitive sciences by first rejecting the claim that linguistics is isolationist in methodology and then by briefly questioning the status of psychology as a connection to other cognitive sciences.

2 A Short History of Mentalism in Linguistics

The standard story concerning the relatively recent history of generative or formal linguistics starts with the idea that it emerged out of the limitations of its predecessors, the motley assortment of views characterized under the umbrella term "structural linguistics". Structuralism, on this view, aimed at producing taxonomies of individual languages and their linguistic structures in isolation. Certain advocates, such as Bloomfield, were skeptical of talk of 'the mental' within linguistics.

> Non-linguists (unless they happen to be physicalists) constantly forget that a speaker is making noise, and credit him, instead, with the possession of impalpable 'ideas'. It remains for linguists to show, in detail, that the speaker has no 'ideas', and that the noise is sufficient. (Bloomfield 1936: 23)

I am not planning to disrupt that narrative here (for that see Joseph 1999; Matthews 2001). Rather I plan to trace a different aspect of the emergence of the study of language under the generative paradigm, namely the understanding of linguistic structure as a reflection of cognitive structure. Even if there was some formal continuity between structuralism and generativism (see Nefdt 2019b), there does seem to have been a marked shift with the advent of the mentalist ontology.

The formal mathematical tradition in linguistics can be argued to have been established by Chomsky (1956, 1957) respectively.[1] Mentalism in linguistics has a slightly different trajectory which can be traced back to

[1] By "formal" here I refer to the mathematical underpinnings of linguistics inspired by Post's developments of canonical production systems in proof-theory. For more information, see Pullum (2011).

Chomsky's (1959) review of B.F. Skinner's *Verbal Behavior* and picked up again in Chomsky (1965).

Mentalism is roughly the view that language can be defined as a mental state or a mental abstraction from a brain-state. In other words, language is a cognitive system. Specifically, generative linguists claim that the object of study is a particular internalized and idealized state of a language user's mind.

> Generative grammar adopted a very different standpoint. From this point of view, expressions and their properties are just data, standing alongside of other data, to be interpreted as evidence for determining the I-language, the real object of inquiry. (Chomsky 1991: 12)

An I-language is a rule-bound internalized conception of the structural descriptions associated with grammatical expressions of a language. It is contrasted with a E-language or external language, often characterized as a set of behavioral dispositions of individual linguistic communities and often marked in terms of socio-linguistic boundaries. Chomskyans have traditionally remonstrated against a possible scientific understanding of language emerging from this latter perspective (see Chomsky 2000; Stainton 2006).

At the time of the Classical Cognitive Scientific Revolution of the 1950s,[2] language proved an essential platform for the study of the mind. The experimental clarity of Behaviorism made it highly influential in American psychology and the strictures of logical positivism reigned in any ontological expansions beyond a parochial conception of parsimony. Psychology was firmly under the sway of the former and philosophy still within the grips of the latter.[3] Thus, neither were in a position to lead a counter-revolution and liberate the concept of mind from the intellectual climate of the time.

---

[2] I avoid the terms "first" and "second" here as Chomsky (1991) and Boeckx (2005) claim the first cognitive revolution occurred in the eighteenth century in the form of the Cartesian representational theory of perception.

[3] For instance, the influential philosopher W.V. Quine even constrained solutions to his infamous indeterminacy arguments to involving only observable behavioral evidence.

Chomsky's (1959) critique of the radical behaviorist proposal of Skinner (1957) on language challenged the stronghold that such accounts had on the study of language. Unlike animal calls, which arguably could be analyzed in terms of stimulus and response, human communication is stimulus independent. Following an insight from Wilhelm von Humboldt on the infinite capacity for linguistic expression based on finite resources, Chomsky extended the mathematics of the time to capture linguistic unboundedness in terms of generative grammars.[4] Again, concepts beyond the reach of radical behaviorism.

If language is such an essential component of what makes us human and its nature can be identified with a kind of mental competence, then linguistics could be the field to banish resistance to the explicit study of the mind. This precise claim was made in Chomsky (1965), which reset the linguistic agenda firmly in favor of mentalism. The now infamous passage from *Aspects of the Theory of Syntax* reads:

> Linguistic theory is concerned primarily with an ideal speaker-listener, in a completely homogeneous speech-community, who knows its (the speech community's) language perfectly and is unaffected by such grammatically irrelevant conditions as memory limitations, distractions, shifts of attention and interest, and errors (random or characteristic) in applying his knowledge of this language in actual performance. (Chomsky 1965: 4)

Not only this, but a language on this view is to be identified with mental competence in that language (later I-language). Linguistics redefined itself as a subset of psychology and in so doing released a respectable mentalism back into the latter. "If scientific psychology were to succeed, mentalistic concepts would have to integrate and explain the behavioral data" (Miller 2003: 142). Talk of "the mind" was moving past the occult into mainstream science. Linguists started to investigate concepts of modularity (under the 'Government and Binding' banner), developmental psychology (within language acquisition studies) and the cognitive

[4] Of course, there were many more components of this result including Bar-Hillel's work on applying recursion theory to syntax, Harris' transformations and Goodman's constructional systems theory. See Tomalin (2006) for a detailed review. See also Pullum and Scholz (2007) for requisite detail on Post's influence somewhat neglected by Tomalin's account.

substrate underlying all human languages (via the Universal Grammar postulate).[5] Pylyshyn described linguistics at the time in the following way:

> [D]espite the uncertainties, none of us doubted that what was at stake in all such claims was nothing less than an empirical hypothesis about how things really were inside the head of a human cognizer. (1991: 232)

The revolution in linguistics therefore played an important role in the establishment of the cognitive sciences. However, it was not the only factor that led to the field emerging as a multi-disciplinary enterprise. It is on to this complicated history that we now move.

3 The Inception of Cognitive Science

Bolstered by the academic successes of formal linguistics across North America and the funding initiatives of the Sloan Foundation, the rigorous study of cognition could finally emerge as a distinct and interdisciplinary field of inquiry.

The Cognitive Science Society was officially founded in 1978 but George Miller, one of the original pioneers, traces the conception of the field to a particular day in 1956.

> I date the moment of conception of cognitive science as 11 September, 1956, the second day of a symposium organized by the 'Special Interest Group in Information Theory' at the Massachusetts Institute of Technology. At the time of course, no one realized that something special had happened so no one thought that it needed a name; that came much later. (2003: 142)[6]

This symposium brought together experts in the diverse fields of Artificial Intelligence (AI), psychology, early cognitive neuroscience, information theory and of course linguistics. Since the inaugural

[5] Although it should be noted that the clinical psychology of the time largely resisted the behavioristic scruples of experimental psychology.

[6] This date is corroborated by Gardner (1985).

conference, certain disciplines have moved out of prominence while others such as philosophy have entered into the field.

> Computer science and psychology have played a strong role throughout. Neuroscience initially was strong, but in the years immediately after the 1956 conference its role declined as that of linguistics dramatically increased. (Bechtel et al. 2001: 2154)

There were a number of early disciplinary influences which ultimately culminated in the emergence of cognitive science in the mid-twentieth century. Some of these influences came from outside of the North American academic milieu while others came from the perceived limitations of approaches available at the time such as informational theory and behaviorism. In the following subsections, I will detail these various developments.

3.1 Influences from Without

Although "cognition" and "the mental" might have been unpopular terms within the scientific lexicon in North America in the 1950s, there were theorists in various fields outside of the USA who embraced mentalistic terminology. I will briefly survey some of the salient figures which prefigured the development of cognitive science.

In Germany, the work on the psychology of language (or *Sprachpsychologie*) of Wundt and the Gestalt psychology of Koffka and others did not shy away from mental categories or "Gestalt" qualities of perception. Importantly, the idea of Gestalten was developed in direct opposition to basic behavioral concepts. "Koffka took his argument a stage further in 1915, arguing for a revision of the concept of "stimulus", which should no longer be seen as a pattern of excitation, but as referring to whole, real objects, in relation to an actively behaving organism" (Sinha 2010: 1278). The Gestalt principles identified a distinctly cognitive dimension of perception not directly reducible to behavioral dispositions. Their position was an interesting mixture of a commitment to physicalism and the autonomy of the mind and its study.

Epstein and Hatfield (1994) describe Gestalt psychology as at once committed to "phenomenal realism" taking mental experience and internal states as real and objective and "programmatic reductionism" which aims all eventual complete explanation at the physiological level.

Another major influence from continental Europe was the genetic epistemological approach of Piaget. The latter was, much like behaviorism, explicitly experimental but unlike behaviorism did not obviate talk of the mental or consciousness. In fact, the so-called "epigenetic" account of development or growth of an organism advocated development as conditioned both internally and externally. For practitioners of this view, understanding knowledge acquisition involved essentially understanding the nature of subject and object and the development of intelligence from infancy. The view set itself apart from psychology at the time by rejecting the claim that knowledge involved some passive reception of external factors and by positing a distinctive aspect of "construction". The model of stimulus and response was thereby modified in terms of the biological analogy of assimilation and association.

> Indeed, no behavior, even if it is new to the individual, constitutes an absolute beginning. It is always grafted onto previous schemes and therefore amounts to assimilating new elements to already constructed structures (innate, as reflexes are, or previously acquired) [...] Similarly, in the field of behavior we shall call *accommodation* any modification of an assimilatory scheme or structure by the elements it assimilates. (Piaget 1976: 18)

Of course, related to this, in the United Kingdom around a similar time, the concept of cognitive schemas was being investigated by Sir Frederic Bartlett. Bartlett criticized the allegedly vacuous accounts of schemas according to which they were merely "storehouses of sensory impressions" and advocated rather that "[s]chemas, are, we are told, living, constantly developing, affected by every bit of incoming sensational experience of a given kind" (1932: 201). Schemas thus involved complex cognitive operations not unlike the accommodation notion of Piaget above. His reconstructive account of memory emphasized the active element in the production of reflective narratives of past events not by merely recording actual features of those events but by recreating and reconstructing them in terms of their meaning and inferences therein involved ("the gist" of the story). The mind, then, actively combines records of actual details with the knowledge contained in pre-existing schemas. This is ostensibly far from a "black-box" notion of the mind.

The above, of course, is only a snapshot of some of the influential research which was being conducted outside of the cognitive program being developed in North America in the 1950s.[7] Sinha (2010) claims many of the above influences skipped a generation and rather informed contemporary cognitive linguistics and second generation cognitive science but it was clear that early cognitive scientists were well aware of these developments and in part inspired by them. As Miller notes, reflecting on the time of the Classical Cognitive Revolution:

> In Cambridge, UK, Sir Frederic Bartlett's work on memory and thinking had remained unaffected by behaviorism. In Geneva, Jean Piaget's insights into the minds of children inspired a small army of followers. And in Moscow, A.R. Luria was one of the first to see the brain and mind as a whole...Whenever we doubted ourselves we thought of such people and took courage from their accomplishments. (2003: 142)

3.2 Influences from Within

There were a number of direct causes which led to the formation of a distinct field designed to explore cognition. Two main types of causes stand out among the others, namely what I call 'development from limitation' and 'development from connection'. The first type is easily seen in the case of behaviorism. Its methodology was restrictive and thereby its results were limited as shown by Chomsky (1959) for the case of language.

Another source of frustration was related to the nature of information theory at the time. The place of Claude Shannon's work on making the notion of information mathematically precise and ushering in the digital era is unquestionable. Shannon's implementation of Boolean logic by means of relays and switches of electrical circuits laid the foundation for

[7] A more full account would include reference to the work of Vygotsky and Luria in the Soviet Union among other things. I thank Anastasia Saliukova for pointing this out to me.

modern computers. As a theory of psychology, however, it had its drawbacks. For instance, the finite Markov processes Shannon used to analyze natural language opened the door for Chomsky's transformational grammar approach which could handle syntactic structures the former could not (see Chomsky 1956). Markov processes of the time mapped well onto behavioral output but for a theory of mental manipulation of grammatical rules it was lacking.[8]

In terms of development by connection, there were a number of emerging fields besides linguistics that had direct application to human cognition. In the early days, the artificial intelligence of Minksy and McCarthy played a central role in cognitive science. Similarly, Newell and Simon's application of computer science to cognitive processes and problem solving set the stage for cognitive psychology to merge with information processing and computer science. More and more advanced techniques were being developed to study the mental realm and the field was primed for the advent of a new interdisciplinary program in the study of mind.

AI and cybernetics certainly provided the conceptual tools and formal rigor required for modeling mental processes. However, although AI was initially a major component of cognitive science, it waned in subsequent years. Part of the reason for this was that AI moved from a model of human imitation inspired by Turing to one based on amplifying human abilities (see Proudfoot and Copeland (2012) for more on this shift). "In AI, it is perfectly legitimate to use methods that have no relation to human thinking in order to attain functional goals" (Chipman 2017: 7). Linguistics provided a further explanatory dimension in its grammatical analysis of complex syntactic patterns. But it was the neuroscience of the time that connected the mind to the brain in interesting ways. The early interdisciplinary work of McCulloch and Pitts on neural networks inspired the later connectionist models of the mind (in the Second Generation Cognitive Science) but the identification of specific areas of the brain and their functions, such as Broca's area associated with

[8] The formal grammars of generative grammar correspond to various automata in accordance with the Chomsky Hierarchy. For instance, context free grammars can be represented by pushdown automata while regular grammars only map onto a more restrictive class of finite state automata.

language, become of utmost importance. Studies on aphasia found their way into linguistics textbooks and defects in mental competence in language were associated with neurophysiological causes thereby cementing the psychological interpretation of language studies. Even the famous competence-performance distinction was in part motivated by neurophysical evidence. As Stabler notes in defense of the distinction,

> The linguistic idealization is apparently grounded in the empirical assumption that the mechanisms responsible for determining how phrases are formed in human languages are relatively independent of those involved in determining memory limitations, mistakes, attention shifts, and so on. (2011: 70)

According to Miller, psychology, computer science and linguistics were to be the core fields within cognitive science. In addition, "[o]ne of the central inspirations for cognitive science was the development of computational models of cognitive performance" (Bechtel et al. 2001: 2155). These two aspects culminated into one of the most influential ideas of the entire movement, later embraced both by linguistics and philosophy, namely the computational theory of mind. However, the later emphasis on performance has also led to a challenge to the competence model of generative grammar mentioned in Sect. 2 and the rise of psycholinguistics within cognitive science (more on this in Sect. 5). In the following section we look at the computational theory of mind and its ramifications for the Classical Cognitive Revolution.

3.3 Language, Representationalism, Computationalism

What linguistics showed possible and what AI and computer science successfully modeled was the idea that cognition could be viewed in terms of the manipulation of mental representations. This is, in essence, the computational theory of mind. The notion that the mind can be understood as a symbol manipulator much like a modern day computer, where the mind is at the software level and the hardware is the brain. Two aspects of this view are important for present purposes, computationalism and representationalism. We'll deal with each in turn.

What is the computational theory of mind? The classical version of the thesis proposed that the mind can be understood as a Turing machine of some sort. There are a number of versions of this claim, computational neuroscience and even connectionism (which diverges considerably from the Turing model) can be viewed as such. Again, the basic idea which underlies such views is that the mind is a computational system. Chalmers (2011: 232) identifies two aspects of this claim.

> First, underlying the belief in the possibility of artificial intelligence there is a thesis of computational sufficiency, stating that the right kind of computational structure suffices for the possession of a mind, and for the possession of a wide variety of mental properties. Second, facilitating the progress of cognitive science more generally there is a thesis of computational explanation, stating that computation provides a general framework for the explanation of cognitive processes and of behavior.

The latter is essential for understanding computationalism and cognitive science in general. The idea of generative grammar assumes this position in defining grammars which operate on syntactic categories in generating all the well-formed expressions of the language. It is a computational system, a symbol manipulation device which operates on mental items. Linguists posited that this computational device was isolated from the rest of the cognitive system. The 'autonomy of syntax' is one explanatory means of isolating the computational aspect of mental language competence. The idea, initially proposed by Chomsky (1957), is that a grammar (or generative grammar) of a language L constitutes a scientific theory of an internalized rule system for generating all and only the grammatical sentences of L. A speaker 'knows' or 'cognizes' this system of rules and produces judgments or linguistic intuitions based on this knowledge of the grammar. But generative linguistics embraces a particular brand of computationalism, namely the representational theory of mind (RTM). RTM is a version of CTM which takes symbolic manipulation to the level of mental representations of a specific sort. One radical expression of this view is captured by the work of Jerry Fodor (1975, 1981,

1987) and the Language of Thought Hypothesis (LOTH) which held that there is a mental language or Mentalese upon which the computational or Turing system operates. This language is compositional, productive and systematic according its advocates. Chomskyans do not generally adhere to the strong claim of LOTH due in part to its semantic nature. They do endorse representationalism of a slightly different order though.

The rules of the language, syntactic rules, are internally represented by speakers of the language. In characteristic fashion, Chomsky states "there can be little doubt that knowing a language involves internal representation of a generative procedure" (1991: 9).[9] Thus, language and generative linguistics provided a viable case for computationalism and representationalism in the rest of the cognitive sciences. All of these properties, however, were challenged in the Second Generative of Cognitive Science. It will nevertheless be my claim that language remains to be a central feature of the field despite the potential paradigm shift.

4 The Relevance of Architecture

The philosophy of cognitive science identifies a distinctively methodological worry within the field, related to its own architecture. "Cognitive science is, of course, a multidisciplinary field of research, but there remains enormous disagreement regarding the relevance and precise role of these various disciplines" (Samuels et al. 2012: 10). For example, so far, I have been using the terms "cognitive science" and "cognitive sciences" interchangeably. But there is a potentially important distinction to be had here. The distinction essentially tracks the difference between cognitive science as a discipline versus the cognitive sciences as a pursuit. It's a matter of union versus intersection.

[9] This view is the subject of the philosophical critique of Devitt (2006) in the philosophy of linguistics and also the more recent so-called 4E approaches to cognitive science represented by embodied, embedded, extended, and enacted cognition. In fact, Menary (2010) argues that perhaps one of the only things connecting these latter views is their mutual rejection of cognitivism or representationalism.

> Thus, one possible definition of cognitive science would include all work in these contributing fields, and everyone working in these fields might be deemed cognitive scientists. That is the union definition of cognitive science. An alternative is the intersection definition of cognitive science. To qualify as cognitive science under this definition, the work must draw on two or more of the contributing fields. (Chipman 2017: 1)

The union approach is multifarious and can include even more fields than have been mentioned here, such as economics and sociology. However, it seems like a bad place to start in defining a field since all of the various parts can have little to nothing to do with one another, hence the more suited term "cognitive sciences". In addition, they would be permitted to pursue independent goals under this interpretation. For instance, AI's movement beyond modeling human thinking would not necessarily remove the field from cognitive science, surely a counter-intuitive situation. This architecture allows for connection but does not require it. Furthermore, a union does not produce an ordered set and the core fields would have no claim to special status. The Fig. 9.1 below.

On the intersectional approach, the characterization of cognitive science aims at finding the core approach(es) within various disciplines aimed at cognition. This strategy seems to be more in keeping with the

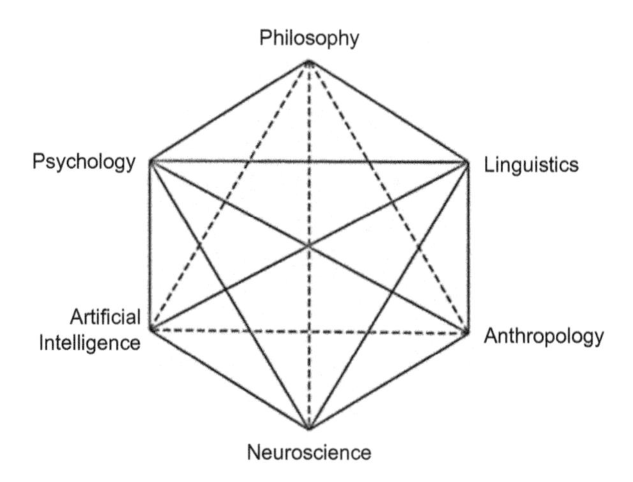

Fig. 9.1 Gardner (1985) *The mind's new science*

original conception of cognitive science. The alternative architecture is one of homing in on the mental and dividing core disciplines from peripheral ones more starkly. The debate then becomes about which disciplines are more beneficial to have at the core than others. In many ways, the Second Generation Cognitive Science makes a strong case for situating psychology as the central field and relegating linguistics to the periphery. This move makes some sense as psychology is precisely the study of the mental processes and it is by historical accident (mainly its long association with behaviorism in North America) that resulted in it not claiming that space in the Classical Revolution. However, psychology does not intersect with all of the relevant fields within the cognitive sciences and advocates might thus favor a union approach for its centrality.

One could of course object that the choice of architecture or core versus periphery makes little difference to the study of mind. I would caution against such a position. One of the main reasons for the development of cognitive science in the first place was that the study of the mind was too complex and broad for a single methodology to capture. An interdisciplinary approach was needed. However, if one understands such an approach as a union of different fields, then shared insights will be hard to come by as theorists continue to operate largely in silos. The resulting field would be interdisciplinary only in name. In order for related goals to be pursued surely a unified approach is necessary. To borrow a metaphor, theorists need to speak each others' languages or at least dialects of the same language. As Sinha (2010) admits, "[t]here can, however, be little doubt that contemporary cognitive science is much less consensual in its fundamental assumptions than was the case a quarter of a century ago" (1266). But some consensus is necessary for the progress of a paradigm, at least for "normal science" (in the Kuhnian sense) to continue. And any discipline or concept which could unify a subject would be advantageous for precisely this reason. The alternative picture suggested by these considerations is outlined in the figure below (Fig. 9.2).

In what follows, I hope to cast some doubt on the claim that psychology should replace language as a core of cognitive science by arguing that language and linguistics better serve the intersectional approach to

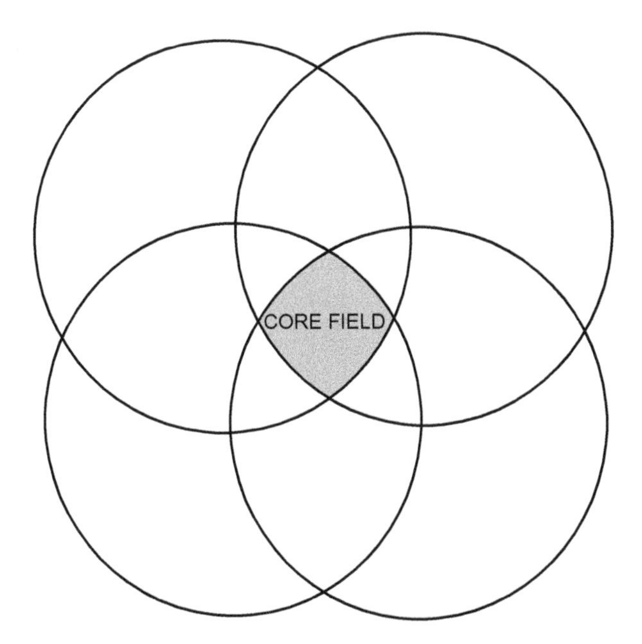

Fig. 9.2 Intersectional approach

cognitive science in addition to some further support from language evolution studies.

5 The Case for Language-Centrality

The argument for language centrality does not preclude psychology or another field from occupying a central position in cognitive science in addition. The aim is merely to offer reasons for reconsidering the move away from language centrality associated with the Second Generation Cognitive Science. The latter term itself was originally coined by Lakoff and Johnson (1999) to describe a new approach to the study of mind, one which relinquished the computational metaphor for an embodied and situated understanding in which cognition is continuous with culture and bio-evolutionary mechanisms. Cognitive linguistics was proposed in order to incorporate an opposing view of language to that of the generative tradition. On this view, any claim about language is subject to evidence from neurobiology, psychology, and cognitive anthropology.[10] Lakoff (1991) calls this the "cognitive commitment" or the idea that one should "make one's account of human language accord with what is generally known about the mind and brain from disciplines other than linguistics" (54). This is certainly a sound principle. However, if we consider the history of linguistics within the classical cognitive scientific tradition we can appreciate adherence to this principle in many forms. In addition, there seems to be a reverse order of explanation within cognitive linguistics, suggested by Lakoff above, which threatens the place of language as a core cognitive phenomenon.

It is not a novel claim to state that generative grammar has had a number of interdisciplinary influences before its official inception and throughout its existence. The more interesting claim is related to how it can be argued that mentalism was the aspect that connected it to these various disciplines, many of which form the core and periphery of cognitive science. This is precisely my claim. Thus, I hope to show below that the methodology of generative linguistics is informed by the methodologies of mathematical logic, computer science, philosophy, psychology and more recently biology. A fact that advocates strongly for its central place within cognitive science.

The history of generative grammar is a broad and complex topic. It is not my purpose to delve into that here (see Newmeyer 1996; Tomalin 2006; Lobina 2017 for some notable attempts). Rather I hope to show that many of the initial and more recent interdisciplinary inspirations were sought for precisely the purpose of understanding the mind.

The first of these insights was adapted for the purpose of explaining linguistic creativity or a related formal property of discrete infinity. Mathematical logic, especially formal proof-theory, offered linguists a way of capturing a universal linguistic property which Chomsky (2000) suggests seemed like a contradiction before such developments, naming how a finite mind can encompass an infinite capacity for expression production and comprehension. Notice that on a Platonist understanding of

[10] This last field has seen a resurgence of interest in the connection between language and thought exemplified by the erstwhile Sapir-Whorf hypothesis. See Reines and Prinz (2009) for an overview of this literature.

linguistics, there would be no obvious cardinality issues since languages could comfortably be considered discretely or even nondiscretely infinite.[11] However, if linguistics is to be a tool for understanding the mind, then an explanation is called for as to how an infinite capacity can be modeled on a finite instrument. The adaptation of the mathematical concept of recursion from logic and early computer science was especially harnessed in linguistics since the time of early syntactic theory. The work of Hilbert, Turing, Post, Kleene, Soare and others provided the basis for understanding recursive and computational systems generally. As Lobina (2017: 38) notes,

> [I]t is important to point out that the results achieved by mathematical logic in the 1930s provided a mechanical explanation for a property of language that had been noticed a long time before; namely, the possibility of producing and understanding an incredibly large, perhaps infinite, number of sentences.

Thus, recursive generative grammars defined in terms of finite rule systems became a mainstay of formal linguistics. A neglected aspect of this methodology is that part of its motivation was the explanation of a linguistic property *qua* mental state.[12]

The next interdisciplinary influence to mention was naturally from analytic philosophy. Due to the linguistic turn, the influence of Wittgenstein and formal logic, first philosophy was mostly considered to be the philosophy of language. A naturalistic understanding of the latter became essential to banishing the recalcitrant debates surrounding mind-body dualism and the emergence of consciousness. Quine's approach to these issues was naturalistic but behavioristic and thus parochial in scope. Advances in the philosophy of mind such as identity theory and the aforementioned LOTH hypothesis provided natural support for a mentalist and internalist understanding of language and even semantics

[11] For instance, Langendoen and Postal (1984) provided an argument for the cardinality of natural language being that of a proper class.

[12] Of course, mentalism in linguistics was more explicitly established after some of these procedural mechanisms were incorporated into the theory.

(traditionally associated with issues of reference and correspondence). Again, the engagement with philosophy (although antagonistic at times) proved to be centered around the mentalism of linguistics and its possible extension to other areas of the philosophy of language such as semantics (see Cappelen 2017 for a negative view of this interaction and Nefdt (2019a) for a response).

In terms of the historical role of philosophy, Chomsky's program benefitted greatly from the work of the philosopher Nelson Goodman on constructional systems theory. But "Chomsky borrowed more from Goodman than merely some of his ideas concerning simplicity measures in constructional systems" (Tomalin 2006: 121). His 1953 paper, 'Systems of syntactic analysis', aimed at connecting concepts from his mentor Harris' linguistic analysis to the work done by Quine and Goodman. Carnap's work on logical syntax also provided useful tools for linguistics especially concerning the development of a formal, in the sense of nonsemantic, computational system that relates to natural language (see Peregrin, this volume, for more). His influence was already felt indirectly in the philosophy of Quine and Goodman from which linguists gained insights and in a more direct way, his work informed Bar-Hillel's first recursive treatment of syntactic structures in 1953.

On the negative side, the aforementioned history also inspired one of generative grammars most controversial claims, namely the autonomy of syntax. Roughly, this claim amounts to an isolation of syntax as a separate purely linguistic computational system which interfaces with semantics and phonology as epiphenomenal features. Jackendoff (2002) goes as far as to claim that this posit, what he calls "syntacto-centricism", is essentially a "scientific mistake" and designs his own theory, the Parallel Architecture, in direct contrast to it. Cognitive linguistics too eschews the autonomy of syntax posit in favor of a more integrated approach in which semantics, pragmatics and general cognitive function can govern syntactic phenomena.

The relationship between psychology and linguistics is perhaps best exemplified by the so-called 'derivational theory of complexity' (Miller and Chomsky 1963) which ultimately led to the development of psycholinguistics, an important albeit peripheral member of the cognitive scientific program. The idea was that the linguistic structures posited by formal theory would find realization in psychological processes and not only this but the effects could be measured. A version of the claim is that "the complexity of a sentence is measured by the number of grammatical rules employed in its derivation" (Fodor et al. 1974: 319). However, this project was largely unsuccessful and yielded no real connection or significant empirical results. The competence and performance distinction proved more problematic than initially thought. Jackendoff (2002) describes the distinction in terms of "hard" and "soft" idealizations. He claims that a soft idealization is a matter of convenience with the ultimate aim of a reintegration of excluded material. The competence-performance distinction based on this latter concept would be a more practical division of labor. "Harder idealizations" are self-contained and severed from the excluded factors in such a way that it is not possible to reintegrate them. He likens the situation in linguistics to that of the distinction hardening over time.

The failure of the derivational theory and the general quest to find formal rules in psychological processes eventually led to a new field being born. Psycholinguistics uses its own methodology and theories to investigate real-time linguistic processing. Although there are a number of interesting more recent proposals linking formal linguistics to real-time processing such as Lobina (2017), Martorell (2018), especially theories under the Minimalist Program (Chomsky 1995). Christiansen and Chater (2016) offer an alternative multilevel representational account of language based on incremental parsing considerations and the "Now-or-never" bottleneck or the fleeting nature of linguistic input, i.e. "[i]f the input is not processed immediately, new information will quickly overwrite it" (Christiansen and Chater 2016: 2). Thus, they advocate that processing constraints be incorporated into the grammar and formal linguistic models themselves, essentially collapsing the distinction between competence and performance.

Another issue that connects generative linguistics to psychology is in the form of the data it uses to construct its theories. Introspective judgments and the judgments of native speakers on linguistic examples form the main source of data for theory construction. However, the reliance on intuitions have also resulted in a number of criticisms, which we cannot delve into here (see Hintikka (1999) for one such criticism, and Maynes and Gross (2013) for an overview of the literature).[13]

The last discipline which has had an impact on generative linguistics also has tentative membership in the cognitive science club, the burgeoning field of evolutionary biology. The inclusion of this discipline has been challenged by Fodor and Piattelli-Palmarini (2010) as having little to no value for cognitive science. Nevertheless, the linguistic enterprise, under generative grammar, has redefined itself as a subfield of biology, so-called "biolinguistics" (Lenneberg 1967) in which evolutionary concerns are paramount. There are also cognitive scientists who similarly claim that the field is continuous with biology (Pinker 1997; Carruthers 2006). The study of Universal Grammar itself has sometimes been claimed to take its cue from biology.

> UG has been studied in a modular fashion, as a mental organ, a coherent whole that can be studied in relative isolation of other domains of the mind/brain. Indeed studies of various pathological cases have shown this mental organ to be biologically isolable. (Boeckx 2005: 46)

And even if biology *simpliciter* seems to be a controversial bedfellow to the other cognitive sciences, contemporary evolutionary biology seems directly relevant to any overarching research agenda. Many studies in language evolution take the view that language and cognition co-evolved (see Bickerton 2016 for a prominent account). This possibility alone, if it were shown to be true, would make a strong additional case for why linguistics and language should occupy a central role in cognitive science.

Lastly, psychology does not embody a similar relationship with other cognitive sciences in terms of history or methodology. Thus, on an intersectional approach, psychology would fare somewhat less favorably as compared with linguistics. Again, the argument here is not to replace psychology but rather to include linguistics within the core of any future cognitive scientific program, be it through the second generation or classical approach.

---

[13] Psychology itself has been less receptive to linguistic theory for numerous reasons; its formalism, data collection techniques and, of course, the many theory changes within generative linguistics did not help matters either. "When Chomsky made significant changes to his theory of grammar, this discouraged many psychologies" (Chipman 2017: 5).

6 Conclusion

In this chapter, I have surveyed the respective histories of linguistics and cognitive science. I have evaluated the early, and some later, disciplinary connections to related fields for the purpose of mounting an argument, based on disciplinary architecture, in favor of keeping language and its formal study within the core fields of cognitive science. This argument took an intersectional model of cognitive science as favorable and thus motivated a picture of the interdisciplinary field as centered around both the study of language and the study of mind.

Acknowledgment I would like to thank audiences at the University of Cape Town and the University of St Andrews respectively for their insights and comments on the content of this chapter. I would especially like to thank Bernhard Weiss, Jack Ritchie, Heather Brookes, and Anastasia Saliukova for their comments and corrections. Any remaining errors are my own.

References

Bartlett, Frederic. 1932. *Remembering: A study in experimental and social psychology*. Cambridge: Cambridge University Press.

Bechtel, William, Adele Abrahamsen, and George Graham. 2001. Cognitive science: History. In *International encyclopedia of the social & behavioral sciences*, 2154–2158. New York: Elsevier Ltd.

Bickerton, Derek. 2016. *More than nature needs: Language, mind and evolution*. Cambridge: Harvard University Press.

Bloomfield, Leonard. 1936. Language or ideas? *Language* 12: 89–95.

Boeckx, Cedric. 2005. Generative grammar and modern cognitive science. *Journal of Cognitive Science* 6: 45–54.

Cappelen, Herman. 2017. Why philosophers should not do semantics. *Review of Philosophy and Psychology* 8 (4): 743–762.

Carruthers, Peter. 2006. *The architecture of mind: Massive modularity and the flexibility of thought*. Oxford: Oxford University Press.

Chalmers, David. 2011. A computational foundation for the study of cognition. *The Journal of Cognitive Science* 12: 323–357.

Chipman, Susan. 2017. An introduction to cognitive science. In *The Oxford handbook of cognitive science*, ed. S. Chipman, 1–22. Oxford: Oxford University Press.
Chomsky, Noam. 1956. Three models for the description of language. *IRE Transactions on Information Theory* IT-2: 113–123.
———. 1957. *Syntactic structures*. The Hague: Mouton Press.
———. 1959. Review of Skinner's verbal behavior. *Language* 35: 26–58.
———. 1965. *Aspects of a theory of syntax*. Cambridge: MIT Press.
———. 1991. Linguistics and adjacent fields: A personal view. In *The Chomskyan turn*, ed. A. Kasher, 3–25. Oxford: Blackwell.
———. 1995. *The minimalist program*. Cambridge: MIT Press.
———. 2000. *New horizons for the study of mind and language*. Cambridge: Cambridge University Press.
Christiansen, Morten, and Nick Chater. 2016. The Now-or-Never bottleneck: A fundamental constraint on language. *Behavioural and Brain Sciences*: 1–72.
Devitt, Michael. 2006. *Ignorance of language*. Oxford: Oxford University Press.
Epstein, William, and Gary Hatfield. 1994. Gestalt psychology and the philosophy of mind. *Philosophical Psychology* 7 (2): 163–181.
Fodor, Jerrold. 1975. *The language of thought*. New York: Thomas Y. Crowell.
———. 1981. *Representations*. Cambridge: MIT Press.
———. 1987. *Psychosemantics*. Cambridge: MIT Press.
Fodor, Jerrold, and Massimo Piattelli-Palmarini. 2010. *What Darwin got wrong*. New York: Farrar, Straus & Giroux.
Fodor, Jerrold, Thomas Bever, and M. Garrett. 1974. *The psychology of language*. New York: McGraw-Hill Publishers.
Gardner, Howard. 1985. *The mind's new science: A history of the cognitive revolution*. New York: Basic Books.
Hintikka, Jaakko. 1999. The emperor's new intuitions. *The Journal of Philosophy* 96 (3): 127–147.
Jackendoff, Ray. 2002. *The foundations of language*. Oxford: Oxford University Press.
Joseph, John. 1999. How structuralist was "American structuralism?". *Henry Sweet Society Bulletin* 33: 23–28.
Lakoff, George. 1991. Cognitive versus generative linguistics: How commitments influence results. *Language & Communication* 11 (1/2): 53–62.
Lakoff, George, and Mark Johnson. 1999. *Philosophy in the flesh*. New York: Basic Books.

Langendoen, Terence, and Paul Postal. 1984. *The vastness of natural languages*. Hoboken: Blackwell Publishers.
Lenneberg, Eric. 1967. *Biological foundations of language*. New York: Wiley.
Lobina, David. 2017. *Recursion: A computational investigation into the representation and processing of language*. Oxford: Oxford University Press.
Martorell, Jordi. 2018. Merging generative linguistics and psycholinguistics. *Frontiers in Psychology* 9: 1–5.
Matthews, Peter. 2001. *A short history of structural linguistics*. Cambridge: Cambridge University Press.
Maynes, Jeffrey, and Steven Gross. 2013. Linguistic intuitions. *Philosophy Compass* 8 (8): 714–730.
Menary, Richard. 2010. Introduction to the special issue on 4E cognition. *Phenomenology ad the Cognitive Sciences* 9: 459–463.
Miller, George. 2003. The cognitive revolution: A historical perspective. *TRENDS in Cognitive Science* 7 (3): 141–144.
Miller, George, and Noam Chomsky. 1963. Introduction to the formal analysis of natural languages. In *The handbook of mathematical psychology*, ed. R. Duncan Luce, R. Bush, and E. Galanter, vol. II. New York: Wiley.
Nefdt, Ryan. 2019a. Why philosophers should do semantics (and a bit of syntax too): A reply to Cappelen. *Review of Philosophy and Psychology* 10: 243–256.
———. 2019b. Linguistics as a science of structure. In *Form and formalism in linguistics*, ed. J. McElvenny, 175–196. Berlin: Language Science Press.
Newmeyer, Frederic. 1996. *Generative linguistics: A historical perspective*. London: Routledge.
Piaget, Jean. 1976. Piaget's theory. In *Piaget and his school*, ed. B. Inhelder et al. New York: Springer.
Pinker, Steven. 1997. *How the mind works*. New York: W. W. Norton & Company.
Proudfoot, Diane, and Jack Copeland. 2012. Artificial intelligence. In *The Oxford handbook of philosophy of cognitive science*, ed. E. Margolis, R. Samuels, and S. Stich, 147–182. Oxford: Oxford University Press.
Pullum, Geoffrey. 2011. On the mathematical foundations of Syntactic Structures. *Journal of logic, language and information* 20 (3): 277–296.
Pullum, Geoffrey, and Barbara Scholz. 2007. Tracking the origins of transformational generative grammar. *Journal of Linguistics* 43 (3): 701–723.
Pylyshyn, Zenon. 1991. Rules and representations: Chomsky and representational realism. In *The Chomskyan turn*, ed. A. Kasher, 231–251. Oxford: Blackwell.

Reines, Maria, and Jess Prinz. 2009. Reviving Whorf: The return of linguistic relativity. *Philosophy Compass* 4 (6): 1022–1032.

Samuels, Richard, Eric Margolis, and Steven Stich. 2012. Introduction: Philosophy and cognitive science. In *The Oxford handbook of philosophy of cognitive science*, ed. E. Margolis, R. Samuels, and S. Stich, 1–17. Oxford: Oxford University Press.

Sinha, Chris. 2010. Cognitive linguistics, psychology, and cognitive science. In *The Oxford handbook of cognitive linguistics*, 1–30. Oxford: Oxford University Press.

Skinner, Burrbus Frederic. 1957. *Verbal behavior*. New York: Copley Publishing Group.

Stabler, Edward. 2011. Meta-meta-linguistics. *Theoretical Linguistics* 37 (1/2): 69–78.

Stainton, Robert. 2006. Meaning and reference: Some Chomskian themes. In *The Oxford handbook of philosophy of language*, ed. E. Lepore and B. Smith, 913–940. Oxford: Oxford University Press.

Tomalin, Marcus. 2006. *Linguistics and the formal sciences*. Cambridge: Cambridge University Press.

10

Linguistics and Brain Science: (Dis-) connections in Nineteenth Century Aphasiology

Els Elffers

1 Introduction

Empirical language-and-brain research started in the nineteenth century. For the first time, autopsies of patients with language disorders gave rise to hypotheses about brain areas where human language capacity was localized. There were earlier ideas about specialized brain areas for mental faculties and there were earlier clinical observations of specific language disorders, distinguished from general cognitive or motor disorders, and caused by brain injury. The combined view that specific language disorders are caused by damage in one or more specialized language centers in the brain came up during the first few decades of the nineteenth century. It gave rise to a host of medical research, theory development and also controversy. Language-and-brain research developed into a genuine

I would like to thank Willem Levelt for his valuable comments on the first version of this article.

E. Elffers (✉)
University of Amsterdam, Amsterdam, Netherlands
e-mail: elffers.els@gmail.com

R. M. Nefdt et al. (eds.), *The Philosophy and Science of Language*,
https://doi.org/10.1007/978-3-030-55438-5_10

research program, which enduringly retained its prominent position on research agendas. In 1865, the term *aphasia* was introduced to indicate language disorders of various types.[1] The term *aphasiology* which refers to the study of language impairment resulting from brain damage, was introduced only in the second half of the twentieth century.

It all started in Paris. During the nineteenth century, the city was the most important center of neurology and psychiatry in Europe. It was the seat of prestigious Academies of Medicine and of Science, and also of large hospitals, such as Charcot's famous Salpêtrière, where numerous patients with neurological and mental deficiencies were cared for. The possibility of autopsy on aphasic patients who eventually died from their brain injuries, and the presence of many brilliant physicians, eager to investigate their brains, together created an excellent climate for epoch-making language-and-brain research. Medical centers in other European countries, especially Great Britain, Germany and Austria, soon followed.

Two early aphasiologists became very famous. Standard introductions in aphasiology all tell about the Parisian physician-scientist Paul Broca (1824–1880), who localized language production in the frontal lobe of the left hemisphere in 1861. In the second place, the German neurologist-psychiatrist Carl Wernicke (1848–1905) is mentioned, who established the existence of a second language area, essential for language comprehension, in the superior temporal lobe in the left hemisphere.[2] Both areas and their related language disorders were named after their alleged discoverers. Nowadays, these terms are still in use, although further

[1] The French physician Armand Trousseau (1801–1867) coined the term *aphasie*. Before 1865, the term *aphémie*, introduced by Broca (1824–1880), was the most current term.

[2] This standard narrative is oversimplified and even partially untrue, as is explained in more scholarly texts about the history of aphasiology (e.g. Tesak and Code 2008, Levelt 2014: ch. 3 and Leblanc 2017). Both Broca and Wernicke had predecessors who prepared their achievements. Moreover, Broca actually observed two similarly damaged brains of aphasic patients in 1861, but only in 1865 did he draw the conclusion that their common damaged area has to be regarded as the seat of articulated language.

scientific developments caused considerable modifications in the neurological as well as the linguistic-pathological phenomena they refer to.[3]

This article discusses questions about the relation between early aphasiology and linguistics, a discipline which was also developing rapidly during the nineteenth century. On the one hand, this relation seems almost unavoidable, given the evident mutual relevance of the disciplines. On the other hand, such a relation might have been hampered, for example by their intellectual and institutional distance.

In the nineteenth century, a genuine interdisciplinary linguistic-aphasiologist research program such as present-day *neurolinguistics* was still far away.[4] But we may ask whether there was, in this period, any contact between the physician-scientists and their linguistic contemporaries, at the meeting-point of emergent aphasiology. Were they aware of each other's research? Did it arouse their interest? Did it influence or contribute to their own research?

Questions like these have been addressed earlier, although they are not discussed in general historiographies of both disciplines.[5] However, they are dealt with in more specialized articles, e.g. Marx (1966), MacMahon (1972), Eling (2009) and Eling and Whitaker (2010). They receive an unanimous negative answer: all authors agree that, in general, neither nineteenth century linguists, nor nineteenth century aphasiologists paid attention to each other's research. For the aphasiologists, they regard this alleged lack of interest as a missed opportunity.

Both Marx and Eling mention one conspicuous exception: the linguist/psychologist Heymann Steinthal (1823–1889). They discuss the

[3] Cf. Tesak and Code (2008: ch. 9). Eling (2018) argues for abandoning the Broca-Wernicke terminology, in light of recent neurolinguistic developments.

[4] Cf. Eling and Whitaker (2010: 580): "By the 1980s, aphasia research had become an interdisciplinary field, combining linguistic (representation), psychological (processes) and neurological (anatomical and physiological mechanism) ideas and models." The term *neurolinguistics* was coined in the late 1940s by Edith Cowell Trager, Henri Hecaen & Alexandr Luria.

[5] In general, historical overviews of both disciplines, e.g., Tesak and Code (2008) and Ivić (1970), contact between linguistics and aphasiology is only discussed as a twentieth century phenomenon, although Tesak and Code mention Steinthal (1823–1889) as an early example of bridge-building between the areas. Leblanc (2017) and Morpurgo Davies (1998), publications entirely devoted to, respectively, nineteenth century aphasiology and nineteenth century linguistics, do not discuss connections with the "other" discipline. Steinthal is dealt with by Morpurgo Davies, but his contributions to aphasiology are not mentioned.

last chapter of his 1871 *Einleitung in die Psychologie und Sprachwissenschaft* ("Introduction to psychology and linguistics"), entitled "Die Sprache als Mechanismus im Dienste der Intelligenz" ("Language as mechanism on behalf of intelligence") and mainly devoted to language pathology. According to Marx and Eling, Steinthal was the only nineteenth-century linguist with interest in and knowledge of aphasiology, and also the only linguistically-informed aphasiologist. Looking backward, they conclude that Steinthal's colleague-aphasiologists were short-sighted in not paying attention to his work.

In the following sections, I will present a somewhat more differentiated picture of nineteenth century interaction between linguistics and aphasiology. In linguistics, interest in aphasiology was almost but not altogether absent. And despite a general lack of linguistic interest in medical circles, aphasiologists occasionally appealed to linguistic ideas. This appeal was scarce and variable, however. I will argue that the use of these ideas does not allow firm conclusions to be made about actual contact between aphasiologists and linguists. Qua content, however, there was some interaction between linguistics and aphasiology, in both directions. Steinthal was not entirely exceptional. Neither was his work entirely uninfluential.

The situation of predominant disconnectedness, however, raises some questions: why was there only a minimal contact between disciplines in a situation of spectacular scientific developments at their intersection? Were scholars short-sighted indeed, or are there more structural explanations? And how can exceptions to this general disconnectedness be explained? A closer look at the intellectual situation in both disciplines will yield some clues to answering these questions.

Special attention will be paid to Steinthal's contribution to aphasiology. His chapter on language disorders has been referred to as a unique and promising example of interdisciplinary neurolinguistics-avant-la-lettre. I will argue that this is an overstatement and that the neglect of Steinthal's aphasiological work is not entirely incomprehensible.

The article's structure is as follows. Section 2 deals with the basic logic of connections between linguistics and brain science. In Sect. 3, earlier views about nineteenth century (dis-) connections between linguistics and brain science will be summarized. In Sects. 4 and 5, the (dis-)

connection issue is discussed in more detail, both from the linguistic and neurological point of view. Steinthal's chapter on aphasiology is the main subject of Sect. 6. In the final Sect. 7, I will summarize the previous results and take a brief glance at the gradual *rapprochement* between linguistics and aphasiology during the twentieth century.

2 Linguistics and Aphasiology: an Asymmetrical Relation

Nowadays, the idea of a narrow connection between linguistics and aphasiology is generally accepted. Neurologists appeal to linguistic concepts in their discussion of disruptions in human language capacity. Data about language disorders are applied by linguists as a test of their theoretical assumptions. Both types of interaction have been amply described in recent literature. A clear explanation is offered, for example, by Avrutin (2001: 5):

> To properly understand the nature of aphasia, it is crucial to understand the structure of language, its internal rules and principles. Only then will it be possible to state what part of the language capacity is disrupted as a result of a particular brain damage. At the same time, research in language impairment can be very important for linguistic research as well. [...] The very presence of such disorder [i.e. specifically related to language] supports the claim that language is an independent system, that is governed by its own rules and principles. Linguistic analysis of the aphasic speech, thus, can provide a theory-based view on the localization of a particular linguistic function: if it can be shown that some language property X is impaired in a population with a damage to area A, it is plausible (although not necessarily true) to claim that this area A is related to X. [...]
>
> Moreover, findings in aphasia research can be used to distinguish between alternative linguistic theories. Suppose we observe that a certain brain damage results in selective impairment of a certain construction S, while leaving another construction, C, intact. Suppose further that there are two competing theories, X and Y, one of which (X) analyses S and C as related, and the other (Y) claiming the opposite. The results obtained from aphasia, in this case, will argue against theory X because, prima facie, two

> linguistically related constructions should be equally impaired as a result to a specific brain damage.

This quotation clearly shows that the relationship between the disciplines is asymmetrical. Aphasiologists necessarily appeal to linguistics, if only as a supplier of concepts and terms "to state what part of the language capacity is disrupted". Linguists reach their analyses independently, and may, in addition, use "findings in aphasia research […] to distinguish between alternative linguistic theories". This appeal to aphasiology is not implied by the logic of linguistic inquiry itself, which is consistent with the fact that only a section of practicing linguists appeals to additional aphasiological evidence. From the nineteenth century origin of aphasiology onward, these *argumentative* appeals to aphasiology by linguists were hardly made until the rise of generative linguistics during the second half of the twentieth century. The strong mentalistic claims of this research program, in combination with its emphasis on linguistic argumentation and theory development, created a fertile breeding ground for the search for psychological and neurological support to linguistic analyses, and, in turn, for the rise of psycholinguistics and neurolinguistics as new and promising disciplines (cf. Eling and Whitaker 2010).

However, this *argumentative pattern* is not the only way in which linguistics may be linked in with aphasiology. As is shown in the above quote from Avrutin (2001), linguists may be actually involved in the analysis of aphasic speech in order to provide a sound theory-based foundation for the localization of language properties. Roman Jakobson's (1896–1982) well-known contributions to aphasiology follow this *foundational pattern*.[6] Finally, there is what might be dubbed the *documentative pattern*. In this case there is no direct theoretical involvement. Attention is paid to aphasiology by linguists in comprehensive overviews of the phenomenon 'language' in, for example, encyclopedias of linguistics or textbooks of general linguistics.[7] The latter two patterns also became current only in the twentieth century.

[6] Cf. Jakobson (1941, 1956).

[7] Malmkjaer (2002^2) and Baker and Hengeveld (eds). (2012) exemplify, respectively, an encyclopedia of linguistics and an introduction into general linguistics that contain extensive chapters on aphasia.

Apart from the documentative pattern, a linguist's degree of involvement in aphasiology depends on his/her philosophy of linguistics. Involvement presupposes the assumption that linguistic concepts fulfill a function in psychological models of language production and comprehension. This, in turn, presupposes the psychological/neurological reality of linguistic concepts, structures and rules. This mentalistic presupposition was and is shared in most philosophies of linguistics. But a minority of linguists denies that the linguist's theoretical constructs directly refer to something "in the head" of language users. They assume that they are "convenient fictions", or that they refer to an abstract "Platonic" or a separate "social" or "linguistic" reality. Others adopt an agnostic, a pluralistic or a "weak- mentalistic" view. For all those linguists, the relation between linguistics and brain science is less evident, and, if acknowledged at all, less direct than for strong mentalists.[8]

In summary, aphasiology is necessarily linked in with linguistics, which provides its conceptual basis. Linguistics is linked in with aphasiology in a more restricted, more incidental and more varied way. The following sections will show how these necessary and possible interdisciplinary interactions took actual shape during the nineteenth century.

3 Separate Worlds?

Interaction between linguistics and aphasiology is generally assumed to start in the twentieth century. The nineteenth century situation of alleged disconnectedness has been dealt with in several publications mentioned above.

Of these articles, Marx (1966) discusses the interaction theme most thoroughly and most systematically. He first deals with the work of a number of prominent nineteenth century linguists (e.g. Humboldt, Schleicher, Müller, Steinthal). His conclusion is that, apart from Steinthal, they hardly pay any attention to the biological basis of language, let alone to aphasia. Subsequently, Marx discusses the work of famous nineteenth

[8] Botha (1993) is the most comprehensive overview of philosophies of linguistics, despite some shortcomings (cf. Pullum and Scholz 1995).

century aphasiologists (e.g. Gall, Broca, Hughlings Jackson, Wernicke), which yields a similar conclusion: their interest in linguistic issues was minimal. The very possibility of these separate descriptions of linguistic and aphasiologist work is significant, according to Marx: "The fact that we could present their development in two separate sections, with little reference to each other already points toward a relatively independent development of their thinking" (Marx 1966: 347). Marx's conclusion is: "We may say [...] that the two groups had a neglectable influence on each other" (Marx 1966: 349).

Marx explains the linguists' lack of interest for aphasiology mainly by an appeal to the dominant nineteenth century view of linguistics as belonging to the humanities, not to the natural sciences, to which brain science belongs.[9] This created an unbridgeable gap for the linguists. The physicians' attitude is associated with their exclusive focus on localization issues, to the detriment of linguistic foundations:

> The heated discussions on the question of localization of language, which ensued everywhere among physicians, very seldom contained an investigation into the basic theoretical questions related to language or a definition of language capacity. (Marx 1966: 348)

The latter diagnosis can also be found in MacMahon (1972: 54):

> [...] the older aphasiological studies were written from a neurological point of view in which the search for centres of language function was a dominant theme, if not an obsession. But in order to localize, neurologists had to know and understand what they were looking for. They took language to be more or less what one finds in school grammar books.

MacMahon emphasizes that this lack of interaction with the achievements of contemporary linguistics was harmful for aphasiology: it prevented the development of "a coherent model of the nature of language

[9] Marx admits that there were nineteenth century linguists who regarded linguistics as a natural science (e.g. Schleicher). Marx supposes, however, that a further development of this view was probably held back by a general fear that attention to the biological basis of language would give rise to "a return to the futile arguments about the origin of language, which they had just recently escaped" (Marx 1966: 348).

with which to discuss aphasia" and led to chaotic classifications and terminologies. He also signalizes, but does not explain, the lack of interest in early aphasiology in linguistic circles, which "meant that earlier errors have gone unnoticed" (MacMahon 1972: 54).

Similar negative conclusions are drawn in Eling (2009) and Eling and Whitaker (2010). The linguists' general non-involvement in aphasiology is observed (but not explained): "Until the latter half of the 19th century, the study of aphasia was mainly performed by physicians and rarely addressed by students of language" (Eling and Whitaker 2010: 571).

As to the nineteenth century physician-scientists, Eling (2009) explicitly thematizes their linguistic ignorance: "Did these scientists have any idea of what they were trying to localize, that is, how did they conceive of language when they were pointing to the spots in the brain where they claimed it was localized"? (Eling 2009: 91). The answer is entirely negative. Eling explicitly refers to Broca and Wernicke as linguistically uninformed scientists; cf. the following citations:

> It is remarkable to see that Broca and Wernicke hardly bothered to explain what language actually is [...]. (Eling 2009: 107)
>
> There is no trace in his papers on aphasia to suggest that Broca was acquainted with linguistics. (Eling 2009: 95)
>
> [...] Wernicke formulated his model [...] not on a sophisticated view of language. (Eling 2009: 102)

Like MacMahon, Eling stresses the harmful effects of the aphasiologists' lack of involvement in linguistics. He regards especially the physicians' exclusive focus on the word level and the neglect of other linguistic levels as a negative effect of their lack of linguistic knowledge:

> Many neurologists always had a rather simple idea of language. In their view all consists of relating words to "objects". [...] The reader looks in vain [in Broca's work] for concepts like 'grammar' or 'word formation'. (Eling 2009: 95)

Eling highly praises Steinthal as an exception to this restricted view: "Apart from Steinthal, these early aphasiologists conceived of language primarily as a process of relating a concept to a word" (Eling 2009: 92). Due to his linguistic scholarship, Steinthal could distinguish, for the first time, aphasic disorders at various linguistic levels.

The physicians' simplistic focus on words is also criticized by Marx (1966: 340), who reports a preference for "simple models of language", that "reduce language to the reception and expression of words".

In summary, earlier analyses of the nineteenth century relation between linguistics and aphasiology present a largely negative picture:

(i) Both groups of scholars were almost entirely ignorant of each other's work.
(ii) Lack of *rapprochement* from the linguistic side is mainly due to the nineteenth century view of linguistics as belonging to the humanities, not to the natural sciences.
(iii) On the physician-scientists' part, lack of *rapprochement* was due to a one-sided focus on localization issues at the expense of serious efforts to develop a coherent and linguistically-informed model of what was localized.
(iv) The lack of linguistic involvement in aphasiology was harmful for the discipline. It led to a chaotic conceptual basis and to a naïve acceptance of simplistic word-directed models.
(v) Steinthal is the only exception. As a linguist with interest in aphasiology and as a linguistically-informed aphasiologist, he was the first to approach aphasia in a linguistically more sophisticated manner.

In the next sections, this picture will be the starting point for further exploration.

4 Nineteenth Century Linguistics and Links with Aphasiology

In this section, I will first take a closer look at the nineteenth century linguistic non-involvement in aphasiology. Attention will be paid to Max Müller (1823–1900), another linguist, who, like Steinthal, had an interest in aphasia, be it more occasionally. Secondly, the non-involvement in aphasiology of the great majority of linguists will be related to the general linguistic state-of-the-art during the nineteenth century.

4.1 Max Müller, a "Modern" Exception

Actually, Steinthal was not the only nineteenth century linguist who got involved in aphasiology. Another exception is the popular German/British linguist Max Müller, who devoted an intriguing passage in his 1887 book, *The science of thought,* to aphasia.[10]

Müller makes a sharp distinction between "rational language", consisting of words that express concepts, and "emotional language", mainly consisting of interjections. He defends this view against opponents who consider the distinction "fanciful and artificial". In this defense, Müller pays attention to the defective language behavior of aphasic patients. He introduces this argumentation saying that it is very fortunate "that physiological observations have been made which confirm in the strongest and most unexpected manner the conclusions at which the students of the Science of Language had arrived in their own way"(Müller 1887: 200).

Müller goes on quoting some long paragraphs from articles published in 1867 by the British aphasiologist John Hughlings Jackson (1835–1911). Hughlings Jackson describes aphasic patients who suffer from injury of a particular region of the left cerebral hemisphere. Their disease "may produce partial or complete defect of *intellectual* language, and not cause corresponding defect of *emotional* or *interjectional* language".

Müller interprets these data as a strong support for the tenability of his distinction between rational and emotional language:

---

[10] In addition, I discovered an informative article about aphasia, published in the 1882 volume of a Dutch linguistic journal for school teachers. It follows the documentative pattern. Its author, Jan Geluk (1835–1919), however, was not a linguist but a head teacher at a village primary school. He was the brilliant son of a farmer, a *uomo universale*, who read widely and published in areas as varied as chemistry, philosophy and history. Although, therefore, his article about aphasia does not fit into the framework of the present article, its publication in this journal shows that the author observed the relevance of linguistic involvement in aphasiology: "I feel that this is a suitable subject for a journal devoted to language and linguistics" (Geluk 1882: 23, transl. E.E.). The article undoubtedly contributed to a wider knowledge of aphasiology among an audience of non-academic linguistic scholarship, characteristic of a period where teacher training colleges were still the "universities of the poor". I did not explore this line of research any further, so I do not know how unique this type of example is.

> [...] so much seems to me firmly established, that if a certain portion of the brain on the left side of the anterior lobe happens to be affected by disease, the patient becomes unable to use rational language; while, unless some other mental disease is added to aphasia, he retains the faculty of emotional language [...]. (Müller 1887: 202)[11]

This evidence shows "that the distinction between emotional and rational language is not artificial or of a purely logical character, but confirmed by palpable evidence in the pathological affections of the brain." (Müller 1887: 202).

Müller's appeal to aphasiological data to support a theoretical-linguistic idea may be the first example of the argumentative pattern explained in Sect. 2 above. Only in the twentieth century do more examples of this pattern appear in linguistic literature, and only in the last few decades of the twentieth century does it become a standardized pattern of reasoning in neurolinguistics.

Steinthal's appeal to aphasiology, which will be discussed in Sect. 6, was of a different type. Although he hinted at the argumentative pattern as well, in his actual approach of connecting linguistics with aphasiology, he followed the other patterns.

4.2 Linguistic Non-involvement in Aphasiology Contextualized

Nineteenth-century linguistics was largely dominated by the new historical and comparative approach to language, which was developed by scholars such as Rasmus Rask (1787–1832) and Franz Bopp (1791–1876). This approach was regarded as the first empirical-scientific way of dealing with linguistic phenomena. The first linguistic chairs at Western European universities, founded in Germany in the first few decades of the nineteenth century, were all occupied by historical-comparatists.[12]

---

[11] Müller briefly refers to the aphasiologist Frederick Bateman (1824–1904) for similar views. Lorch and Hellal (2016: 120) discuss a lecture given by Müller in 1873, in which the argument about aphasia is already presented, including a reference to Bateman, alongside Hughlings Jackson. Remarkably, Marx (1966) discusses Müller, but does not mention his involvement in aphasiology.

[12] Cf. Morpurgo-Davies (1998: ch. 1) and Karstens (2012).

In this approach, words and sounds were regarded as the central linguistic elements. Languages were conceived as autonomous organisms. They were studied in the form of linguistic products: written or printed material. The law-like historical development of words and sounds and the reconstruction of proto-languages were the central foci of interest. Syntax and sentences as functional units of speech were a neglected area. The prominent linguist August Schleicher (1821–1868) even excluded syntax from linguistics, because of its dependency on the human will. This property was regarded as incompatible with the ideal of linguistics as a genuine science. Wilhelm Wundt (1832–1920) characterized this period as the period of *negative syntax* (cf. Noordegraaf 1982).

In this research climate, no attention was paid to language in use: actual processes of sentence production and understanding. Questions about the psychological and neurological foundations of these processes remained undiscussed. So the non-involvement of most nineteenth century linguists in the spectacular aphasiological developments, happening simultaneously with the spectacular rise of comparative-historical linguistics, is hardly surprising.

The incorporation of linguistics in the humanities, mentioned by Marx (1966) might have been relevant too, but this view was not shared by all linguists; e.g. Schleicher and Müller regarded linguistics as a natural science. This multiply ambiguous philosophical issue was probably less influential than the lack of connection of daily linguistic practice with psychological aspects of language.[13] Conceived in terms of institutions, journals etc., however, the distance between the humanities and the sciences certainly hampered disciplinary contact, not only from the linguists' but also from the physicians' point of view.[14]

Alongside historical and comparative linguistics, there was a less prominent current, sometimes called "the other side of 19th-century linguistics": general grammar, a descendant of seventeenth- and eighteenth-century French *grammaire générale*. This approach focused on synchronic sentence analysis in terms of alleged universal categories of language and

[13] Cf. note 9. Cf. Reill (1994) for ambiguities and complications with respect to the dichotomy 'natural sciences – humanities'.

[14] Cf. Lorch and Hellal's (2016: 11) remark about this period: "[...] there typically tended to be little cross-referencing among those working in disparate fields of humanities and sciences".

thought. Especially in nineteenth-century Germany, this was a productive research program, resulting in a considerable growth of synchronic-syntactic insights. At the new linguistic university departments, however, this type of research was completely overshadowed by the successes of historical-comparative linguistics, and the program was frequently dismissed as "non-empirical" because of its philosophical roots and its association with prescriptive school grammar. Prominent grammarians of this type, such as Karl Becker (1775–1851) and Simon Herling (1780–1849) operated outside academia and exerted their main influence in educational circles.[15]

By the end of the nineteenth century, general grammar became integrated into the much broader, thoroughly theoretical but also empirical discipline of general linguistics. The influential work of Wilhelm von Humboldt (1767–1835) was crucial as a starting point for this development.

Although general grammar focused on sentences and their underlying thought processes, the conceptualization of these thought processes was still largely based upon philosophical reflection and connected with general philosophical issues, as was most early cognitive psychology-avant-la-lettre. Therefore, these grammarians' non-involvement in aphasiological research is understandable: interest in neurological aspects of language presupposes a data-oriented approach to language psychology. Only in a predominantly empirical psychological research program are language-pathological and neurological phenomena relevant. Such a program was still lacking in most nineteenth century general grammar.

This precisely explains Steinthal's exceptional status. As we will see in Sect. 6, Steinthal belonged to "the other side of 19th-century linguistics", but, unlike his predecessors, he wanted to develop a new and more scientific psychological foundation of linguistics. This triggered his interest in language pathology.

[15] Becker was originally a physician, later a school teacher; Herling was a gymnasium professor. Due to its minor role in contemporary academic linguistics, nineteenth- century general grammar has long been neglected in linguistic historiography. Nowadays, for example, Jan Noordegraaf has paid due attention to "the other side of nineteenth century linguistics" (and coined this expression). See, e.g., Noordegraaf (1982, 1990).

Müller's involvement in aphasiology has to be explained in a different way. Müller belonged to the academic historical-comparative mainstream, but his area of interest was much broader; he was, e.g., also an orientalist, a theologist and a philosopher. Müller took a warm interest in the question of the origin of language, a widely-debated and controversial issue in those days.[16] His ridiculing characterizations of two popular answers to this question became well-known: the "bow-wow theory" (language originated in onomatopoeia) and the "pooh-pooh theory" (language originated in interjections). Both theories were rejected by Müller for the reason that onomatopoeia and interjections are marginal language phenomena, irrelevant to what he considers the kernel of human language: words that represent concepts (Müller 1887: 200). In this argumentation, a sharp distinction between rational language (words) and emotional language (interjections and sound-imitative cries) is essential. Therefore, Hughlings Jackson's data were a welcome support.[17]

5 Nineteenth Century Aphasiology: What Precisely Was Localized?

I observed in Sect. 2 that discussing aphasia presupposes some linguistic theory, if only to furnish concepts in terms of which language disorders can be described. Section 3 dealt with the linguistic ignorance of nineteenth-century aphasiologists. The paradox is only apparent, because of the physicians' alleged focus on words and lack of attention to other linguistic levels. If 'word' was the only linguistic concept they appealed to, they could avoid linguistics, because this concept is well-known outside linguistics too.

[16] There were even doubts about the meaningfulness of the question itself. In 1866, the Linguistic Society of Paris banned any existing or future discussion of the origin of language. Müller (1887: 207) criticizes this ban and vehemently argues against the equation of the origin-of-language question with "problems of the *perpetuum mobile* or the squaring of the circle".

[17] Müller followed developments in aphasiology, and especially work of British physician-scientists, from the 1860s onward. Vice-versa, British aphasiologists were acquainted with the -highly controversial- equation of language and thought by Müller, who was "a recognized public intellectual" and whose lectures at the Royal Institution of Great Britain "attracted huge popular interest" (Lorch and Hellal 2016: 111).

This section mirrors the previous one. I will first discuss some exceptions: cases of occasional appeal to linguistic concepts other than 'word', by aphasiologists preceding Steinthal. Secondly, the general absence of aphasiologists' involvement in linguistics will be connected with properties of the intellectual climate, in linguistics as well as in brain science.

5.1 Occasional Appeals to Linguistic Concepts

Due to the nineteenth century aphasiologists' idea that, in language, "all consists of relating words to 'objects'" Eling (1999: 95, transl. E.E.), the architecture of human language capacity was mainly conceived of as a storehouse of words: sound images, connected, sometimes via concepts, with sensory or internal impressions, or directly with external objects. Language disorders were described as disturbances in this storehouse or in its function within the mechanism of language production or understanding. In this approach, a classification of types of aphasia in terms of damage at various linguistic levels, more and more usual in later periods, could simply not be made.

This does not mean that nineteenth century physicians did not classify at all. On the contrary, some of them made rather complex typologies of aphasia, not in terms of linguistic notions, but in terms of the language modalities that could be damaged, such as volitional speech, reading aloud, repeating spoken words, understanding written or spoken language etc. All these modalities were related to alleged locations in the neural language circuit, which was represented in intricate diagrams.[18]

However, linguistically-based classifications were not entirely absent. In the first place, there were incidental classifications in terms of word classes. Jean-Baptiste Bouillaud (1796–1881), an early Paris aphasiologist, described differences between language disorders in terms of traditional word categories such as noun and verb. He claimed that

> ...this organ [= "the cerebral organ responsible for articulated language"] is composed of several portions each of which presides over the formation

[18] A famous "diagram-maker" was Ludwig Lichtheim (1845–1928), Wernicke continued Lichtheim's approach.

> and the memory of one of the words, such as the substantive, adjective, verb etc., which taken together compose speech. (Bouillaud 1825: 289, quoted and translated in Head 1926: 16)

Furthermore, there were some incidental appeals to the distinction between interjections and other word classes. In the previous section, we discussed Hughlings Jackson's view of the difference between "intellectual language" and "emotional or interjectional language". This distinction was not entirely the original idea of Hughlings Jackson; a similar distinction could be found in the work of some other aphasiologists (cf. note 11). It was often observed that almost speechless aphasic patients preserved some rudimentary, mainly emotional language. This language certainly did not consist merely of interjections, but Hughlings Jackson regarded interjections as prototypical examples.[19]

For Hughlings Jackson, "intellectual language" was not only characterized by the use of other word classes than interjections, but, more importantly, by the capacity to "propositionalize" i.e. to make statements (Hughlings Jackson 1874). Emotional language was assumed to consist of psychologically unanalyzable cries or automatized utterances. In intellectual language, "original ideas are being encoded into newly generated and novel referential utterances" (Tesak and Code 2008: 58). However, despite this clear awareness of the relevance of the sentence level for an adequate characterization of language disorders, Hughlings Jackson did not elaborate this idea any further. His actual work focused on words and was silent about the psychological basis of sentence formation.[20]

The above examples show that 'word' was not the only linguistic concept appealed to by nineteenth century aphasiologists. There were

[19] Hughings Jackson's "emotional or interjectional language" could include, for example "cursing and swearing, automatic rote-learnt serial verbal activities like automatic counting, nursery-rhymes, prayers and the recitation of arithmetic tables" (Tesak and Code 2008: 57–58). For the early aphasiologist Franz Joseph Gall (1758–1828), "natural language" (equivalent to Hughlings Jackson's "emotional or interjectional language") was thought to consist of interjections, alongside gestures (cf. Marx 1966: 337).

[20] In the words of Marx (1966: 340): "The interrelationships of words did not receive the attention which he knew they deserved". According to Marx, this lack of clarity contributed to Hughlings Jackson's lack of impact on his contemporaries. However, the aphasiologist William Henry Broadbent (1835–1907) included a separate center for "propositioning" in his diagram of the language circuit and borrowed this idea from Hughings Jackson (cf. Eadie 2015).

incidental references to word classes and to propositions/sentences. However, none of these few exceptions proves that nineteenth century aphasiology was actually linked in with contemporary linguistics, because, just like 'word', the other linguistic concepts applied in aphasiology were also available outside linguistics.

In the first place, concepts such as 'noun' and 'verb' belonged to traditional school grammar, so their application is not a strong indication of involvement in contemporary academic linguistics.

In the second place, attention to interjections as semi-linguistic emotional utterances without intellectual meaning was common among linguists, but also among philosophers and psychologists. As was explained in the previous section, many nineteenth century scholars working in these disciplines were involved in the issue of the origin and evolution of language in primitive man. A popular theory was the "interjections theory", the one Müller nicknamed "pooh-pooh theory". In this context, interjections were considered prototypes of a larger class of emotional utterances and gestures at the threshold of language.[21] Aphasiologists typically appealed to the concept 'interjection' in this wider sense, which was current in and outside linguistics. A plausible source for Hughlings Jackson is the work of the philosopher Herbert Spencer (1820–1903), a scholar very influential to Hughlings Jackson and an adherent of the "interjections theory".[22]

In the third place, the concept 'proposition' as applied by Hughlings Jackson, belonged to logic rather than to linguistics, although it was not uncommon in general grammar. Hughlings Jackson was well acquainted with the work of John Stuart Mill (1806–1873). In his own work, he

---

[21] In Elffers (2007) I argue that phonological, syntactic and semantic features that were attributed to interjections from Antiquity onward made them suitable candidates for "primitive words" when, many centuries later, language evolution became a subject of discussion. It is interesting to observe that, nowadays, different ideas about interjections led to new interpretations of interjections in aphasic speech. Interjections are no longer mainly regarded as utterances of individual emotions. Instead, their intersubjective, conversational functions are emphasized. Interjections, preserved in aphasic speech, are consequently interpreted as attempts to display the patient as a competent conversation partner, despite his/her defective language competence (Heyde 2011).

[22] Révész (1946: 35–45) discusses the "interjections theory" as a member of the larger class of "biological" theories of the origin of language. Spencer is mentioned among its advocates. Spencer's influence to Hughlings Jackson is well documented, cf., for example, Jacyna (2011).

refers to Mill's magnum opus *A system of logic* (Jacyna 2011: 3126; see also Lorch and Hellal 2016: 119).

Although we cannot exclude the possibility that the scarce use of linguistic concepts by aphasiologists was based on interaction with contemporary linguistics, non-linguistic sources appear to be more plausible.

5.2 Aphasiological Non-involvement in Linguistics Contextualized

Was the dominant focus on localization issues the main cause of the nineteenth-century aphasiologists' lack of attention to linguistic foundations, as was suggested in several publications discussed in Sect. 2 above? There are indications that this was indeed the case. Historical overviews of the relevant period all show a type of "normal science" focused on localization. Scholars read each other's papers about localization and gave their positive or negative reactions, based upon their own brain research.

For questions about the basic architecture of the human language capacity, there was not such a steady exchange of ideas. Simple word-directed models prevailed, but deviations such as Bouillaud's and Hughlings Jackson's views did not elicit many reactions. Both scholars met criticism, but this was almost exclusively directed against their ideas about the localization of language in the brain (cf. Tesak and Code 2008: 45 and Leblanc 2017: 177). So the linguistic-psychological basis remained discontinuous and underdeveloped.[23]

Lorch (2008) presents an example which shows the lack of attention to this basis in a very concrete way. In 1868, Broca presented a paper at a meeting in Norwich. Its main subject was not localization -although that was discussed as well-, but a refinement of the taxonomy of language defects, which presupposes a refinement of the model of human language capacity. "Broca was trying [...] to determine the nature of the mental,

[23] Wernicke was exceptional in his explicit argumentation against Steinthal's and Broadbent's (cf. note 20) emphasis on sentences/propositions as basic units of speech. Cf. Levelt (2014: 79): "Wernicke could not always suppress his disdain for philosophers and linguists (in particular for Steinthal) who consider the sentence as the basic unit of speech. Words are the hard core of language and speech; the clinician should not go beyond them [...]."

linguistic and motoric components of their [the patients'] difficulties" (Lorch 2008: 1664).

During the discussion after Broca's lecture, his ideas about localization were given a lively airing, but

> ...little attention appears to have been paid to this major aspect of Broca's paper. He called for a refinement in clinical observation as well as signalling a significant development of the theoretical distinctions between speech, language memory and thought. In the reading of the contemporary British neurological literature there appears little if any uptake of this proposal. (Lorch 2008: 1664)

However, the emphasis on brain observation, characteristic for the intellectual climate of early aphasiology, is not the only explanatory factor for the physicians' general non-involvement in linguistics. Institutional distance between sciences and humanities, mentioned in Sect. 4.2, was an impediment in both directions. Moreover, for the aphasiologists, a very great distance had to be bridged, namely the distance to general grammar. Mainstream historical-comparative linguistics was, as was explained above, as focused on words as the aphasiologists themselves were. General grammar offered insights beyond the word level, but this program was overshadowed by the widely known successes of historical-comparative linguistics and hardly visible in academia.

Historians of aphasiology may regret the neglect of linguistic issues by early aphasiologists, but MacMahon's and Eling's suggestion, reported in Sect. 3, that their linguistic ignorance could have been avoided by studying contemporary linguistics seems unjustified. Even if the physicians had actually consulted contemporary mainstream linguistics, this would hardly have changed their naïve focus on words.

6 Steinthal as Linguist and Aphasiologist

As a scholar, Heymann Steinthal was unique in his period. He studied theology, philosophy and general linguistics. Bopp taught him comparative linguistics, but the works of Humboldt were his main source of linguistic inspiration. One of Humboldt's main statements was that language is *energeia,* not *ergon*. Therefore, it should be studied as ongoing processes, not as products such as written texts. Steinthal's work can be conceived as one great attempt to put this device into practice.

Steinthal's most important innovation was his lifelong project of establishing linguistics on a firm and advanced psychological foundation. To realize this, he became a self-taught psychologist. The main foundation was provided by the work of Johann Friedrich Herbart (1776–1841), one of the first designers of a scientific psychology. Steinthal reinterpreted the linguistic concepts of general grammar and Humboldt's linguistic theory in terms of Herbart's framework of psychical entities (percepts, representations) and operations (associations, apperceptions).

Steinthal's first book consisted of a criticism of general grammar, especially Becker's variety, and a plea for a radical replacement of its logical-philosophical basis by a Herbartian psychological basis (Steinthal 1855). Herbartian psychology was also applied to wider linguistic issues such as the origin of language, its development (ontogenetic and phylogenetic) and language typology. This, however, required a considerable expansion and detailing of Herbart's theory and conceptual apparatus, which was all brought about by Steinthal himself. As a result, the first part of Steinthal's 1871 *Einleitung in die Psychologie und Sprachwissenschaft* ("Introduction to psychology and linguistics") contains an almost purely psychological "introduction" of 270 pages before language comes in. The linguistic part, including the last chapter on aphasia, accounts for 120 pages. Not surprisingly, Steinthal is regarded as the founder of psycholinguistics, due to "his grand endeavour to provide the science of language with a psychological basis" (Levelt 2014: 50).

Was he also the founder of neurolinguistics, as Eling (2006: 1076) claims, basing himself upon Steinthal's contributions to a linguistically-oriented aphasiology?[24] In the following subsections I will first discuss the precise content of claims about Steinthal as the nineteenth century pioneer of linguistics-aphasiology bridge-building. Next, Steinthal's chapter on aphasia and its reception will be dealt with.

[24] In Eling's words: "[...] to the best of my knowledge, Steinthal may be considered the first neurolinguist".

6.1 Neurolinguistics-avant-la-lettre?

In discussions about Steinthal's involvement in aphasiology, he is sometimes presented as a bridge-builder from the linguistic bank to the neurological bank. Marx (1966) includes Steinthal in his section about linguists and regards his position of "the first of the linguists to examine the reports on aphasia" as a direct consequence of his turning to psychology, in which "he saw the method for examining the biological language capacity of man" (Marx 1966: 348). According to Marx, Steinthal was "disappointed to find them [the reports on aphasia] insufficient", due to psychological and linguistic deficiencies. He quotes from Steinthal's chapter on aphasia (in translation):

> Over exercitions to find its location [the center of articulated speech], observation of the psychological manifestations of the disease has been neglected. The clinical descriptions are much too incomplete and are unaccurately recorded. Our physicians have as yet no clear concept of what the function of language is. (Marx 1966: 334)[25]

This quotation suggests that Steinthal followed the foundational pattern, trying to furnish a better linguistic basis for clinical description, but Marx does not present further details.[26]

Eling and Whitaker (2010: 571) mention Steinthal as an exceptional contributor of linguistic insights to earlier aphasiology. They fill the gap left by Marx, by discussing Steinthal's classification of types of language disorders (*aphasia* vs. *akataphasia*) in terms of linguistic levels:

---

[25] Cf. Steinthal (1881^2: 464). This is an early report of the alleged connection, often observed in recent publications (see Sect. 3), between a strong focus on location and a neglect of accurate observation of aphasic behavior. Steinthal also paid early attention to the logical priority of precise observation: "If one wants to localise, then one must know first exactly what one can localise" (Steinthal 1874: 50, cited in Tesak and Code 2008: 76).

[26] Marx (1966: 348) suggests that the "linguistic deficiencies" Steinthal observed "kept the work on aphasia from influencing or contributing to linguistic theory. Other linguists after Steinthal ignored it". Given the general linguistic climate discussed in Sect. 4.2, which sufficiently explains this ignoring attitude, I have some doubts about this additional explanation, which presupposes at least some knowledge of aphasiology in linguistic circles.

> Steinthal proposed that there could be a number of disorders at different stages of this process of unpacking [of the message to be conveyed, E.E.], resulting in disorders that are not modality specific (input-output, auditory-visual), but are related to the level of representation. For Steinthal, aphasia is a disorder at the word level, and akataphasia is a disorder at the sentence level. (Eling and Whitaker 2010: 577)

Tesak and Code (2008: 73, 179) also mention Steinthal's criticism of the physicians' "superficial and linguistically shallow coverage of language". They refer to his own "more exact analysis of aphasic language" (resulting in the differentiation between aphasia and akataphasia), and to his attempts "to establish linguistics as a central discipline relevant for aphasiology".

Alongside the foundational pattern, reported in these citations, the argumentative pattern is also followed by Steinthal, according to Eling and Whitaker (2010: 577): Steinthal "developed a psycholinguistic theory and applied it to language disorders [...] as a test of its validity" (cf. also Eling 2009: 102 and Levelt 2014: 48).

In most expositions about Steinthal and aphasiology, he is regarded as a bridge-builder in the other direction as well, namely as a linguistically-informed aphasiologist, be it an aphasiologist with a new, non-clinical and purely psycholinguistic approach (cf. Tesak and Code 2008: 73). Given his psychological and process-oriented approach to linguistics, Steinthal could easily join the medical aphasiologists in the program of theory development about the mental organization of languages, the neural circuit of centers and pathways and the possible disturbances in each of them. In this respect, his assumptions conform to widely accepted ideas (cf. Eling 2006: 1078).

In summary, Steinthal is unanimously regarded as a scholar who, due to his double expertise as a linguist and a psychologist, was in a unique position to connect aphasiology to linguistics. This would result in

(i) his criticism of "the physicians": they should focus on an accurate and linguistically sound analysis of aphasic phenomena, instead of ignoring this task in favor of localization issues.

(ii) a more linguistically sophisticated observation of language disorders, resulting in a classification in terms of linguistic levels, in particular a distinction between *aphasia* (in a narrow sense) at the word level and *akataphasia* (which was later called *agrammatism*) at the sentence level.

Steinthal defended his views on aphasia not only in his 1871 book, but also during a meeting of the Berlin Society of Anthropology, Ethnology and Prehistory in 1874. On that occasion, he contrasted his views with those of the "Messrs Physicians" again. According to Tesak and Code (2008: 76–77) he "insisted that imprecise descriptions of symptoms will not do" and "he was appalled by their lack of understanding of language and communication, contending that their observation must therefore be superficial and without theoretical value". In a discussion of the same meeting, Eling (2006: 1081) reports that Steinthal claimed that the physicians, when analyzing aphasic phenomena, "should do so with a more sophisticated notion in the back of their minds of how the language system operates". The physicians, in turn, accused "the gentlemen of the language side" (Steinthal and his colleague and friend Moritz Lazarus (1824–1903)) of "thinking too lightly about the complex phenomena" and insisted that they "should make the effort to study the phenomena of aphasic speech in the patients themselves".

This was indeed what Steinthal had not done. According to a footnote in his 1871 chapter on aphasia, all his cited cases are taken from three unnamed inaugural dissertations of the Berlin medical faculty, published between 1867 and 1869.[27] At the Berlin meeting, Steinthal refers to this chapter in his answer to "the physicians", saying that "he had already extensively described the phenomena and how they could be interpreted" (Eling 2006: 1081). In the next subsection, Steinthal's description and interpretation of the phenomena will be the central issue.

[27] Levelt (2014: 48) observes that this is unchanged in the 1881 edition of the book and that, due to to these scant sources, neither Broca nor Wernicke are referenced by Steinthal.

6.2 The Chapter on Aphasia

Die Sprache als Mechanismus im Dienste der Intelligenz stretches to 35 pages and consists of 56 numbered sections of unequal length. After a few summarizing sections about general psychology of language, the text is devoted to "some pathological phenomena of language capacity".[28] Steinthal motivates his turn to language pathology in terms of the rather modern argumentative pattern: he claims that the scientific value of pathology "for the physiology of body and mind", is its function as imaginary experiment in situations where real experiments are forbidden or impossible. However, this theme does not return in the rest of the chapter; no imaginary experiment is presented.

Steinthal's text has a rather unbalanced structure and content. Notwithstanding claims of Eling and others, it is not in any way presented as a thorough plea for and presentation of a new and linguistically informed approach to aphasia. The distinction between *aphasia* (in a narrower sense) and *akataphasia* is not prominently introduced as an example of such a new approach. Actually, akataphasia turns up only in the very last sections of the text as one of six types of language disorder, and its status as a disorder at the sentence level is ambiguous.

It is hard to characterize the chapter in general terms. As a tentative guide, I have drawn up a scheme of the content (Table 10.1). The division in parts (1–4) and the labels attached to them are mine, but the division is supported by Steinthal, in that he inserts a blank line between them, except between (3) and (4), where the division, however, is marked by an explicit announcement of a thematic change.

After the first part, already described above, the second part contains detailed descriptions of a wide variety of cases of pathological behavior, described, as in other works about aphasia, in terms of the damaged modalities (speaking, reading, writing etc.). Besides aphasia and *anarthria* (a purely motor speech disorder, discussed alongside semi-normal deficiencies such as stammering and stuttering), Steinthal pays considerable attention to disorders in semi- or non-linguistic modalities, such as writing (*agraphia*), music (*amusia*), purposeful tasks (*apraxia*) and

[28] My translation from German, as in all following cases.

Table 10.1 Scheme of content of Steinthal's chapter on aphasia

	Subject	Number of sections	Content
1	Thought and language	5	Summary of psychology of language Motivation of turn to pathology
2	Aphasia, apraxia, amusia, agraphia, asemia, anarthria	16	Case studies and interpretation
3	Brain and nervous system Symbolic and practical actions. Word classes and aphasia	16	Arguments against language-specific brain center Criticisms of "the physicians"
4	Aphasia	18	Case studies and interpretation Aphasia vs. akataphasia Other classifications

symbols in general (*asemia*). As far as aphasia is discussed, the loss of words is the central issue, just like in works of other aphasiologists. All disorders and their subcategories are tentatively interpreted in terms of the speaker's language-and-thought circuit, sometimes through common sense reasoning, sometimes through an appeal to Herbartian psychology of language.

After this wide-ranging part (2), Steinthal focuses in part (3) on the thought-and-language circuit itself, including its anatomic neural counterpart as proposed by "the physicians". Steinthal accepts their ideas about the physical-motoric speech center, related to anarthria, but has doubts about its psychical counterpart, related to aphasia. The reason is the physicians' alleged "incomplete and inaccurate observation of aphasic behaviour" and their ignorance about the "function of language". The passage cited by Marx on p. 260 above is presented in this context.

What inaccuracies Steinthal has in mind is not immediately revealed, however. He approvingly refers to Ferdinand Finkelnburg's (1832–1896) view of aphasia as only one manifestation of asemia. Finkelnburg's ideas about the neural counterpart of the "symbolic center" give rise to a lengthy exposition about the nervous system in general. Then Steinthal returns to Finkelnburg and criticizes his assumption of a symbolic center. His main argument is based upon a comparison between cases of asemia, cases of apraxia and some in-between cases. His conclusion is that "symbolic and practical acts are inseparable":

> With respect to the physiological mechanism, there cannot be a difference between my lifting my arm for greeting or for work, or my kneeling at the service of God or of my manual work. And we observed how, similarly, in case of aphasia, mimicry and practical movements are disrupted. [...] Therefore, there cannot be a specific centre for symbolism. (Steinthal 1881^2: 468–469)

Given this inseparability, there is no difference, in case of disorders, between, on the one hand, asemia and aphasia, and, on the other hand, apraxia. Steinthal's conclusion is that "it is improbable that there has to be assumed a specific brain location for speech, nor for symbolism, separated from general intelligence" (Steinthal 1881^2: 469). The conclusion is supported by the observation that, also psychologically, practical and symbolic actions are similar: the underlying psychical act is apperception.[29]

This denial of language-specific disorders, and, therefore of a specific speech center, runs counter to the basis of all thought about aphasia. Marx (1966: 334) explicitly criticizes this "refusal to accept the factual clinical findings of aphasia studies [...] on purely logical grounds".[30]

Steinthal's criticism of "the physicians" becomes more clear now. They should have observed the inseparability of symbolic and practical actions. This would have prevented their undue search of specific symbolic or linguistic neural centers. He concludes:

---

[29] Apperception is the process of assimilation of new ideas into the conglomerate of ideas acquired earlier. It is distinguished from association, the connection of ideas. On this issue, physicians are in error as well, according to Steinthal. In many cases, they mistake apperception for association: "Apparently, our physicians do not see clearly the insufficiency of the category "association"" (Steinthal 1881^2: 469).

[30] Steinthal was not the only "non-localizationist" of his time; there were others, more influential than Steinthal, for example Hughlings Jackson, Sigmund Freud (1856–1939), Wilhelm Wundt (1832–1920) and Adolf Kussmaul (1832–1902). These scholars, however, did not deny language-specific disorders, but adopted more "holistic" views of the neural seats of language (cf. Levelt 2014: 88–92).

> The physicians have to make clear for what, to what degree, or whether at all there can be a specific locally determined brain organ for psychical functions. To establish this, it is before all necessary, that they observe the psychical phenomena more accurately and analyze their form and content more carefully [...]. (Steinthal 1881[2]: 473)

The criticism of inaccurate observation is also applied to aphasiologists' assumptions about separate brain locations for word classes (noun, verb etc., cf. p. 254 above). This idea is hardly compatible with the observation that proper nouns are always damaged first in progressive aphasia, while verbs and adjectives are less vulnerable and reappear first during recovery. Steinthal explains this in terms of differences in strength of the associative relation of word classes to their content: the better entrenched, the more damage-proof. Nouns refer to external objects; these can also be mentally represented without the help of words. In the case of movements and qualities, the referents of verbs and adjectives, mental representations are so divergent that words are necessary for their unification. In such cases, words become linked with representations very tightly. In a similar way, objective relations are better entrenched and therefore less vulnerable than subjective, symbolic relations. This explains that aphasia can appear without other cognitive disturbances (Steinthal 1881[2]: 469–472).

Steinthal's general conclusion of part 3 is that "psychology is the most necessary preparation for brain physiology" (Steinthal 1881[2]: 473).

Part (4) of the chapter resembles part (2) in its presentation of a great wealth and variety of pathological case descriptions and attempts to interpret them as disorders in the psychical and/or motoric language mechanism. Now the focus is only on cases of anarthria and aphasia, no longer on apraxia and other non-linguistic disorders. In a series of six types of cases, the third type is labeled *akataphasia. Aphasia* is described as the mental incapacity to reproduce sound images of words; *akataphasia* refers to the more severe incapacity to reproduce the associated semantic representation. In this case, the speaker's transition from percept to representation is hampered, which means that sentence formation is hampered. Some examples are presented of what is nowadays called, respectively, *agrammatism* (lack of finite verbs and conjunctions) and *paragrammatism* (lack of grammatical structure) (Steinthal 1881²: 478). In a later section, the distinction between *aphasia* and *akataphasia* is summarized as follows:

a) The sound image of a word is not remembered: aphasia in the narrower sense
b) The image and meaning of a word is not remembered, the representation, the sentence form is not constructed: akataphasia. (Steinthal 1881²: 485)

These summarizing definitions are immediately followed by a presentation of the distinction as related to various parts of language, respectively: the multitude of semantic representations, and the methods (laws, rules) and means (particles, grammatical forms) to connect these representations in sentence formation.

In the last sections of the chapter, Steinthal presents a fine-grained classification of language disorders. The basic divisions are modality-specific and described in terms of the alleged location of disturbances in the language circuit: the organic mechanism (stammering, stuttering, anarthria), the psychical mechanism (aphasia and akataphasia) and the semantic content (mental illness).

6.3 Steinthal: Aphasiologist *Tout Court*

I observed that Steinthal was intensively involved in aphasiology, but not as a linguist with a "message" for aphasiologists, nor as an aphasiologist pleading the importance of linguistics for their discipline. This picture, presented in recent literature, seems to be an example of anachronistic "ancestor hunt", maybe induced by the situation at present: despite the rise of interdisciplinary neurolinguistic research, the linked-in situation is not perfectly smooth. In present-day aphasiology, there is still some controversy between predominantly "technical" brain researchers and researchers who pay full attention to linguistic analyses of aphasic behavior. The latter group is inclined to make an example of Steinthal.[31]

[31] For example, according to Tesak and Code (2008: 73), Steinthal's criticism of "the physicians" is echoed in "the contemporary critique of some brain imaging research". As late as 1969, a Dutch

For Steinthal, the great challenge was: trying to understand aphasic (and asemic etc.) behavior in all its multiple varieties. Detailed case descriptions and questions about their interpretation fill most pages of his aphasiological chapter. His primary explanatory framework is his own psycholinguistic theory, based on Herbart's views.

The chapter on aphasia has by no means a programmatic character in favor of a more linguistically-informed aphasiological approach. Only in part 3 is there programmatic criticism of "the physicians", but this mainly concerns their alleged neglect of *psychological* reflection about the inseparability of symbolic and practical behavior. Even in the most "linguistic" examples of the criticism -the word classes issue-, the focus is on a psychological aspect: entrenchment.

In his discussion of aphasia, Steinthal focused on the loss of words and described disorders in modality-specific terms, just like other aphasiologists. The distinction *aphasia-akataphasia* was undoubtedly a valuable innovation in a period of naïve word-directedness in aphasiology.[32] It was, however, only briefly discussed and hardly illustrated by examples. It was introduced only very late in a multifarious chapter on language (and related) pathology, amidst other, modality specific distinctions. It played no prominent role as an example of an augmented linguistic involvement. Even its status as early linguistic level-related distinction of language disorders is ambiguous, because akataphasia was primarily introduced as a psychological failure at the word level: the speaker's transition from percept to representation is hampered, therefore, word meaning is not remembered. Steinthal's subsequent equation of this with sentence-formation can only be understood against the background of a current assumption of nineteenth century logicians, psychologists and general grammarians: the idea that, psychologically, not only the combination of two sentence elements, but also the use of one single word

interdisciplinary aphasiological research group was organized, which merely consisted of physicians of various specialisms and some speech therapists. The non-participations of linguists was severely criticized by the linguist Uhlenbeck (1969).

[32] However, Huglings Jackson's somewhat similar attention to the propositional level arose almost simultaneously. Hughlings Jackson did not elaborate his idea, but neither did Steinthal.

embodies a subject-predicate relation, namely the classification of the word's referent into the category indicated by its meaning.[33]

6.4 Reception of Steinthal's Aphasiology

In all recent literature, the reception of Steinthal's ideas about aphasia is characterized as negative: reactions were either dismissive or non-existent. His approach was regarded as too unempirical, his basic psychological categories as too introspective and too complicated, compared to the physicians' clear-cut idea of words as the only elements of speech and a corresponding clear-cut model of the central nervous system.[34] Eling (1999: 98) concludes: "Unfortunately, it was too much to ask for researchers in the sciences to take this armchair theory seriously. Although Steinthal became a well-known linguist, his views of language disorders and aphasia hardly came further than his own textbook" (transl. E.E.).

The latter comparison is somewhat flawed, because Steinthal's linguistic production was substantial, and widely surpasses the single chapter devoted to aphasiology. As additional reasons for the lack of positive reactions, the unbalanced and untransparent structure of Steinthal's chapter on aphasia can be mentioned, as well as his non-localizationism, which was a minority view, and moreover, supported by Steinthal with very debatable arguments.[35]

---

[33] Cf. Elffers (1991: ch. 9). In Steinthal's case, this classification was conceived of as the speaker's apperception of a percept into a representational category (cf. Nerlich 1992: 58). In general, the degree to which general grammarians and their psychologistic successors, including Steinthal, equated meaning with *the speaker's* thought processes preceding talk, can hardly be overestimated (cf. Knobloch 1988: ch. 6). In the quotation on p. 261, Eling and Whitaker unjustifiedly present Steinthal's aphasiology in terms of stages in *the listener's* process of unpacking the message. Actually, this process was almost non-existent in Steinthal's psycholinguistics.

[34] A striking example of such a physician is Wernicke, who used to ridicule Steinthal's views (cf. Levelt 2014: 48–49, 70, 72 and note 23 above). Although Herbart's psychology can be regarded as a major step in the emancipation of psychology from philosophy, it was still considered "philosophical" by natural scientists, due to its philosophically-oriented conceptual apparatus and its introspective method.

[35] The fact that this debatable argumentation was partially based upon a thorough study of apraxia provides a positive counterbalance: in historiography of apraxia research, Steinthal figures as the originator and name-giver of this area (cf. Binkofski and Reetz 2008: 67).

This is not the whole story, however. One important counterbalance was the work of the aphasiologist Adolf Kussmaul (1822–1902), who valued Steinthal positively. Taking into account Steinthal's distinction *aphasia-akataphasia*, Kussmaul (1877) introduced the term *agrammatism* for disorders at the sentence level. This term has survived until today.

According to Marx, Kussmaul's own theories were not valued very highly by his colleagues, for the same reasons as in Steinthal's case:

> One reason that their [Steinthal's and Kussmaul's] work was mostly unacceptable for their contemporaries was that their formulations seemed to endanger the clear-cut simple view of a central nervous system, composed of sensory motor reflex arcs with a precise division of the two components. Another reason may well have been the nature of their psychologies. Neurologists [...] wanted to keep their young discipline from a domination by philosophy. The psychology of [...] Kussmaul, like Steinthal's, was still mainly philosophical and introspective. Marx (1966: 349)

However, Kussmaul widely surpassed Steinthal by elaborating his theory in a systematic and comprehensive textbook (Kussmaul 1877), which included a separate linguistics section and a transparent three-stage speech production model, which pays due attention to syntax.[36]

7 Conclusion and Look Ahead

For aphasia studies, the nineteenth century was too early to be linked in with linguistics. Mainstream linguistics was not in a suitable state to help the aphasiologists go beyond their restricted focus on words, nor to be interested in their results for the benefit of their own theories. General grammarians had not yet made the empirical turn necessary to take an interest in aphasiology. Aphasiologists were too focused on brain research to look further than mainstream linguistics, if they looked at all. Nevertheless, there are, apart from the well-known case of Steinthal,

[36] According to Willem Levelt (p.c.) "Kussmaul's text was for decades the standard source on language disorders, going through reprint after reprint." Cf. also Levelt (2014: 82–87).

some more examples of early border crossings between linguistics and aphasiology.

Only in the twentieth century, did structuralist linguists, such as Jakobson, and subsequently generative linguists gradually develop a more systematic interest in aphasia, in line with their increasing focus on synchronic linguistics. Simultaneously, aphasiologists became more aware of their unavoidable dependence on linguistics, in principle but also in practice: they needed the new linguistic approaches for structural insights beyond the word level.

Notwithstanding recent historiography, Steinthal's work on aphasia was not a genuine breakthrough to a situation of linguistics linked in with aphasiology. In his chapter on aphasia, Herbartian psychology is more prominent than linguistics. The chapter has not at all the character of a programmatic plea for a new linguistically-informed approach to aphasiology. His discussion of akataphasia, the predecessor of agrammatism, was innovative, but it remained unelaborated and was not the result of such a program. The suggestion that Steinthal was a direct predecessor of present-day neurolinguists has to be rejected as an example of "ancestor hunt".

References

Avrutin, Sevey. 2001. Linguistics and agrammatism. *Glot International* 6: 1–35.

Baker, Ann, and Kees Hengeveld, eds. 2012. *Linguistics*. Chichester: Wiley-Blackwell.

Binkofski, Ferdinand, and Kathrin Reetz. 2008. Apraxia. In *Cognitive neurology. A clinical textbook*, ed. S.F. Cuppa et al. Oxford: Oxford University Press.

Botha, Rudolph. 1993. *Twentieth-century conceptions of language. Mastering the metaphysics market.* Oxford: Blackwell.

Bouillaud, Jean-Baptiste. 1825. Recherches cliniques propres à démontrer que la perte de la parole correspond à la lesion des lobules antérieurs du cerveau, et à confirmer l'opinion de M. Gall, sur le siège de l'organe du langage articulé. *Archives Générales de Medicine* 8: 25–45.

Broca, Paul. 1865. Sur le siège de la faculté du langage articulé. *Bulletin de la Société d'Anthropologie de Paris* 6: 377–393.

Eadie, Mervyn. 2015. William Henry Broadbent (1835–1907) as a neurologist. *Journal of the History of the Neurosciences. Basic and Clinical Perspectives* 24: 137–147.

Elffers, Els. 1991. *The historiography of grammatical concepts. 19th and 20th-century changes in the subject-predicate conception and the problem of their historical reconstruction*. Amsterdam: Rodopi.

———. 2007. Interjections – On the margins of language. In *Linguistische und epistemologische Konzepte – diachron*, ed. S. Matthaios and P. Schmitter, 115–138. Münster: Nodus.

Eling, Paul. 1999. Heymann Steinthal: de eerste neurolinguïst. *Stem-Spraak- en Taalpathologie* 8: 94–99.

———. 2006. The psycholinguistic approach to aphasia of Chajim Steinthal. *Aphasiology* 20: 1072–1084.

———. 2009. Aphasia: Where and how. *Sartoniana* 22: 91–111.

———. 2018. De afasie van Broca en Wernicke behoren tot de geschiedenis. *Neuropraxis* 22: 167–171.

Eling, Paul, and Harry Whitaker. 2010. History of aphasia: From brain to language. In *History of neurology*, ed. Stanley Finger, Francois Boller, and Kenneth Tyler, 571–580. Edinburgh: Elsevier.

Geluk, Jan. 1882. Over spraakgebreken in 't algemeen, en over aphasie in 't bijzonder. *Noord en Zuid* 5: 16–30.

Head, Henry. 1926. *Aphasia and kindred disorders of speech*. Vol. 1. Cambridge: Cambridge University Press.

Heyde, Cornelia. 2011. The structure and function of interjections in English aphasic conversation. Two case studies. *TABU conference paper*. Groningen.

Hughlings Jackson, John. 1874. On the nature of the duality of the brain. *Medical Press and Circular* 1 (19): 41–63.

Ivić, Milka. 1970. *Trends in linguistics*. The Hague-Paris: Mouton.

Jacyna, Stephen. 2011. Process and progress: John Hughlings Jackson's philosophy of science. *Brain* 134: 3121–3126.

Jakobson, Roman. 1941. *Kindersprache, Aphasie und allgemeine Lautgesetze*. Frankfurt: Suhrkamp.

———. 1956. Aphasia as a linguistic problem. In *Psycholinguistics. A book of readings*, ed. S. Saporta, 419–426. New York: Holt, Reinhart & Winston.

Karstens, Bart. 2012. Bopp the builder: Discipline formation as hybridization: The case of comparative linguistics. In *The making of the humanities*, ed. R. Bod, J. Maat, and T. Weststeijn, vol. II, 103–127. Amsterdam: Amsterdam University Press.

Knobloch, Clemens. 1988. *Geschichte der psychologischen Sprachauffassung in Deutschland von 1850 bis 1920*. Tübingen: Niemeyer.
Kussmaul, Adolf. 1877. Die Störungen der Sprache. Versuch einer Pathologie der Sprache. In *Handbuch der Speziellen Pathologie und Therapie. H. Anhang*, ed. H. Ziemssen. Leipzig: F.C.W. Vogel.
Leblanc, Richard. 2017. *Fearful Asymmetry: Bouillaud, Dax, Broca, and the Localization of Language, Paris, 1825–1879*. Montreal: McGill-Queen's University Press.
Levelt, Willem. 2014. *A History of psycholinguistics. The pre-Chomskyan era*. Oxford: Oxford University Press.
Lorch, Marjorie. 2008. The merest logomachy: The 1868 Norwich discussion of aphasia by Hughlings Jackson and Broca. *Brain* 131: 1858–1870.
Lorch, Marjorie, and Paula Hellal. 2016. The Victorian question of the relation between language and thought. *Publications of the English Goethe Society* 85: 110–124.
MacMahon, Michael. 1972. Modern linguistics and aphasia. *International Journal of Language and Communication Disorders* 7: 54–63.
Malmkjaer, Kirsten, ed. 2002[2]. *The linguistics encyclopaedia*. London: Routledge.
Marx, Otto. 1966. Aphasia studies and language theory in the 19th century. *Bulletin of the History of Medicine* 40: 328–349.
Morpurgo Davies, Anna. 1998. Nineteenth-century linguistics. In *History of linguistics*, ed. G. Lepschy, vol. IV. London: Longman.
Müller, Max. 1887. *The science of thought*. London: Longmans.
Nerlich, Brigitte. 1992. *Semantic theories in Europe: 1830–1930. From etymology to contextuality*. Amsterdam: Benjamins.
Noordegraaf, Jan. 1982. Traditie en vernieuwing in de taalwetenschap. Twee 'problemen'. In *Studies op het gebied van de geschiedenis van de taalkunde*, ed. L. van Driel and J. Noordegraaf, 81–109. Kloosterzande, Duerinck-Krachten.
———. 1990. National traditions and linguistic historiography. The case of general grammar in the Netherlands. In *Understanding the historiography of linguistics. Problems and projects*, ed. W. Hüllen, 287–302. Münster: Nodus.
Pullum, Geoffrey, and Barbara Scholz. 1995. Review Botha, R. 1993 Twentieth-century conceptions of language. Mastering the metaphysics market. *Language* 71: 157–160.
Reill, Peter. 1994. Science and the construction of the cultural sciences in late Enlightenment Germany: the case of Wilhelm von Humboldt. *History and Theory* 33: 345–366.
Révész, Geza. 1946. *Ursprung und Vorgeschichte der Sprache*. Bern: Francke.

Steinthal, Heymann. 1855. *Grammatik, Logik und Psychologie. Ihre Prinzipien und ihr Verhältniss zu einander*. Berlin: F. Dümmler.

———. 1874. Diskussionsbeiträge. *Verhandlungen der Berliner Gesellschaft für Antropologie, Ethnologie und Urgeschichte* 1874: 47–50, 131–134, 138, 139, 140.

———. 1881[2].[1871]. *Einleitung in die Psychologie und Sprachwissenschaft*. Berlin: F. Dümmler.

Tesak, Juergen, and Chris Code. 2008. *Milestones in the history of aphasia. Theories and protagonists*. Hove/New York: Psychology Press.

Uhlenbeck, Eugenius. 1969. Studie van taalfunctiestoornissen te Leiden zonder taalfunctiedeskundigen. *Forum der Letteren* 1969: 55–56.

Wernicke, Carl. 1874. *Der aphasische Symptomencomplex: eine psychologische Studie auf anatomischer Basis*. Breslau: Cohn & Weigert.

11

Epistemic Transfer Between Linguistics and Neuroscience: Problems and Prospects

Giosuè Baggio

1 Introduction

Among the new sciences to have emerged in the early twentieth century, linguistics and neuroscience play a special role in the quest for understanding the human mind in scientific terms. The potential for linguistics and neuroscience to jointly illuminate mental structures, both from a formal and a material perspectives, may have been clear before the advent of cognitive science (Levelt 2012). But it is only in the past few decades that fundamental challenges have become visible and that conceptual tools for addressing them have been developed and deployed. Here, I focus on one such challenge, that of *effective integration* between those two disciplines (Embick and Poeppel 2015; Poeppel and Embick 2005), as seen through the conceptual lens of *epistemic transfer*: the idea that knowledge of language, as produced by linguistics, may be turned over 'as

G. Baggio (✉)
Department of Language and Literature, Norwegian University of Science and Technology, Trondheim, Norway
e-mail: giosue.baggio@ntnu.no

R. M. Nefdt et al. (eds.), *The Philosophy and Science of Language*,
https://doi.org/10.1007/978-3-030-55438-5_11

is' to neuroscience and used to generate new knowledge of how language is learned, represented, and processed in the brain. I will argue that, to the extent that integration is a realistic target, unidirectional epistemic transfer (UET) from linguistics to neuroscience may not be the best way to achieve it. I will show that some successful cases of UET actually exist, but other attempts have not led to the desired outcomes. I will introduce two new concepts—*deep competence* and *epistemic pooling*—and suggest ways in which they may be used to address the integration challenge. But let us start with a catalog of transferables: possible candidates for pure UET drawn from current theoretical approaches to language.

1.1 Structuralism

The notions of distinctive feature and of closed systems of linguistic representation based on finite feature sets are lasting contributions of structuralism in linguistics, with important potential consequences for the neurobiology of language. In fact, I will argue below that (apparently) successful instances of UET between linguistics and neuroscience involve precisely distinctive features, in particular in phonology. Traditionally, features are *binary*, that is, they can take either of two values, + or −, specifying whether a given unit of representation, such as a phonological segment, has (+) or lacks (−) a particular feature (Chomsky and Halle 1968; Trubetzkoy 1969; Jakobson and Halle 1971). The actual feature sets being postulated may differ across theories, but there remains significant convergence, at least on the ideas of feature subsets describing the voicing, manner, and place of articulation of phonemes, and of phonemes as unique feature combinations: e.g., the consonants /p/ and /b/ can be distinguished by a single feature, VOICE (whether the vocal cords vibrate or not with the articulation of the segment), which is present for /b/ and absent for /p/. Closed systems of binary features have been used in parametric analyses of syntax and grammatical acquisition and variation (Chomsky and Lasnik 1993) and in some decompositional models of semantics, in combination with other representations (Katz and Fodor 1963; Jackendoff 1983; Pustejovsky 1995). In all these cases, where distinctive features (or equivalent) have been described and applied, it is tempting to assume that linguistics has successfully identified the 'parts list' (Ben Shalom and Poeppel 2008) of a given system of representation (phonology, syntax, semantics), and that it now falls to neuroscience to identify corresponding 'parts' in the brain: neurons, circuits, areas, or networks that respond preferentially (show sensitivity) to individual distinctive features, or components in neural signals that do so.

1.2 Formal Linguistics

A second set of transferables can be identified within areas of generative grammar and logical semantics—for lack of a better term, to distinguish them from classical structuralism, I will refer to these fields jointly as 'formal linguistics'. Theoretically, the constructs posited by formal linguistics, that could be candidates for pure UET, are quite different from distinctive features: they are not representational devices, but *structure-building operations*, such as Merge or Unification in syntax (Chomsky 1995; Shieber 1986/2003) and Functional Application in logical semantics (Heim and Kratzer 1998; Chierchia and McConnell-Ginet 2000). These operations are typically at the heart of deductive or derivational formalisms, assumed to model (by design or incidentally) human linguistic competence. In spite of this, these theories often have a characteristic procedural vein, such that core structure-building operations may be accompanied by explicit algorithms for constructing the relevant 'objects', for example syntactic trees, interpretable logical forms, models etc. But even when algorithms or some other computational implementation are not provided, it may still be possible to begin a search for the neural correlates of such operations, or at least to evaluate some of the theory's processing consequences. UET here requires linking hypotheses that establish some kind of relationship between complexity of the derivation—*within* the linguistic theory—and complexity of the process that is (actually or possibly) executed by the brain. Formal linguistics may be assumed to have identified syntactic or semantic composition operations that are necessary to explain human linguistic competence or, to put it in more mundane terms, to make the theory do (tentative) explanatory work. The job of neuroscience then is to find the neural footprints of such operations in language processing experiments.

1.3 Usage-Based Approaches

A third class of approaches in linguistics has recently coalesced into the view that language is not only learned by using it—it is also *shaped by* usage (Christiansen and Chater 2016). In these theories, the focus is shifted from the formal structures and operations that *constitute competence* (e.g., in structuralism and formal linguistics: distinctive features, syntactic and semantic composition etc.) to the processes that *construct competence*: knowledge of a language is said to be built from experience, which is usually characterized as embodied, enactive, interactive or social etc. This shift should not mislead. Structuralism and formal linguistics may come equipped with theories of learning and performance, as much as usage-based accounts have distinctive views of structure in phonology, syntax, and semantics. Moreover, there are some theories that blend elements of these different traditions, e.g., the theory of language learning by Yang (2002, 2004) and the theory of syntax by Culicover and Jackendoff (2005). However, these approaches diverge sharply on the issue of the *balance between storage and computation* in the language system. The explanatory strategy used by generative grammar and formal semantics is to assume powerful structure-building operations plus a finite set of building blocks—the 'parts list' of the language: what can be computed is computed, and only what must be stored is stored. In usage-based models, language is seen as a network of constructions, or learned form-meaning pairings, ranging from morphemes to phrasal and sentential constructions: in this account, what can be stored is stored, and only what must be computed is computed. More precisely, usage-based accounts place the burden of processing on memory systems in brains and machines, eliminating or minimizing any form of computation other than learning. This view seems to be partly shared by construction grammar (CxG), distributional semantics, connectionist and deep learning network models, and language models as developed in NLP research. The kind of criticism that is often levied, explicitly or implicitly, against neurolinguistic projects built upon generative grammar, logical semantics, and related approaches is that "neuroscience has so far failed to enlighten what we consider to be the core of our linguistic powers, namely the specifically human ability to variably combine meaningful units into rule-conforming higher-order structures" (aka composition; Pulvermüller et al. 2013; Pulvermüller 2010). The relevant candidates for UET are *constructions* (or, rather, whatever the theory says is stored as a result of learning) and *learning mechanisms*. The former are directly transferred from linguistics, and the latter are lifted from learning theory, statistics, machine learning etc. The task for neuroscience is to find evidence for constructions in the brain and to describe language learning in neural systems with specific inductive biases and constraints.

1.4 Theoretical Pluralism

Considering structuralism, formal linguistics, and usage-based approaches, I have identified three types of candidates for UET between linguistics and neuroscience: distinctive features, structure-building operations, and constructions and learning mechanisms involved in their acquisition. The rest of the chapter is an assessment of recent attempts at searching for the neural correlates or underpinnings of these different linguistic 'objects'. The goal is to evaluate the success of each attempt and to identify specific challenges in cases of partial success or failure. The conclusion, perhaps not surprisingly, will be that UET is hard, even when the transferables can be reliably classified as knowledge, and that, so far, only few attempts at pure UET have been successful. However, even some failed cases of UET can help to generate new knowledge, sparking discovery in the neurobiology of language: of such cases too, I will provide representative examples.

A plea for theoretical pluralism is in order, before we move on. To anyone working in the field of linguistics, or in neighboring fields, it may be clear that language can be rationally reconstructed in multiple ways. Some may be provably equivalent (a respectable industry in linguistics is consecrated to proving equivalence theorems for given grammatical formalisms), others may not, and may therefore be counted as competitors;

in that case only, some theories may prevail over others. Even then, different formalisms or theories may gradually converge on similar results, or may blend in ways that reveal complementarity or compatibility of formalisms. A glaring example in contemporary linguistics is perhaps Jackendoff's evolving view of the 'architecture of language', which incorporates essential elements of all three approaches above. What makes integration of different linguistic ideas possible, in Jackendoff's case, is his general orientation toward cognition and processing. How likely is it that either structuralism, formal linguistics, or usage-based models have hit on 'The One Right Theory of Language in the Brain', given that none of them has been informed by neuroscientific concepts from the outset of theory building, and that none of them makes systematic use of neuroscience results in guiding theory development? What is more likely, at this very early stage of inquiry, is that grains of truth may be found in the best theories from each of the approaches above. The challenge is to find them and tie them together in theoretically coherent ways.

2 Three Scenarios for Epistemic Transfer

In what follows, as anticipated above, I will consider a particular form of epistemic transfer between linguistics and neuroscience, discussing alternative models later, in Sect. 3: knowledge transfer that is *unidirectional* (between linguistics and neuroscience, but not the other way around—whether or not the latter is possible or currently attested, I will leave open here) and *conservative* (it leaves knowledge generated in linguistics largely 'as is', unmodified prior to and during transfer). Additionally, I will accept as largely unproblematic the facts that conservative UET (i) involves *true knowledge* in the source domain (linguistics), as opposed to either conjectures, beliefs, or constructs with a weaker epistemic status than knowledge; that is not the same as assuming that current theoretical linguistics only generates true knowledge, or that the knowledge that is transferred is only knowledge of the object language(s) under study; (ii) that knowledge may effectively be transferred from linguistics to neuroscience in one or several steps, that, however, do not alter the structure or the epistemic status of the source (linguistic) knowledge; (iii) that knowledge transfer fulfills certain epistemic purposes in the target domain, where it may be useful or necessary to build theories and models, to contribute to guiding or constraining experimental design etc., as a result of transfer.

Applying a key distinction from current work in the philosophy of science (Herfeld and Lisciandra 2019), and elaborating on point (i), knowledge in the source domain can be of three kinds: (A) knowledge about some or all languages, or about human language as a cognitive capacity; (B) knowledge of linguistic formalisms, theories, or models, paired with knowledge that such constructs may be applied to some or all natural languages with some degree of success; (C) knowledge of how to use or apply a set of formal or empirical methods, techniques, or tools from various areas of linguistics (e.g., computational linguistics) to advance research or knowledge in the target domain (neuroscience). Type-A knowledge could provide candidates for *realist interpretations* of linguistic theories in the target field (e.g., that construct X is 'neurally real'); type-B knowledge stems from theories that may explain aspects of linguistic competence, but where a realist interpretation in the target domain is limited by the meta-theoretic nature of the knowledge involved; type-C knowledge only provides formal tools that may serve primarily data analysis and occasionally theory or model development in the target domain, but where issues of realism, as is normally understood in the philosophy of science, may not arise. The distinction between these three types of linguistic knowledge that can be transferred to areas of brain science (e.g., cognitive neuroscience, neurology, and neurobiology) is used below to describe three possible scenarios for conservative UET.

2.1 Scenario 1: Phonological Features, Syntactic Parameters, Constructions

The first scenario considered here includes all cases in which type-A knowledge is available from linguistics, and its transfer to neuroscience leads to discoveries that allow realist interpretations of the relevant linguistic 'objects' and of some of their immediate consequences. Distinctive features in phonology are probably the most successful example of type-A knowledge transfer in the neurobiology of language. Mesgarani et al.

(2014) studied how speech is represented in areas of the superior temporal gyrus (STG), and specifically whether local neural responses are sensitive to phonemes, to distinctive phonological features, or to low-level spectrotemporal parameters (power in specific frequency bands over time). They recorded cortical surface potentials from STG using implanted electrode arrays in epileptic patients, while they were listening to natural speech samples in English. The authors found that cortical signals from some electrodes were selective for particular phonemes. Yet, the majority of STG electrodes were sensitive not to individual phonemes, but to particular *groups of phonemes*. This raises the possibility that the neural code in STG represents properties shared by different phonemes, e.g., distinctive features. For instance, among all electrodes that were sensitive to plosives, some responded to place of articulation or voicing features. In general, the data showed grouping of phonemes based on shared features, "suggesting that a substantial portion of the population-based organization can be accounted for by local tuning to features at single electrodes" (Mesgarani et al. 2014; Grodzinsky and Nelken 2014).

The representation of speech sounds in STG, and generally in the temporal lobe, is complex and still not completely understood, but it most likely involves distinctive features (for recent research and reviews, see Chang et al. 2010; Hickok and Poeppel 2016; Hullett et al. 2016; Tang et al. 2017; Hamilton et al. 2018; Jasmin et al. 2019). The Mesgarani et al. (2014) data do not show that distinctive features are the only, or the most fundamental, representations of speech, but they do show that features are 'neurally real', at least in the sense that they can be put into strict, systematic correspondence with local cortical responses. If distinctive features (the classical structuralist notion of Jakobson, Trubetzkoy etc.), constitute type-A knowledge—knowledge of language, and not just knowledge of a possible way of coherently reconstructing language—, then these results may count as a successful case of pure UET.

As noted in Sect. 1, closed systems of distinctive features have also been used in parametric syntax and decompositional semantics. However, in those areas, there are few studies that systematically assess the 'neural reality' of distinctive features or other units of representation in a linguistic formalism. For instance, research in neuropsychology has (indirectly) addressed some of the basic conceptual features posited by decompositional semantic theories, e.g., the distinctions between living and non-living things, or animate and inanimate entities (Patterson et al. 2007). In those cases, it is impossible to speak of UET between linguistics and neuroscience, because these distinctions are transferred individually, and not as part of a closed system of features. In parametric syntax, one challenge to transfer to neuroscience is that features (parameters: e.g., pro-drop, head directionality etc.) apply to *whole languages* (and define them), and not to individual linguistic segments, as is the case for distinctive features in phonology. This renders it difficult or impossible to search for the 'parts list' of syntax in the brain (if one accepts a parametric theory) by means of standard language processing experiments. But what is possible is to design stimuli that violate lawful ways of setting some parameters, such as strings where function words (e.g., negation) occupy a fixed linear-order position (e.g., as the *n*-th word). This pseudo-property effectively lies *outside* the space of principles and parameters that define human languages; such sentences belong instead to an 'impossible language' (Moro 2016). It has been shown that such strings, compared to actual sentences (e.g., in German or Japanese), *disengage* (result in reduced activations in) classical language networks, including the left inferior frontal gyrus (LIFG; Musso et al. 2003), and are moreover harder to learn by children (Nowak and Baggio 2017). Although this is a less clear case than the phonology case, discussed earlier on, it illustrates an alternative approach to type-A epistemic transfer. Here, one is not looking for a correspondence between 'parts lists'—in the theory and in the brain—, but rather for evidence that neural language processing is constrained in ways compatible with the parametric theory (Moro 2019).

The existence of constructions, if not the framework of CxG as such, is now largely accepted in linguistics. The fact that some sentence types have meanings that may not be derivable compositionally, i.e., are not transparently related to the syntactic structure of the sentence, is an argument for 'listing' those constructions as lexical entries, that is, stored form-meaning pairings (Goldberg 1995, 2003, 2006). These should include, at least, idioms and widely attested phrasal constructions, such as sound + motion constructions, beneficiary dative constructions, resultatives, 'X let alone Y' constructions, and many more (Culicover and Jackendoff 2006). Yet, it is not clear whether the radical notion that language is "constructions all the way down" (or up, i.e., that the smallest and the largest structures of language, whatever they are, are the same kind of form-meaning pairings) could serve as a basis for type-A epistemic transfer. Moreover, there is little experimental research in neuroscience on accepted or less controversial grammatical constructions, which could count as type-A knowledge. For instance, Cappelle et al. (2010) report MEG data consistent with the constructionist notion that phrasal verbs (e.g., 'rise up', 'break down' etc.) are not composed syntactically and semantically, but are retrieved as single lexical units. Research using fMRI has shown that idioms engage largely the same regions as literal language (Bohrn et al. 2012; Rapp et al. 2012), but it is not clear whether such activations can be taken as evidence for retrieval of idiom form and meaning, as opposed to also composition, especially if one takes into account factors such as idiom decomposability or semantic transparency (Vulchanova et al. 2019). Finally, the embodied learning and representation mechanisms assumed by most theories in the CxG camp have been challenged by contrary evidence from studies on single words, phrases, and sentences (Pavan and Baggio 2013; Papeo et al. 2014; Ghio et al. 2016). In sum, usage-based theories, in particular CxG, have generated true type-A knowledge that, importantly, does not also follow from either generative syntax or formal semantics—constructions are real, whether or not one also accepts a broad view of language as "nothing but constructions". However, such knowledge has not yet been successfully transferred to neuroscience. Whether constructions are also 'neurally real', and which phrasal and sentential structures are retrieved as lexical units, are important topics for future work in psycho- and neurolinguistics.

2.2 Scenario 2: Syntactic and Semantic Composition

In the previous section, I have discussed three instances of type-A knowledge, i.e., knowledge about language(s), as such. Distinctive features in phonology, syntactic parameters, and constructions are fairly widely accepted in theoretical linguistics, even if one does not endorse in toto the formalisms they are embedded in and the assumptions they make.

One hallmark of type-A knowledge is that few theoretical alternatives exist for these particular constructs that are (a) not merely notational variants and (b) empirically equally successful or more. In this section, I consider type-B knowledge in linguistics: knowledge of linguistic theories (and not directly of language as such) and of their application to modeling specific natural language phenomena. In these cases, there may be well-specified formal constructs, just like type-A knowledge, but conditions (a) and (b) are met. Merge and Unification seem to be good examples, in that these structure-building operations are formally quite different (they are not reciprocal notational variants), yet they seem to have much the same empirical coverage of syntactic structures (they are *universal* operations, in that sense). Functional Application is another example, the issue being whether or not all semantic composition boils down to it (aka 'Frege's conjecture'; Heim and Kratzer 1998), or whether additional composition operations must be introduced. Systems with those operations are not merely notational variants of a system with only Functional Application, but the latter may be sufficient to guarantee universal generation of logical forms, given appropriate lexical and syntactic inputs (Heim and Kratzer 1998; Chierchia and McConnell-Ginet 2000). Therefore, type-B knowledge is *meta-theoretic knowledge* of ways of formally reconstructing aspects of syntax and semantics. This invites caution when transferring such knowledge to neuroscience. The question of the 'neural reality' of Merge, Unification, or Functional Application may not be answerable, and should be replaced with the question (given relevant criteria of theoretical coherence, empirical adequacy etc.) of what may be the best account of syntactic and semantic composition in the brain, and what is the role of specific formal operations (Merge, Unification, Functional Application etc.) in such endeavor, in particular at the computational and/or algorithmic levels of analysis (Marr 1982; Baggio et al. 2012a, b, 2015, 2016; Martin and Baggio 2019).

Questions about the 'neural reality' of structure-building operations in syntax and semantics have been frequently raised in the literature (Townsend et al. 2001; Marantz 2005; Embick and Poeppel 2015; Poeppel and Embick 2005). Recently, ECoG and fMRI experiments have provided data supporting (or at least consistent with) Merge (Nelson et al. 2017; Zaccarella et al. 2017), but none of those studies offers arguments or evidence against Unification (Snijders et al. 2008) or other syntactic structure-building operations with similar coverage and processing predictions. In fact, cortical activations to syntactic structures, against various baselines, typically involving areas of the left inferior frontal gyrus and of the left temporal cortex, can be explained equally well by neurocognitive models assuming Merge (Friederici et al. 2017) and by Unification-based accounts (Hagoort 2003). Merge and Unification may (or may not) be teased apart in other domains than syntax, such as semantics (Hagoort et al. 2009; Baggio and Hagoort 2011; Baggio 2018). Yet, as models of core syntactic structure-building operations, neither Merge nor Unification have, so far, led to successful UET. We *are* making progress in identifying the cortical networks underlying syntactic structure building, but this knowledge has not been produced as a result of transferring type-B linguistic knowledge about Merge, Unification etc. to neuroscience. Rather, it results from the implementation of certain experimental designs, which are not strictly based on one specific contemporary theory of syntax, but may be interpreted in the framework of several such theories.

The case of semantic composition is slightly different from the perspective of UET. MEG studies have sought to identify a neural correlate of semantic composition, in the sense of Functional Application, but (so far) without success (Pylkkänen 2019). Pylkkänen and collaborators found stronger MEG responses from the left anterior temporal lobe (LATL) in 'minimal' NPs (e.g., 'red boat') *vs.* word lists containing the same critical noun ('cup, boat'), at about 200–250 msec after noun onset (Bemis and Pylkkänen 2011, 2013; Blanco-Elorrieta and Pylkkänen 2016; for EEG data, see Flò et al. 2020, Olstad et al. 2020 and Fritz and Baggio 2020). This early LATL response was initially linked to structure-building operations (syntactic composition), but further research by the same lab has shown that it may reflect, instead, combinatorial processes sensitive to specific conceptual relations between words (e.g., the specificity of the noun relative to the modifying adjective, as in 'spotted fish' *vs.* 'spotted trout'; for a review and critical discussion, see Westerlund and Pylkkänen 2017, Pylkkänen 2019). So far, no evidence has been produced that Functional Application (or some other, formally equivalent meaning composition operation) corresponds to a specific type of neural response: there is no evidence, in other words, that Functional Application is 'neurally real' (Olstad et al. 2020). Still, this does not entail that there is no such thing as a (semantic) composition operation in the brain, or that those attempts at type-B knowledge transfer have been sterile. In fact, even though the 'neural fingerprints' of meaning composition were not found, knowledge *was* generated about conceptual combination and the role of the ATL in on-line semantic processing, as a result of attempts at UET between linguistics and neuroscience. Perhaps not the success that was hoped for, but not a failure either.

For our purposes, one pressing philosophical question is what distinguishes cases of successful UET for type-A linguistic knowledge (e.g., sensitivity to phonological distinctive features; Mesgarani et al. 2014) from cases in which attempted transfer of type-B knowledge leads to novel empirical results, but not necessarily to tighter integration of linguistics and brain science (e.g., studies showing that syntactic or conceptual-semantic processing have specific neural correlates, without, however, providing sharp evidence for the 'neural reality' of one particular formal operation, as envisaged by linguistics, such as Merge, Unification, Functional Application etc.). I cannot attempt a definite answer here. This is a deep problem our field is likely to struggle with for some years to come. But it seems the distinction between type-A and type-B knowledge should be part of the answer. There is a difference between knowing that phonemes, across languages, are defined by distinctive features, that the grammar of languages varies parametrically within limits, and that some types of construction exists (type-A), *vs.* knowing that syntactic or semantic composition can be formalized as Merge or Unification or Functional Application (type-B). This is precisely the difference between knowledge of language and knowledge of ways of (formally) reconstructing language. Attempts at either type-A or type-B transfer can lead to generation of new knowledge in neuroscience, but only the former can, and perhaps only in exceptional circumstances, establish the 'neural reality' of the relevant linguistic 'objects'. Type-B epistemic transfer can, at most, provide formal frameworks within which experimental data may be interpreted, without however adjudicating between basic constructs (e.g., Merge *vs.* Unification). What is at stake here, given the distinction between type-A and type-B knowledge, is precisely the issue of *scientific realism* for neurolinguistics and the neurobiology of language: in what conditions may one take a *realist stance* on particular linguistic constructs in the context of theories of language in the brain?

2.3 Scenario 3: Computational Linguistics and Neural Network Models

The third scenario for UET between linguistics and neuroscience relates to the use of methods, techniques, and tools from computational linguistics and neighboring fields, primarily as aids in the analysis of experimental behavioral and neural data, and secondarily as support in theory and model development in the neurobiology of language. The type-C knowledge involved in these cases is knowing how certain methods, techniques, or tools work when applied to relevant data types, the kinds of inferences they support, and their limitations (given performance benchmarks). Often, type-C knowledge includes type-A and/or type-B knowledge from linguistic theory (e.g., NLP models of parsing and interpretation necessarily make abundant use of linguistic notions). But what gets transferred in this case is knowledge of the method as such. That is what is used to generate new knowledge in the target area, usually to discover patterns in neural data. Recently, some authors have argued for the need for tighter integration or interplay between the neuroscience of language and computational linguistics and NLP/AI research (Hasson et al. 2018; Hasson and Nusbaum 2019). Recent studies have used methods from computational research on parsing in the analysis of MEG data from language processing experiments (Brennan and Pylkkänen 2017) or in classifying known ERP responses to sentences (Michalon and Baggio 2019). Other work has used language models and distributional semantic models to investigate the effects of word predictability and semantic similarity on neural signals (Frank and Willems 2017). All these studies, via modeling, have revealed patterns in M/EEG or fMRI data that standard statistical analyses would be blind to: for example, that responses from the LATL reflect a predictive left-corner parsing strategy (Brennan and Pylkkänen 2017); that P600 effects in ERP data can index a conflict between the assignment of grammatical roles (e.g., subject and object) by autonomous, parallel meaning and grammar systems (Kim and Osterhout 2005; Michalon and Baggio 2019); that word predictability and semantic similarity are correlated with activations in different brain networks, though they leave similar traces on ERP signals (Frank and Willems 2017). In these cases, type-C epistemic transfer is largely successful: new knowledge about language processing in the brain is generated, strictly as a result of the application of computational linguistics methods to neural data. However, it is clear that the methods used are 'just' tools for discovering or modeling patterns in data, and are not *representations* (theories or models) of the relevant biological systems at work: the brain is not (or does not contain) a classifier, or a probabilistic language model, or a dependency parser. In type-C epistemic transfer, questions of realism do not really arise. Rather, instrumentalism is the default interpretation of the 'objects' (methods) transferred from linguistics to brain science.

It is important to emphasize the limitations of this approach, particularly vis-à-vis the detection of patterns in data that do not correspond to (and are not correlated with) any construct or component in the models being applied. Although methods from computational linguistics, NLP, machine learning, and AI are often extolled as excellent discovery tools, greatly extending the human capacity to find patterns in data, they are only as powerful as the formal linguistic assumptions they embody, from a representational and computational perspectives. Recent work by Lyu et al. (2019) illustrates this point well. Their goal was to study "the neural mechanisms underpinning semantic composition" (in the sense above), through a combination of experimental M/EEG data and computational linguistics tools. Composition, qua Functional Application, is guided by syntax; therefore, functions and arguments are *independent*—there are no relations or constraints between them that can facilitate or hinder, allow or block composition.[1] As noted above, research so far has failed to find a neural correlate of semantic composition. Lyu et al. used topic modeling, an NLP technique, to capture statistical relationships between sub-

[1] This holds only for arguments of the appropriate logical type. Obviously, constraints exist on which functions may be applied to which logical argument types. For examples, see the extensive literature on type mismatches, type shifting, and coercion (Pustejovsky 1995; Heim and Kratzer 1998).

jects, verbs, and objects in sentences, based on corpus data: some verbs constrain their object noun more than others, e.g., 'eat' selects something edible as a direct object noun, while 'want' places fewer constraints on object nouns. The main result is that activation in left perisylvian cortex is best explained by a model that takes these constraints into account, and not by a model that only captures the contribution of the object noun, independent of the context in which it occurs. The time course of the effect, the networks engaged, and the context-sensitive nature of the process all support current models of semantic activation in the brain (see Hagoort et al. 2009; Baggio and Hagoort 2011; Baggio 2018). This is an important finding, but it does not reveal either the "neural mechanisms" or the "neural dynamics of semantic composition", because composition is not context-sensitive in the way most topic models are, and requires instead independent representations of functions and arguments (Martin and Doumas 2019). Computing conditional probabilities in topic models (in general, in language models) is quite different from composition in formal or distributional semantics, which involves logical or algebraic operations on constituent meanings. If compositional dynamics *were* present in the data, topic models would simply be blind to them, precisely because these models do not embody the representations and operations formally involved in syntactic or semantic composition.

Another example of the limitations of current computational models as applied to neural data is provided by recent research by Rabovsky et al. (2018), describing a connectionist model of the N400 component and effect in ERPs. More specifically, they model semantic processing as the probabilistic updating of sentence meaning representations in a connectionist network. Their simulations reproduce as many as 16 known instances of the N400 effect qualitatively, and other data points (e.g., discourse-level N400 effects) seem fully within the model's reach. This impressive performance sets a new benchmark for future algorithmic-level models of semantic processing and for neural-level models of the N400. Their long-term success must, however, be measured by their capacity to reproduce *all and only* the N400 effects reported in the literature. It is now largely accepted that the N400 reflects aspects of lexical semantic access *and* integration, and recent studies show that the N400 has complex functional and spatio-temporal characteristics (see Baggio and Hagoort 2011 for a theory and Nieuwland et al. 2019 for supporting data from a large-scale study). However, the challenge posed by N400 results that were originally taken to back up the integration account remains. It could be instructive to train and test the model on stimuli from studies on quantifiers, event structure, negation, and other 'hard cases' for simple activation theories of the N400 (Urbach et al. 2015; Ferretti et al. 2007; Baggio et al. 2010; Kuperberg et al. 2010). These would be important steps toward assessing the model's empirical adequacy more thoroughly. Perhaps the most probative question is not whether Rabovsky's model is powerful enough, but whether it is 'too powerful'. The representations that are dynamically encoded and updated in the network are "implicit", in that they lack any predefined formal structure, unlike traditional models in linguistic semantics. Moreover, there are no bounds on the kinds of semantic contents that the network may come to represent. Such constraints, the authors argue, may well be "an impediment to capturing the nuanced, quasiregular nature of language" (Rabovsky et al. 2018). Yet, their model may overgenerate for precisely this reason: it may predict N400 effects where none are found experimentally. Studies have reported sustained negative ERPs different from the N400—in latency, duration, and scalp distribution—in reaction to critical words in sentences that trigger 'updates' of event representations in imperfective, aspectual coercion, and light verb constructions (see Baggio et al. 2008; Paczynski et al. 2014; Wittenberg et al. 2014). Their model should also be trained and tested on stimuli from these studies. If semantic updates in their model are specific to the processes underlying the N400, there should be no differences in activation levels, in the model, before and after presentation of the words that elicit sustained ERPs. These considerations show that applications of type-C knowledge to experimental data should still meet basic requirements of empirical adequacy to lead to effective knowledge transfer and generation of new knowledge in neuroscience: the chosen probabilistic update algorithm can only be considered a theory of the mechanisms behind the N400, if the model captures all and only known instances of the effect.

3 Deep Competence and Epistemic Pooling

In Sect. 2, I have argued that type-A and type-C linguistic knowledge may lead to production of new knowledge in the neuroscience of language, as the *direct result* of transfer. Typically, this new knowledge concerns the 'neural reality' of linguistic constructs (type-A) and patterns in behavioral or brain data (type-C). The problem for type-A transfer is to identify 'true' linguistic knowledge among the conjectures, hypotheses, or other constructs generated by theoretical linguistics: that would be knowledge (i) invariant across formalisms (equal alternatives are mere notational variants), (ii) widely accepted in the linguistic community, and (iii) such that there are no competitors with equal or greater explanatory scope and power—these are all hallmarks of a 'mature' theory, according to modern philosophy of science. The challenge for type-C transfer is to select appropriate data analytic methods among those currently available, and to resist realist interpretations of the constructs that are being applied in data analysis or modeling.

As for type-B transfer, different and seemingly harder problems arise, particularly if we consider knowledge from generative grammar, logical semantics, and related theoretical frameworks. Currently, much research in the neuroscience of language operates with formal linguistics as/in its background, but whatever results may be generated, they are *not* strictly and directly the outcome of knowledge transferred from that background. As an example, several fMRI studies have converged on the description of a left-hemispheric perisylvian network for basic structure building, regardless of whether these studies assumed Merge or Unification (quite different operations formally) as part of their background, as if one's preferred syntactic theory did not matter for the purposes of designing and conducting neuroimaging experiments that may reliably identify language- or syntax-relevant brain regions. There seem to be, in other words, factors (conceptual or empirical) preventing our best theories in formal linguistics to make direct contact with experimental neural data. Why have we not found any robust brain correlates, let alone mechanisms, of syntactic and semantic composition *as defined by formal linguistics*? Why, when we look for those correlates, do we invariably find brain responses with very different characteristics from those that formal linguistics has ascribed to syntactic and semantic composition—i.e., graded, probabilistic, context-dependent, even predictive, when the latter are discrete, algebraic, context-independent, and bottom-up?

At present, this is one of the most obvious puzzles in the neurobiology of language, and it concerns primarily type-B knowledge from formal linguistics. Some authors have argued that, effectively, generative syntax has already been integrated within the neuroscience of language (e.g., Marantz 2005, Friederici 2017), but many seem to share a more pessimistic outlook. Here, I suggest two strategies for dealing with this predicament, based on the concepts of *deep competence* and *epistemic pooling*, both meant to support type-B transfer. None of them is a proven strategy, but both are fairly straightforward next steps in the search for alternatives to type-B UET.

3.1 Deep Competence

One epistemological obstacle to further integration between theoretical linguistics and neuroscience is a particular notion of linguistic competence as (a) knowledge of a formal deductive or derivational system for the generation and interpretation of linguistic structures, (b) distinct from processing and 'performance' (Jackendoff 2002; Baggio et al. 2012a). The problem is whether and how a derivational system (e.g., a formal theory of logico-semantic composition based on the λ-calculus) could be transferred to a target domain where the goal is to develop processing theories, as in psycholinguistics, neuroscience, computer science or AI. Transferring that knowledge 'as is' (conservative transfer) and qua deductive system would miss the point, especially if the goal is to apply it to study physical machines, such as brains, where computation is not (all) deductive. The strategy used by 'strict competence' (Steedman 1992), according to which processing algorithms are determined by the competence theory and, therefore, computation is deductive, is not viable: the well-documented early failure of the so-called derivational theory of complexity in syntax (DTC) testifies to that (Fodor and Garrett 1966, 1967). Here is the fork in the road: either we drop the requirement or desideratum that competence theories or models are or involve deductive formalisms, effectively adapting them to demands from the target domain (constraint-based formalisms in linguistics may be seen as instances of this approach, e.g., Jackendoff 2007), or we adopt a 'weak competence' stance, such that the structures being computed (syntactic trees or graphs, logical forms, models etc.) are those sanctioned by one's (deductive) competence theory, but computation is not deductive. Either way, we are compelled to take processing seriously, and examine how a theory of competence—whether deductive or based on some other formalism—may be implemented algorithmically and neurally.

Deep competence is precisely the idea that what matters for the study of linguistic (syntactic and semantic) competence is not what formalisms one adopts at the top level of analysis (Marr and Poggio 1976; Marr 1982), but what aspects and elements of one's top-level (linguistic) theory are implemented algorithmically and neurally. So, linguistic competence is not just *declarative* knowledge (e.g., knowledge of the syntactic and semantic 'facts' of a language), but also *procedural* knowledge (e.g., of how to compute form and meaning, resolve ambiguities etc.), so long as it can be plausibly and tractably implemented in models of brain processes. One important advantage of this perspective is that now linguistic theories need not be accepted or rejected wholesale as competence theories. We can distinguish, within the top-level computational theory (the formal linguistic theory as such), which aspects are implementable, and are therefore part of deep competence, and which elements support the competence theory 'from the outside', providing *constraints on the computation*: these may affect what the system does (or what it can do), but are not part of what it 'knows' (deep competence). Similarly, Poggio (1981) and Marr (1982) have argued that a computational-level theory of vision requires a precise characterization of the "properties of the visible world that constrain the computational problem". Key aspects of the structure of the physical world should be "incorporated as critical assumptions of the computation", which is required to make computational problems "well defined and solvable". The same would apply to theories of syntax and semantics. The role that some formal notions play in the theory would be restricted to the computational level, without however being part of what the system effectively computes. Notions of truth and entailment, e.g., truth conditions tout court, may be an example. Those constructs are necessary to make (some versions of) semantic theory work, also qua theory of competence, but they cannot be reasonably assumed to be implemented in the brain, if only because the logical formalisms in which such notions are typically specified lack the properties (e.g., decidability) that would make them fully implementable algorithmically (Frixione 2001; Partee 2018).

How could these ideas facilitate or help clarify the nature of the problems involved in type-B epistemic transfer? Deep competence may be seen as a filter on transfer: only those formal constructs that are implementable, algorithmically and neurally, will be transferred to the target domain—we may assume that the goal of transfer to neuroscience is to establish 'neural correlates' of (or mechanisms associated to) some of those constructs, e.g., Merge, Move, Unification, Functional Application etc. The intuitive notion of 'implementability' should be broken down into appropriate technical concepts from computational complexity theory, the most consequential of which may be *tractability*. Syntactic and semantic composition, as implemented in the brain, are constrained by its finiteness—the fact that there are limited space (memory) and time resources for computation. Therefore, the functions describing composition should be computationally tractable, i.e., a subset of polynomial-time computable functions, plus some super-polynomial-time computable functions for which one or more input parameters are small in practice (van Rooij 2008). There is little formal work applying these specific computational constraints to modeling language processing, but the importance of this approach, as a preliminary step in epistemic transfer, should be clear: knowledge about composition in the brain (let us assume there is such a thing) can be generated *as a direct result of transfer* only if tractable composition operators are considered.[2] Note, however, that there is no guarantee that any particular tractable operator is the true one. For instance, some solutions may have unpleasant consequences for

[2] Among the computational problems whose combinatorial search spaces may be super-polynomial, van Rooij (2008) mentions one that is different from composition, but sufficiently closely related to deserve attention: i.e., the Input is a surface form s, a lexicon D, and a lexical-surface form mapping mechanism M, and the Output is a set of lexical forms U generated by D from which M can create s (Barton et al. 1987; Ristad 1993).

the system's implementational *architecture* (e.g., for the number, nature, and interrelation of the proposed algorithms; for discussions about syntax, and Merge in particular, see Frampton and Gutmann 1999, Petersson and Hagoort 2012, Trotzke and Zwart 2014). So, tractability is surely necessary, but not sufficient: other factors may constrain algorithmic and neural implementation, such as plausibility of the resulting architecture, learnability and evolvability of the relevant functions etc. (van Rooij and Baggio 2020). These computational considerations may guide UET between formal linguistics and neuroscience, in particular when type-B knowledge, or 'theoretically loaded' linguistic knowledge more generally, is concerned.

3.2 Epistemic Pooling

Deep competence suggests a fairly straightforward strategy for narrowing the gap between linguistic constructs and neural reality: only constructs that are provably implementable (e.g., tractable functions or operators) may count as candidates for type-B UET. This move is designed to reduce the initial set of transferables a priori and strictly on computational grounds, so that empirical or experimental research can proceed with testing fewer but formally better specified ideas.

The second strategy I sketch out here is compatible with the first one, in that it too emphasizes the need to work at the algorithmic-and-architectural level of analysis to achieve better integration between linguistics and neuroscience. However, this second proposal modifies the very idea of epistemic transfer. As before, I consider epistemic transfer, both UET and its new modified version, in the context of Marr's (1982) multi-level framework for the analysis of information processing systems: epistemic transfer corresponds to devising algorithmic- and neural-level analyses that effectively implement a given computational (i.e., linguistic) theory. According to the traditional 'functionalist' approach, one starts at the top of the scheme, from the computational level, and works one's way 'down' to the algorithmic and neural levels—in that order. Because stating the computational problem is a prerequisite for algorithmic analyses, this is obviously the correct approach. But it is much less clear that the same kind of relationship of logical priority should hold between the algorithmic and neural levels of analysis. If we accept this notion (and here I argue that we needn't and shouldn't), we run against two problems. The first is the curse of multiple realization, meaning that, much as the same function may be computed by different algorithms, the same algorithm may be executed by different physical machines, possibly with different architectures, components etc. This problem can only be addressed, or at least mitigated, by taking into account neural constraints, for example, from experimental research or, in any case, from the 'outside' of one's processing theory. A second issue is that, ironically, formalism can be an 'iron cage'. The workings of even the smallest physical components of a system, at the lowest of its levels of organization, may be described algorithmically: single 'neurons' are a classic case (McCulloch and Pitts 1943), but the same applies, for example, even to ion channels (e.g., see van Dongen 1996). This blurs the line between algorithmic- and neural-level descriptions, which may well be a virtue—the blurrier the line, the smaller the gaps to be filled—, but still poses the problem of integrating information from experimental measurements, which are typically continuous, dynamic, noisy etc., different from algorithmic model outputs. These issues point to the challenges of anchoring one's algorithmic theory in neurobiology strictly from the top down.

Epistemic pooling is a special case of transfer, which is not unidirectional and does not just involve one source (linguistics) and one target (neuroscience) domains. In epistemic pooling, knowledge is transferred from two domains into a third one: in this case, from linguistics and neuroscience into a field of inquiry equidistant from both (i.e., not a branch of linguistics or a branch of neuroscience), autonomous but capable of assimilating linguistic and neuroscientific knowledge, and emphasizing theory at Marr's intermediate level of analysis, that is, the level of representations, algorithms, and architectures. A computationalist psycho-/neurolinguistics may fit this rough characterization reasonably well (for an early programmatic statement, see Arbib and Caplan 1979). Developing verisimilar algorithmic-level analyses of the problems or tasks involved in language processing requires pooling of knowledge from linguistic theory (specifying what functions are computed) and neuroscience (constraining the implementation of the relevant computations; for example, how much representational redundancy and architectural parallelism may be involved, see Michalon and Baggio 2019 for a case study). This strategy is effectively a middle ground between a functionalist approach to modeling (it saves the mathematically necessary computational→algorithmic link) and the newer neuroscience-oriented bottom-up stances (it reverses the algorithmic→neural link, allowing neural-level constraints to co-determine algorithmic-level analyses; Hasson et al. 2018). There currently is a paucity of research using this specific approach, particularly on topics involving type-B linguistic knowledge about composition(ality) and related issues. However, greater theoretical emphasis and analytic efforts placed on intermediate levels of analysis are likely to stimulate pooling of knowledge from both linguistics and neuroscience, and perhaps eventually greater integration between the two.[3]

4 Conclusion

Language has a puzzling 'dual nature': as the object of theoretical linguistics, it is a system defined by characteristic formal structures and operations; as the object of human neuroscience, it is a complex, integrative function implemented via specific neurophysiological mechanisms in the brain. Convergence between linguistics and neuroscience is often viewed as a desirable goal, but so far achieving it has proved exceedingly hard. Because the concepts, methods, and background assumptions of linguistics and neuroscience are so different, their results, although fundamentally about the same object of inquiry—i.e., language as a human mental capacity—, are not immediately integratable. Moreover, and for similar reasons, neither discipline seems able to adapt to the other to an extent sufficient to enable integration. Here, I have argued that this onerous task

[3] A precursor of this approach is the co-evolutionary stance adopted by Churchland, Sejnowski, and others. I share with that proposal the notions that (1) algorithmic analyses are formally independent of implementational theories (e.g., because of multiple realizability), but discovery of the algorithms underlying cognition or behavior will depend on understanding aspects of the structure and function of relevant brain systems (Churchland and Sejnowski 1988, 1992), and (2) one goal of neuroscience is the discovery of such algorithms, as opposed to the study of how independently-assumed algorithms could be physically or biologically realized (Sejnowski et al. 1988).

falls to disciplines operating in areas *between* linguistics and neuroscience—primarily psycholinguistics, but also newer research endeavors (e.g., in computational linguistics, AI etc.) whose theoretical constructs fit the bill of Marr's algorithmic theory, that is, interfacing computational-level and physical-level descriptions of the given cognitive systems at work. This perspective confers on psycholinguistics the status of a 'bridge science'. Although it used to be a 'frontier science', driving research forward in early modern linguistics and brain sciences (aphasiology and neurology) (Levelt 2012), it may now be best construed as a semi-autonomous field, which can achieve the goal of re-connecting linguistics and neuroscience only by building on high-quality results from both disciplines, as is implied by the concept of epistemic pooling introduced here.[4]

I have argued that integration between linguistics and neuroscience may occur via unidirectional, conservative epistemic transfer, but only if the linguistic knowledge involved is either directly about language (type-A) or about methods applicable to behavioral or neural data analysis and data modeling (type-C). Knowledge of ways of formalizing linguistic structures and operations (type-B) is unlikely to generate new knowledge about how language is represented and processed in the brain, as is testified by many recent problematic attempts at identifying neural correlates or mechanisms of syntactic and semantic composition operations *as defined by formal linguistic theories*. Greater emphasis on tractable, plausible composition functions, and on neural constraints on their algorithmic implementation, may help build the missing bridges between type-B linguistic knowledge and human neurobiology. As I have suggested here, this move requires that we revise our concepts of linguistic competence (*deep competence*) and epistemic transfer (*epistemic pooling*), so as to refocus theoretical efforts on algorithmic-level analyses. This has two effects. The first is that 'mature' theories of language in the brain will be

[4] This account of psycholinguistics is an application of a broader view of human cognitive psychology as a semi-autonomous field, whose primary epistemic purpose is to integrate results from the formal and biological sciences relevant to understanding human behavior. Philosophically, this also entails that one may deconstruct and reconstruct the (metaphysical) category of mind in precisely the ways that serve best the bridging between the formal sciences (logic, linguistics, probability, decision and game theory etc.) and the biological sciences (neuropsychology, neurobiology, neurogenetics etc.).

algorithmic. They will be neither versions of current linguistic theories, whose core formal constructs are replaced with 'neurally plausible' ones (e.g., vectors instead of symbols), nor giant computer simulations of the brain 'doing language'. The second is that none of the currently dominant epistemological stances results in a viable methodology for the present era of language science: functionalism and rationalism (formal linguistics) will at best give us computational-algorithmic models untethered to neural reality, while physicalism and empiricism (neuroscience, usage-based linguistics) might be ultimately unable to extract from learning models a complete account of linguistic capacities. What may be more likely to succeed, given the arguments articulated in this essay, is an approach that weds minimal but strong structural assumptions on the search for the algorithms that compute the linguistic functions of interest (e.g., tractability analyses) with neurobiological constraints on implementation, taking seriously the idea that characteristics of the brain co-determine, together with the computational problem as such, the algorithms, representations, architectures, and mechanisms that constitute the language system functionally.

Acknowledgments I am grateful to Yosef Grodzinsky, Richard Lewis, Andrea E. Martin, James McClelland, Andrea Moro, Iris van Rooij, and Mathew Walenski for inspiration, input, and feedback on the ideas presented here.

References

Arbib, Michael, and David Caplan. 1979. Neurolinguistics must be computational. *Behavioral and Brain Sciences* 2 (3): 449–460.

Baggio, Giosuè. 2018. *Meaning in the brain*. Cambridge, MA: MIT Press.

Baggio, Giosuè, and Peter Hagoort. 2011. The balance between memory and unification in semantics: A dynamic account of the N400. *Language and Cognitive Processes* 26 (9): 1338–1367.

Baggio, Giosuè, Michiel Van Lambalgen, and Peter Hagoort. 2008. Computing and recomputing discourse models: An ERP study. *Journal of Memory and Language* 59 (1): 36–53.

Baggio, Giosuè, Travis Choma, Michiel van Lambalgen, and Peter Hagoort. 2010. Coercion and compositionality. *Journal of Cognitive Neuroscience* 22 (9): 2131–2140.

Baggio, Giosuè, Michiel van Lambalgen, and Peter Hagoort. 2012a. Language, linguistics and cognition. In *Philosophy of linguistics*, ed. R. Kempson, T. Fernando, and N. Asher, 325–355. Amsterdam: Elsevier.

———. 2012b. The processing consequences of compositionality. In *The Oxford handbook of compositionality*, ed. W. Hinzen, E. Machery, and M. Werning, 655–672. Oxford: Oxford University Press.

———. 2015. Logic as Marr's computational level: Four case studies. *Topics in Cognitive Science* 7 (2): 287–298.

Baggio, Giosuè, Keith Stenning, and Michiel van Lambalgen. 2016. Semantics and cognition. In *The Cambridge handbook of formal semantics*, ed. M. Aloni and P. Dekker, 756–774. Cambridge: Cambridge University Press.

Barton, Edward, Robert Berwick, and Eric Ristad. 1987. *Computational complexity and natural language*. Cambridge, MA: MIT Press.

Bemis, D.K., and L. Pylkkänen. 2011. Simple composition: A magnetoencephalography investigation into the comprehension of minimal linguistic phrases. *Journal of Neuroscience* 31 (8): 2801.

Bemis, Douglas, and Liina Pylkkänen. 2013. Basic linguistic composition recruits the left anterior temporal lobe and left angular gyrus during both listening and reading. *Cerebral Cortex* 23 (8): 1859–1873.

Ben Shalom, Dorit, and David Poeppel. 2008. Functional anatomic models of language: Assembling the pieces. *The Neuroscientist* 14 (1): 119–127.

Blanco-Elorrieta, Esti, and Liina Pylkkänen. 2016. Composition of complex numbers: Delineating the computational role of the left anterior temporal lobe. *NeuroImage* 124: 194–203.

Bohrn, Isabel, Ulrike Altmann, and Arthur Jacobs. 2012. Looking at the brains behind figurative language—A quantitative meta-analysis of neuroimaging studies on metaphor, idiom, and irony processing. *Neuropsychologia* 50 (11): 2669–2683.

Brennan, Jonathan, and Liina Pylkkänen. 2017. MEG evidence for incremental sentence composition in the anterior temporal lobe. *Cognitive Science* 41: 1515–1531.

Cappelle, Bert, Yury Shtyrov, and Friedemann Pulvermüller. 2010. Heating up or cooling up the brain? MEG evidence that phrasal verbs are lexical units. *Brain and Language* 115 (3): 189–201.

Chang, Edward., Jochem Rieger, Keith Johnson, Mitchel Berger, Nicolas Barbaro, and Robert Knight. 2010. Categorical speech representation in human superior temporal gyrus. *Nature Neuroscience* 13 (11): 1428.

Chierchia, Gennaro, and Sally McConnell-Ginet. 2000. *Meaning and grammar: An introduction to semantics*. Cambridge, MA: MIT Press.

Chomsky, N. (1995). The Minimalist Program. Cambridge MA: The MIT press.

Chomsky, Noam, and Morris Halle. 1968. *The sound pattern of English*. New York: Harper & Row.

Chomsky, Noam, and Howard Lasnik. 1993. The theory of principles and parameters. In *Syntax: An international handbook of contemporary research*, ed. J. Jacobs et al., vol. 1, 506–569. Berlin/New York: Walter de Gruyter.

Christiansen, Morten, and Nick Chater. 2016. *Creating language: Integrating evolution, acquisition, and processing*. Cambridge, MA: MIT Press.

Churchland, Patricia, and Terry Sejnowski. 1988. Perspectives on cognitive neuroscience. *Science* 242 (4879): 741–745.

———. 1992. *The computational brain*. Cambridge, MA: MIT Press.

Culicover, Peter, and Ray Jackendoff. 2005. *Simpler syntax*. Oxford: Oxford University Press.

Culicover, P. W., and Jackendoff, R. (2006). The simpler syntax hypothesis. *Trends in Cognitive Sciences, 10*(9), 413–418.

Embick, David, and David Poeppel. 2015. Towards a computational(ist) neurobiology of language: *Correlational*, *integrated* and *explanatory* neurolinguistics. *Language, Cognition and Neuroscience* 30 (4): 357–366.

Ferretti, Todd, Marta Kutas, and Ken McRae. 2007. Verb aspect and the activation of event knowledge. *Journal of Experimental Psychology: Learning, Memory, and Cognition* 33 (1): 182.

Fló, E., Cabana, Á., and Valle-Lisboa, J. C. (2020). EEG signatures of elementary composition: Disentangling genuine composition and expectancy processes. Brain and Language, 209, 104837.

Fodor, Jerrold, and Merill Garrett. 1966. Some reflections on competence and performance. In *Psycholinguistics papers*, ed. J. Lyons and R.J. Wales, 135–154. Edinburgh: University of Edinburgh Press.

———. 1967. Some syntactic determinants of sentential complexity. *Perception & Psychophysics* 2: 289–296.

Frampton, John, and Sam Gutmann. 1999. Cyclic computation, a computationally efficient minimalist syntax. *Syntax* 2 (1): 1–27.

Frank, Stefan, and Roel Willems. 2017. Word predictability and semantic similarity show distinct patterns of brain activity during language comprehension. *Language, Cognition and Neuroscience* 32 (9): 1192–1203.

Friederici, A. D. (2017). Language in Our Brain: The Origins of a Uniquely Human Capacity. Cambridge MA: The MIT press.
Friederici, Angela, Noam Chomsky, Robert Berwick, Andrea Moro, and Johan Bolhuis. 2017. Language, mind and brain. *Nature Human Behaviour* 1 (10): 713–722.
Fritz, Isabella, and Giosuè Baggio. 2020. Meaning composition in minimal phrasal contexts: Distinct ERP effects of intensionality and denotation. *Language, Cognition and Neuroscience* 35 (10): 1295–1313.
Frixione, Marcello. 2001. Tractable competence. *Minds and Machines* 11 (3): 379–397.
Ghio, Marta, Matilde Vaghi, Daniela Perani, and Marco Tettamanti. 2016. Decoding the neural representation of fine-grained conceptual categories. *NeuroImage* 132: 93–103.
Goldberg, Adele. 1995. *Constructions: A construction grammar approach to argument structure.* Chicago: University of Chicago Press.
———. 2003. Constructions: A new theoretical approach to language. *Trends in Cognitive Sciences* 7 (5): 219–224.
———. 2006. *Constructions at work: The nature of generalization in language.* Oxford: Oxford University Press.
Grodzinsky, Yosef, and Israel Nelken. 2014. The neural code that makes us human. *Science* 343 (6174): 978–979.
Hagoort, Peter. 2003. How the brain solves the binding problem for language: A neurocomputational model of syntactic processing. *NeuroImage* 20: S18–S29.
Hagoort, Peter, Giosuè Baggio, and Roel Willems. 2009. Semantic unification. In *The cognitive neurosciences*, ed. M. Gazzaniga, 4th ed., 819–836. Cambridge, MA: MIT Press.
Hamilton, Liberty, Erik Edwards, and Edward Chang. 2018. A spatial map of onset and sustained responses to speech in the human superior temporal gyrus. *Current Biology* 28 (12): 1860–1871.
Hasson, Uri, and Howard Nusbaum. 2019. Emerging opportunities for advancing cognitive neuroscience. *Trends in Cognitive Sciences* 23 (5): 363–365.
Hasson, Uri, Giovanni Egidi, Marco Marelli, and Roel Willems. 2018. Grounding the neurobiology of language in first principles: The necessity of non-language-centric explanations for language comprehension. *Cognition* 180: 135–157.
Heim, Irene, and Angelica Kratzer. 1998. *Semantics in generative grammar.* Oxford: Blackwell.

Herfeld, Catherine, and Chiara Lisciandra. 2019. Knowledge transfer and its contexts. *Studies in History and Philosophy of Science Part A* 77: 1–10.

Hickok, Gregory, and David Poeppel. 2016. Neural basis of speech perception. In *Neurobiology of language*, ed. G. Hickok and S. Small, 299–310. London: Academic.

Hullett, Patrick, Liberty Hamilton, Nima Mesgarani, Christoph E. Schreiner, and Edward Chang. 2016. Human superior temporal gyrus organization of spectrotemporal modulation tuning derived from speech stimuli. *Journal of Neuroscience* 36 (6): 2014–2026.

Jackendoff, Ray. 1983. *Semantics and cognition*. Cambridge, MA: MIT Press.

———. 2002. *Foundations of language: Brain, meaning, grammar, evolution*. Oxford: Oxford University Press.

———. 2007. A parallel architecture perspective on language processing. *Brain Research* 1146: 2–22.

Jakobson, Roman, and Morris Halle. 1971. *Fundamentals of language*. The Hague: Mouton.

Jasmin, Kyle, Cesar Lima, and Sophie Scott. 2019. Understanding rostral-caudal auditory cortex contributions to auditory perception. *Nature Reviews Neuroscience* 20 (7): 425–434.

Katz, Jerrold, and Jerrold Fodor. 1963. The structure of a semantic theory. *Language* 39 (2): 170–210.

Kim, Albert, and Lee Osterhout. 2005. The independence of combinatory semantic processing: Evidence from event-related potentials. *Journal of Memory and Language* 52 (2): 205–225.

Kuperberg, Gina, Arim Choi, Neil Cohn, Martin Paczynski, and Ray Jackendoff. 2010. Electrophysiological correlates of complement coercion. *Journal of Cognitive Neuroscience* 22 (12): 2685–2701.

Levelt, Willem. 2012. *A history of psycholinguistics: The pre-chomskyan era*. Oxford: Oxford University Press.

Lyu, Bingjiang, Hun Choi, William Marslen-Wilson, Alex Clarke, Billi Randall, and Lorraine K. Tyler. 2019. Neural dynamics of semantic composition. *Proceedings of the National Academy of Sciences* 116 (42): 21318–21327.

Marantz, Alec. 2005. Generative linguistics within the cognitive neuroscience of language. *The Linguistic Review* 22 (2–4): 429–445.

Marr, David. 1982. *Vision: A computational investigation into the human representation and processing of visual information*. San Francisco: Freeman.

Marr, David, and Tomaso Poggio. 1976. From understanding computation to understanding neural circuitry. *MIT AI Memos* 357: 1–22.

Martin, Andrea, and Giosuè Baggio. 2019. Modelling meaning composition from formalism to mechanism. *Philosophical Transactions of the Royal Society B: Biological Sciences* 375 (1791): 1–7.

Martin, Andrea E., and Leonidas Doumas. 2019. Predicate learning in neural systems: Using oscillations to discover latent structure. *Current Opinion in Behavioral Sciences* 29: 77–83.

McCulloch, W. S., and Pitts, W. (1943). A logical calculus of the ideas immanent in nervous activity. *The Bulletin of Mathematical Biophysics*, *5*(4), 115–133.

Mesgarani, Nima, Connie Cheung, Keith Johnson, and Edward Chang. 2014. Phonetic feature encoding in human superior temporal gyrus. *Science* 343 (6174): 1006–1010.

Michalon, Olivier, and Giosuè Baggio. 2019. Meaning-driven syntactic predictions in a parallel processing architecture: Theory and algorithmic modeling of ERP effects. *Neuropsychologia* 131: 171–183.

Moro, Andrea. 2016. *Impossible languages*. Cambridge, MA: MIT Press.

———. 2019. The geometry of predication: A configurational derivation of the defining property of clause structure. *Philosophical Transactions of the Royal Society B: Biological Sciences* 375 (1791).

Musso, Mariacristina, Andrea Moro, Volkmaar Glauche, Michel Rijntjes, Jürgen Reichenbach, Christian Büchel, and Cornelius Weiller. 2003. Broca's area and the language instinct. *Nature Neuroscience* 6 (7): 774.

Nelson, Matthew, Imen El Karoui, Kristof Giber, Xiaofang Yang, Laurent Cohen, Hilda Koopman, Sydney Cash, Lionel Naccache, John Hale, Christophe Pallier, and Stanislas Dehaene. 2017. Neurophysiological dynamics of phrase-structure building during sentence processing. *Proceedings of the National Academy of Sciences* 114 (18): E3669–E3678.

Nieuwland, Mante, Dale Barr, Federica Bartolozzi, Simon Busch-Moreno, Emily Darley, David Donaldson, Heather Ferguson, Xiao Fu, Evelien Heyselaar, Falk Huettig, Matthew Husband, Aine Ito, Nina Kazanina, Vita Kogan, Zdenko Kohút, Eugenie Kulakova, Diane Mézière, Stephen Politzer-Ahles, Guillaume Rousselet, Shirley-Ann Rueschemeyer, Katrien Segaert, Jyrki Tuomainen, and von Grebmer Zu Wolfsthurn, Sarah. 2019. Dissociable effects of prediction and integration during language comprehension: Evidence from a large-scale study using brain potentials. *Philosophical Transactions of the Royal Society B* 375 (1791).

Nowak, Iga, and Giosuè Baggio. 2017. Developmental constraints on learning artificial grammars with fixed, flexible and free word order. *Frontiers in Psychology* 8: 1816.

Olstad, A. M. H., Fritz, I., and Baggio, G. (2020). Composition decomposed: Distinct neural mechanisms support processing of nouns in modification and predication contexts. *Journal of Experimental Psychology: Learning, Memory, and Cognition* 46 (11): 2193–2206.

Paczynski, Martin, Ray Jackendoff, and Gina Kuperberg. 2014. When events change their nature: The neurocognitive mechanisms underlying aspectual coercion. *Journal of Cognitive Neuroscience* 26 (9): 1905–1917.

Papeo, Liuba, Angelika Lingnau, Sara Agosta, Alvaro Pascual-Leone, Lorella Battelli, and Alfonso Caramazza. 2014. The origin of word-related motor activity. *Cerebral Cortex* 25 (6): 1668–1675.

Partee, Barbara. 2018. Changing notions of linguistic competence in the history of formal semantics. In *The science of meaning: Essays on the metatheory of natural language semantics*, ed. D. Ball and B. Rabern, 172–196. Oxford: Oxford University Press.

Patterson, Karalyn, Peter Nestor, and Timothy Rogers. 2007. Where do you know what you know? The representation of semantic knowledge in the human brain. *Nature Reviews Neuroscience* 8 (12): 976.

Pavan, Andrea, and Giosuè Baggio. 2013. Linguistic representations of motion do not depend on the visual motion system. *Psychological Science* 24 (2): 181–188.

Petersson, Karl, and Peter Hagoort. 2012. The neurobiology of syntax: Beyond string sets. *Philosophical Transactions of the Royal Society B: Biological Sciences* 367 (1598): 1971–1983.

Poeppel, David, and David Embick. 2005. Defining the relation between linguistics and neuroscience. In *Twenty-first century psycholinguistics*, ed. A. Cutler, 103–118. Mahwah: Lawrence Erlbaum Associates.

Poggio, T. (1981). Marr's computational approach to vision. *Trends in Neurosciences, 4*, 258–262.

Pulvermüller, Friedemann. 2010. Brain embodiment of syntax and grammar: Discrete combinatorial mechanisms spelt out in neuronal circuits. *Brain and Language* 112 (3): 167–179.

Pulvermüller, Friedemann, Bert Cappelle, and Yury Shtyrov. 2013. Brain basis of meaning, words, constructions, and grammar. In *The Oxford handbook of construction grammar*, ed. T. Hoffmann and G. Trousdale, 397–416. Oxford: Oxford University Press.

Pustejovsky, James. 1995. *The generative lexicon*. Cambridge, MA: MIT Press.

Pylkkänen, L. (2019). The neural basis of combinatory syntax and semantics. Science, 366(6461), 62–66.

Rabovsky, Milena, Steven Hansen, and James McClelland. 2018. Modelling the N400 brain potential as change in a probabilistic representation of meaning. *Nature Human Behaviour* 2 (9): 693.

Rapp, Alexander, Dorothee Mutschler, and Michael Erb. 2012. Where in the brain is nonliteral language? A coordinate-based meta-analysis of functional magnetic resonance imaging studies. *NeuroImage* 63 (1): 600–610.

Ristad, Eric. 1993. *The language complexity game*. Cambridge, MA: MIT Press.

Sejnowski, Terry, Christof Koch, and Patricia Churchland. 1988. Computational neuroscience. *Science* 241 (4871): 1299–1306.

Shieber, Stuart. 1986/2003. *An introduction to unification-based approaches to grammar*. Stanford: CSLI Publications.

Snijders, Tineke, Theo Vosse, Gerard Kempen, Jos Van Berkum, Karl Petersson, and Peter Hagoort. 2008. Retrieval and unification of syntactic structure in sentence comprehension: An fMRI study using word-category ambiguity. *Cerebral Cortex* 19 (7): 1493–1503.

Steedman, Mark. 1992. *Grammars and processors*. Technical report MS-CIS-92-52, Department of Computer and Information Science, University of Pennsylvania.

Tang, Claire, Liberty Hamilton, and Edward Chang. 2017. Intonational speech prosody encoding in the human auditory cortex. *Science* 357 (6353): 797–801.

Townsend, David, Thomas Bever, and Matthew Crocker. 2001. *Sentence comprehension: The integration of habits and rules*. Cambridge, MA: MIT Press.

Trotzke, Andreas, and Jan-Wouter Zwart. 2014. The complexity of narrow syntax: Minimalism, representational economy, and simplest merge. In *Measuring grammatical complexity*, ed. F.J. Newmeyer and L.B. Preston, 128–147. Oxford: Oxford University Press.

Trubetzkoy, Nicolai. 1969. *Principles of phonology*. Berkeley: University of California Press.

Urbach, Thomas, Katherine DeLong, and Marta Kutas. 2015. Quantifiers are incrementally interpreted in context, more than less. *Journal of Memory and Language* 83: 79–96.

Van Dongen, Anne. 1996. A new algorithm for idealizing single ion channel data containing multiple unknown conductance levels. *Biophysical Journal* 70 (3): 1303–1315.

Van Rooij, Iris. 2008. The tractable cognition thesis. *Cognitive Science* 32 (6): 939–984.

Van Rooij, Iris, and Giosuè Baggio. 2020. Theory before the test: How to build high-verisimilitude explanatory theories in psychological science. PsyArXiv preprint, https://doi.org/10.31234/osf.io/7qbpr.

Vulchanova, Mila, Evelyn Milburn, Valentin Vulchanov, and Giosuè Baggio. 2019. Boon or burden? The role of compositional meaning in figurative language processing and acquisition. *Journal of Logic, Language and Information* 28 (2): 359–387.

Westerlund, Masha, and Liina Pylkkänen. 2017. How does the left anterior temporal lobe contribute to conceptual combination? Interdisciplinary perspectives. In *Compositionality and concepts in linguistics and psychology*, ed. J.A. Hampton and Y. Winter, 269–290. New York: Springer.

Wittenberg, Eva, Martin Paczynski, Heike Wiese, Ray Jackendoff, and Gina Kuperberg. 2014. The difference between 'giving a rose' and 'giving a kiss': Sustained neural activity to the light verb construction. *Journal of Memory and Language* 73: 31–42.

Yang, Charles. 2002. *Knowledge and learning in natural language*. Oxford: Oxford University Press.

———. 2004. Universal grammar, statistics or both? *Trends in Cognitive Sciences* 8 (10): 451–456.

Zaccarella, Emiliano, Lars Meyer, Michiru Makuuchi, and Angela Friederici. 2017. Building by syntax: The neural basis of minimal linguistic structures. *Cerebral Cortex* 27 (1): 411–421.

Part IV

Linguistics and the Humanities

12

Linguistics Meets Hermeneutics: Reading Early Greek Epistemological Texts

Anna Novokhatko

1 Introduction

1.1 General Considerations

In 1988 Jan Ziolkowski forced a group of leading American philologists to formulate definitions of philology. A special issue of *Comparative Literature Studies* brought together talks delivered at a conference at Harvard University in 1988 entitled "What is philology?"[1] (Ziolkowski

[1] Originally published as a special-focus issue of *Comparative Literary Studies* vol. 27, no. 1, 190. This epistemological question became hot once again during the last decade – several volumes focused on the sense and capacities of philology have been published, such as König 2009 and Pollock, Elman and Chang 2015. For a detailed bibliography, see Pollock, Elman and Chang 2015. On the term 'philology', see the arguments in Ziolkowski 1990b. Onexcellents accounts on philology, see Momma 2013, 1-27 and Hamilton 2018.

A. Novokhatko (✉)
Classical Philology, Aristotle University Thessaloniki, Thessaloniki, Greece

Albert Ludwig University of Freiburg, Freiburg im Breisgau, Germany
e-mail: anna.novokhatko@altphil.uni-freiburg.de

R. M. Nefdt et al. (eds.), *The Philosophy and Science of Language*,
https://doi.org/10.1007/978-3-030-55438-5_12

1990a). Richard Thomas described there philology as "a component of textual criticism and editing, the writing of commentaries, stylistic and metrical studies, as well as those modes of interpretation and literary history wherein the notions of "affection", "respect", or "close proximity" to the text are maintained. At the same time it draws from history, archaeology, palaeography, epigraphy, historical linguistics, anthropology, the study of religion, and critical theory, for all of these potentially aid in the quest for facts and truths about literary texts" (Thomas 1990: 69).

Establishing a text is far from being a mechanical act. Editing cannot proceed without, nor should it simply precede, interpretation and commentary. Thus comprehending a text may involve an examination of the individual words and a determination of their meanings on the basis of earlier and later evidence. Further, philology incorporates restoring to these words as many of their original nuances and as much of their original power as we can manage. Therefore philology has often enough been identified with literary criticism and theory, or with some of its components, such as grammar and textual criticism. From the outset, however, it was formed and came to be understood through detailed interpretation of poetic texts.

The separation of linguistics from philology, which occurred in European thought only gradually, was signaled in the nineteenth century works of August Schleicher (1821–1868),[2] and then later in a much-quoted statement from Ferdinand de Saussure's (1857–1913) *Course in General Linguistics* published posthumously in 1916: "Quant à la philologie, nous sommes déjà fixés: elle est nettement distincte de la linguistique, malgré les points de contact des deux sciences et les services mutuels qu'elles se rendent."[3]

While the objective of modern canonical philology founded by Friedrich August Wolf (1759–1824), was mainly to examine, edit and comment on texts, linguistics dealt with signs, sounds and speech.[4] The

[2] On Schleicher's distinction, see more in Koerner 1997: 170.

[3] Saussure 1922 (1916): 21. Saussure's most influential work, *Cours de linguistique générale*, was published posthumously in 1916 by his former students Charles Bally and Albert Sechehaye, on the basis of notes taken from Saussure's lectures in Geneva. On the separation of linguistics from philology in the 1910s and 1920s worldwide, see Chang 2015.

[4] It was in Halle in the years 1783–1807, that Wolf first laid down the principles of the field he would call "philology". He defined philology as the study of human nature as exhibited in antiquity (*Altertumswissenschaften*). Its methods included the examination of the history, writing, art, and other examples of ancient cultures. It combined the study of history and language (*Sprachstudium*)

main separation between literary and textual criticism and linguistics is clear: linguistics deals with language itself (including non-written language such as those of some tribes in Northern America or Africa) while literary and textual criticism (it is best to avoid 'philology' as a term here, as philology should be understood in a broader sense) deal with the interpretation (literary criticism) and edition (textual criticism) of texts. However, interpretation not supported by linguistic data and statistics, and not proved by linguistic analysis of the text structure, can easily become subjective and misleading (although here again we should recall the on-going debates on data-oriented versus theory-oriented linguistics).[5] Thus with the so-called scientific turn of the 1960s, attempts were made to "mathematize" both linguistics and literary studies.[6] The description of literary texts with linguistic and structuralist methods promised exactness to the field of literary criticism, part of this being the application of exact methods to literary texts and to linguists more generally (the 1970 foundation of the journal in German studies with the speaking title *Die Zeitschrift für Literaturwissenschaft und Linguistik*, as well as the De Gruyter *Journal of Literary Semantics* in English studies, and, somewhat later, the British *Language and Literature, Journal of the Poetics and Linguistics Association* are noteworthy here).

In parallel another fusion was attempted, and a branch of linguistics called 'text linguistics' was developed, this dealing with texts as communication systems (De Beaugrande and Dressler 1981 and Janich 2008). Its original aims lie in uncovering and describing text grammars. Text linguistics takes into account the form of a text and the way in which it is situated in an interactional context. The main criteria of textuality are coherence (the use of deictic, anaphoric and cataphoric elements or a logical tense structure, as well as presuppositions and implications connected to general world knowledge) and cohesion (the purely linguistic grammatical and lexical linking within a sentence that holds a text

and textual criticism (*Wortkritik*), through interpretation (*Hermeneutik*), in which history and linguistics coalesce into an organic whole. For an overview of early philological thought in Germany, see Chang 2015: 313–317.

[5] See Hogg 1994 for more on this distinction.

[6] On the 'scientific turn' and the paradigm change for literary and linguistic studies in the 1960s, see Fix 2010: 21–22.

together and gives it meaning). Both the author of a (written or spoken) text as well as its addressee are taken into consideration in their respective (social and/or institutional) roles in the specific communicative context. In general 'text linguistics' is an application of discourse analysis at the much broader level of text, rather than just a sentence or word.[7]

It is noteworthy that the narratological approach, developed firstly in Russian formalist circles and then later by French structural linguists, can serve as a good example of applying linguistic methods and tools to the analysis of literary texts (Garrett 2018) and Liveley 2019. Time and space become principles of narrative structure: the past, the present and the future afford different items for inspection and accommodate different interventions on each other. Text interpretation is being fulfilled through reference to temporality and spatiality. Temporal dimensions within narrative structure, grammatical tenses manifested by the use of specific forms of verbs, particularly in their conjugation patterns, as well as syntactic structures, and adverbs of time and space, should be analysed. When the present mirrors the past, the overall temporality is regressive; when the past mirrors the future, then the overall temporality is progressive, and so on.

Furthermore, the developing field of the so-called "digital humanities" has grown dramatically over recent decades. The most important branch here is corpus and computational linguistics: a way of approaching language and literary texts through new methods of analysis that involve the use of large electronic corpora. In the place of earlier hand-made, limited corpora, there is an enormous amount of online accessible, electronically documented texts, in the majority of cases compiled according to certain questions or topics. The electronic storage and the electronic accessibility allow for the provision of a comprehensible and complete analysis of seemingly incomprehensible amounts of texts, in contrast to earlier times when only accidental findings could occur. Thus, for example, collocations (the usual connections of words) and co-occurrences (an above-chance frequency of occurrence of two terms from a text corpus alongside each other in a certain order) that used for ages to be judged mainly in

[7] On the linguistic text analysis, see Brinker-Colfen-Pappert 2014.

terms of intuition can now be demonstrated to be accurate in terms of quantitative numbers.[8]

Such a stream of linguistic research provoked the next turn. In his 1982 article "The Return to Philology" the deconstructivist Paul De Man (1919–1983) questioned the compatibility of linguistics and hermeneutics. "It is by no means an established fact that aesthetic values and linguistic structures are incompatible. What is established is that their compatibility, or lack of it, has to remain an open question" (De Man 1986: 25). He rephrases the notion of philology as "an examination of the structure of language prior to the meaning it produces" (De Man 1986: 24). He points out that literary criticism becomes important as soon as the approach to literary texts ceases being based on historical and aesthetic considerations, but moves closer to linguistics. Texts should be read "closely as texts" and not move "at once into the general context of human experience or history" (De Man 1986: 23). The interest in semiology of critics such as Roman Jakobson and Roland Barthes, argues De Man in his other essay "The Resistance to Theory", thus "reveals the natural attraction of literature to a theory of linguistic signs" (De Man 1986: 8).[9]

To sum up, despite the multiple directions and methods in which linguistic and literary studies have developed, for those working with texts philologically there is a mutual desirability of close interaction of linguistic practice, especially with interpretative analysis.[10]

1.2 To the Question of the Juxtaposition of Linguistics and Hermeneutics in Classics

Applying these aforementioned debates to the classics, the remarkable growth of purely linguistic studies on Ancient Greek and Latin in the last decades revealed a clear need for a linguistic approach to ancient texts. Novel and fruitful results for textual understanding were brought forward through novel treatment of the same well-known but also at times

[8] Chronopoulos-Maier-Novokhatko 2020.

[9] On the application of linguistic concepts to literary criticism and the "blurring boundaries" between literary criticism and cognitive linguistics, see recently also Allan 2020.

[10] For instance, Koerner (1997: 173): "I fail to see that their relationship needs to be a troubled one".

unknown material.[11] This process created new conditions for disciplinary and methodological interaction between linguistics and textual and literary criticism.

The process started more than a century and a half ago. The German philologist Georg Curtius (1820–1885) published the monograph "Die Sprachvergleichung in ihrem Verhältniss zur classischen Philologie" (1845) in which he tried to demonstrate the usefulness of 'Sprachwissenschaft' (i.e. historical-comparative grammar) to classical philology. Recent linguistic theories such as narratological analysis or the speech-act theory developed during the last decades are now being applied to Greek and Latin texts, thus proving the correctness of Curtius' claim.

Within the digital humanities, mentioned above, the developing field of the so-called digital classics, the digital study of Ancient Greek and Latin texts from editorial, hermeneutic, but mainly linguistic perspective, is noteworthy.[12] It became possible, to mention only a few examples, for scholars to observe through a machine the insecure use of particles and prepositions or punctuation marks ignored over hundreds of years. Linguists make use of these materials for the determination of use frequencies, word fields, semantic networks, and ultimately for the compilation of dictionaries. They are used in pragmalinguistics as well for text comparisons employing intertextual references. Again, linguistics and hermeneutics find themselves in a close cooperation and mutual dependence.

Thus classics, par excellence a data-oriented discipline dealing with material texts (literary or epigraphic), represents the chance for reflection of the on-going debates of the compatibility of linguistic and literary approaches to the same material. In what follows, the process and results of such compatibility will be analysed as a case study.

---

[11] See Allan and Buijs 2007 with bibliography.

[12] See e.g. the claim "reinvention of philology" in Bamman and Crane 2010.

2 The Earliest Attestations of Greek Epistemological Thought

In this paper linguistic and hermeneutic methods will be applied to specific pre-classical Greek texts, texts where hermeneutic methods are also questioned, though the questions we pose of them may be considered anachronistic.

The passages I selected for analysis engage in the first epistemological and gnoseological critical considerations and construct early approaches for the theoretical discussion of the methodology of criticism and hermeneutics in terms of concepts and vocabulary up until the first half of the fifth century BCE.[13] The early Greek criticism and textual exegesis include various disconnected but not necessarily clearly distinguishable directions, and should be regarded together in the context of the theories of perception and epistemology of the pre-Socratic tradition.[14] I will try to read the passages applying a mainly pragmatic approach, i.e. those where the transmission of meaning depends not only on the structural and linguistic knowledge of the author and his audience such as grammar and vocabulary, but also on the context of the utterance, and on pre-existing knowledge of those involved. Early Greek epistemological considerations are thus interpreted as a tool to analyse cognitive mechanisms of a multi-layered process of textual exegesis. It is significant that the same syntactic formulas and structural models are used when the author introduces or summarizes a self-referential or critical judgment as an innovative form, as this evolved in the late Archaic period. Thus a distance between the author and his product is emphasized and a coherent message is transmitted to the audience, framed in an always syntactically recognizable construction.

From the beginning of the fifth century BCE onward scholarly notions, concepts and vocabulary were developing at an intense pace. Proto-terms for scientific methodology, such as "preciseness", "error", "research", "evidence",

[13] On the emergence of epistemology and early epistemological reflections on the interaction between mortal and divine in the poetry of Hesiod, Xenophanes and Parmenides, see Tor 2017. See also Gerson 2009: 1–26.

[14] For early Greek poetological considerations, see Lanata 1963 and Nagy 1989.

"attestation" and many others, come more and more into use. This is a crucial step toward the borrowing of terminology for philological and hermeneutic studies and the creation of a methodology. Here in order to start *ab ovo* the Archaic and late Archaic texts of the 6th c. and early 5th c. BCE will be examined.

2.1 Examining, Self-Examining and Searching

The lyric poet Archilochus (c. 680 – c. 645 BCE) was the earliest known Greek poet to compose almost entirely describing his own emotions and experiences. He claimed to have combined the activities of soldier and poet, saying that he was the servant of the god of war Ares and of the Muses, having knowledge (ἐπιστάμενος) of their lovely present (fr. 1 IEG).[15]

The following epistemological aphorism written in iambic trimeter and ascribed to Archilochus constituted the starting point for Isaiah Berlin's 1953 essay on the fox and the hedgehog.

> fr. 201 IEG *the fox knows many things* (πόλλ' οἶδ') *but the hedgehog (knows) one big thing* (ἓν μέγα)

Our knowledge of cultural and anthropological function of the fox and the hedgehog in Archaic Greek world is not sufficient to draw conclusions as to what exactly the allegory should mean and also as to what the fox and the hedgehog actually "know". And we are automatically biased by Berlin's interpretation of this verse and think of a broad variety of experience (fox) versus a system built around of one big idea (hedgehog) in the world history of thought. Without knowing the exact meaning of Archilochus' verse we can be sure that the nature, capacities and limits of knowledge are questioned by Archilochus. The verb of knowledge is used (οἶδα). One kind of knowledge (cunning, experimental, searching?) is contrasted to the other kind (defensive, spiny, conservative?) through the

[15] Cf. the similar use of the participle πιστάμενος by the statesman and poet Solon (c. 638-558 BCE) referring to a poet "who has skill (πιστάμενος) in the measure of desirable wisdom" (fr. 13, 52 IEG). Here and below all translations from Ancient Greek are my own.

adversative conjunction "but" (ἀλλ') and through the juxtaposition of "many" (πόλλ') on the one side and "one" (ἓν) on the other. The other knowledge, that of the hedgehog, is emphatically marked as significant at the end of the verse, "something which is one and big" (ἓν μέγα).[16]

Archilochus further repeatedly brandishes his own knowledge or skills in general and in particular his capacity to create or perform poetry. The vocabulary of self-referential knowledge is thus well represented in his fragments: the poet knows to sing dithyramb (οἶδα, fr. 120 IEG), knows a remedy (οἶδα, fr. 67 IEG), knows how to love a friend (ἐπ]ίσταμαι, fr. 23, 14 IEG), knows one big thing (ἐπίσταμαι, fr. 126 IEG).[17] The verb "I know" (οἶδα or ἐπίσταμαι) is always used in the 1st person singular of the speaker himself. Both verbs mean "to know / to be skilled" and the difference at this early stage is not clear.[18] In Classical Greek due to the development of scientific and epistemological thought the difference will grow, the verb ἐπίστασθαι being opposed to εἰδέναι and meaning "to have exact (scientific) knowledge" (cf. Arist. *Met.* 1, 2, 10 982b20–23: *Therefore as in order to escape ignorance people studied philosophy, it is clear that they pursued scientific knowledge* (τὸ ἐπίστασθαι) *through (general) knowledge* (τὸ εἰδέναι), *and not for any practical reason*). In Archaic Greek this is hardly the case, and the verbs should rather be understood synonymously.

Both the lyric poet Theognis of Megara and the philosopher and poet Xenophanes of Colophon were probably active in the second half of the 6th c. BCE. In Theognis' elegies we read:

---

[16] See Bowra 1940 and Swift 2019, 385–387 on this fragment. One of the interpretations of this juxtaposition might be suggested by Solon's reference to a fox (before 558 BCE): "Each one of you walks in the tracks of the fox, but the mind of all of you is empty; for you look at the language and the words of a cunning man, and you do not see at all the work accomplished" (Solon fr. 11 West). On the craft (μῆτις) of a fox, see also Detienne et Vernant 1978: 41–46, on Archilochus' fragment esp. p. 43, n. 64.

[17] Cf. the 6th c. BCE lyric poet Anacreon who in the 1st person singular claimed "to sing graceful songs and to know (οἶδα) how to speak graceful verses" (fr. 57 PMG). Campbell 1983: 255 and Bossi 1990: 178 with bibliography.

[18] Calame 2019, 56–57.

> *I cannot understand* (οὐ δύναμαι γνῶναι) *the mind the citizens have* (νόον ἀστῶν ὅντιν' ἔχουσιν), *for I do not please them either acting well or badly* (vv. 367–368 IEG)

Here the context is ethics, and the verb "to know/understand" (γνῶναι) has to be explained as an admonition to comply with aristocratic ideals. However, although the verb is employed in a moral and not scholarly context, the act of self-estimation, self-examination and of reflection on one's own achievements remains significant.

Further down in Theognis' text the same reference to the perception of others' mind occurs (vv. 1013–1016 IEG):

> *Oh blessed, fortunate and happy is he, who went down to Hades' black house without experiencing contest, before he had had the chance to crouch down in front of his enemies, before he had been transgressed by force, and before he could examine his friends, and examine their real mind.* (ἐξετάσαι τε φίλους, ὅντιν' ἔχουσι νόον)

This text is the earliest attestation of the verb "to examine" (ἐξετάζειν) in Greek, which will become *terminus technicus* for "testing, examining and reviewing".[19] Again the verb is employed in an ethical context, but the claim of scrutinizing and examining is noteworthy.

The much-discussed manifesto by Theognis on the poet's duties (vv. 769–772 IEG) is formulated with the use of a variety of epistemological vocabulary.[20] The poet (or the one who shares knowledge/skill), claims Theognis, even if he were to have any greater knowledge (εἴ τι περισσόν εἰδείη) should not become jealous with his skill, but search things (τὰ μὲν μῶσθαι), reveal things (τὰ δὲ δεικνύεν), create/make things (ἄλλα δὲ ποιεῖν). What use would it be for him if he alone has knowledge (μοῦνος ἐπιστάμενος)?

---

[19] See the scholarly usage of this verb referring to the correctness of words and phrases in Plato's *Theaetetus* 184c: δι' κριβείας ξεταζόμενον "precisely scrutinized". See further the fixed meaning of the verb as "to test, examine" already in 1st–9th cent. CE Greek (Lampe 1961 s.v.).

[20] See a thorough analysis of these verses in Woodbury 1991. See also Van Groningen 1966: 297–299 and Ford 1985: 92–93.

The density of the vocabulary of knowledge and perception (εἰδείη, σοφίης, ἐπιστάμενος) is striking in these verses. Even more striking is the list of poet's duties with the aim of his sharing knowledge with others. Theognis uses a rare verb for "searching" (μῶσθαι) attested in Epicharmus and in tragedy and explained by the scholiasts through a more common verb for "searching" ζητεῖν (Tor 2017: 127). The object of searching should remain his knowledge, the things that he understands (εἰδείη). The second verb means "to show/display/illumine/represent" (δεικνύεν). As the next stage (the sequence emphasized through the particles μὲν... δὲ... δὲ) after searching is "to find/not to find", "to show" here should come close to "to discover/find out", so that the third duty becomes "to make/create/invent/compose".[21] Thus the complete process of epistemological search and the creation of an intellectual product is represented: the poet looks for knowledge, reveals it and then composes verses out of his discovery.

Human knowledge is, however, limited, claims Theognis at another point, again employing the vocabulary of thinking (vv. 141–142 IEG):

> *We mortals have vain thoughts* (μάταια νομίζομεν)[22] *knowing nothing* (εἰδότες οὐδέν)*; the gods, however, bring everything to pass according to their mind/understanding.* (κατὰ σφέτερον... νόον)

Exactly this question of the relativity of human intellectual capacities is made a theme in Xenophanes' poetry, although in different terms. The vainness of human knowledge itself is not contradicted, but Xenophanes focuses on the process of cognition and acquiring knowledge, on searching and finding out:

DK 21 B18 = D53 Laks-Most:

> *not from the beginning did the gods show* (ὑπέδειξαν) *everything to people,*
> *but they find out better* (ἐφευρίσκουσιν ἄμεινον) *in process of time whilst searching* (ζητοῦντες)

[21] "δείκνυμι can then signify the declaration and publication of what is sought and found" (Woodbury 1991: 485, n. 8).

[22] On the verb νομίζομεν cf. the commentary by Van Groningen 1966: 56: "Le verbe signifie dans ces cas: 'avoir l'habitude de penser'".

Xenophanes in his verses emphasized rational critical thought and a distance between a rational approach and the mythological tradition of poetry. The gods did not reveal (ὑπέδειξαν) things to mortals: Xenophanes uses the same verb but with the prefix ὑπο- that occurred in Theognis without a prefix (δεικνύεν) in the passage, discussed above, on the claim what it is that somebody competent in poetry should know.[23] Further, the pair ζητεῖν 'to search' and ἐφευρίσκειν 'to find out' are both important principles of knowledge and cognition that would soon come to belong to the working vocabulary of science and scholarship (both were to be scientific terms already rooted in the language of this period and indicating epistemological activity).[24] The opposition of the first line (mythological tradition, stated in terms "the gods show to mortals") to the second (rational critical thought) expressed both through the negation "not… but" (οὔ …ἀλλὰ) and through the temporal deixis "not from the beginning" and "in process of time", also remains worth noting.

Xenophanes' way of thinking was connected in the later Ancient tradition to the discourses on the contrast between critical reason and the imagination. The Peripatetic philosopher Aristocles (1st c. CE) credits Xenophanes along with some members of the so-called Eleatic school in Magna Grecia[25] – Parmenides of Elea (*fl.* late sixth or early fifth century BCE), Zeno of Elea (c. 490 – c. 430 BCE) and Melissus of Samos (*fl.* 5th c. BCE) – with having argued that one should trust reason alone (αὐτῶι δὲ μόνον τῶι λόγωι) while perceptions (τὰς μὲν αἰσθήσεις) and appearances/imaginations (τὰς φαντασίας) should be rejected (DK 21 A49 (= R10 Laks-Most = Aristocl. *Philos.* 7)). With this in mind, Xenophanes' fragment containing the argument that people search and find out the essence of things, should be seen not as a casual statement at random, but a decisive methodological principle.

---

[23] On the use of the verb δείκνυμι without a prefix on divine disclosure and on the unique use of ὑποδείκνυμι on the divine action here in Xenophanes, see Tor 2017: 117–118. On the prefix ὑπο- and its connotations (such as indirect and cryptic manner), see Lesher 1991: 237, n. 19 and Lesher 1992: 149 and 153.

[24] On the use of both words in early authors, see Lesher 1991: 242–243 and Lesher 1992: 154–155. See the detailed discussion of Xenophanes' fragment DK21 B1= D53 Laks-Mostin Lesher 1991; Lesher 1992: 149–155; and Lesher 2013: 86–87. See also Heitsch 1994: 19–20 and Tor 2017: 127–128.

[25] On Xenophanes himself belonging not to the Eleatics, but to the Ionians, perhaps having had "his Eleatic period", see Lesher 1992: 3–7.

The (Attic) verb ζητεῖν 'to search/seek/inquire', or rather its Ionic (Homeric) equivalent δίζησθαι occurs in a critical sense in Heraclitus' fragment (Heracl. DK22 B101 = D36 Laks-Most: *I searched out myself* (ἐδιζησάμην ἐμεωυτόν), which is contemporary to Xenophanes.[26] The first attested use of the verb in the 1st person emphasized by the medium voice and the reflexive 1st person pronoun is striking. The reference to self-examination reveals an epistemological context of the discovery of one's own nature, as suggested by many other fragments by Heraclitus also.[27] This fragment is transmitted by Plutarch (ca. 45–125 CE), who in his "Reply to the philosopher Colotes" ridicules the Epicurean Colotes for attacking Socrates and ignoring self-reflexivity. Heraclitus, says Plutarch, having accomplished something great and important, said "I searched out myself", and the Delphic inscriptions contained the famous line "know yourself", which gave instructions to Socrates in his puzzling and inquiring.[28] Heraclitus' engagement in self-discovery is attested in another fragment as well where the same Delphic maxim might be quoted: "All people have a share in knowing themselves (γινώσκειν ἑωυτοὺς) in being sound of mind" (Heracl. DK22 B116 = D30 Laks-Most).

Heraclitus' contemporary Eleatic philosopher Parmenides, mentioned above, wrote a poem *On Nature*, surviving in fragments which can be interpreted only with difficulty. Here questions on being and becoming and knowledge about them are posed.[29] The same verb δίζησθαι is used for the act of philosophical inquiry, linked to the acquisition of divine knowledge. Thus the being/what-is is called by Parmenides one, unoriginated (ἀγένητον), imperishable, and continuous. "For what origin (γένναν) could you inquire (διζήσεαι) for it?" (DK28 B8, 3–6 = D8, 8–11 Laks-Most) The verb is thus employed with the meaning of searching for/investigating the origins of nature itself.

[26] On the verb δίζησθαι employed in the earliest texts for the act of consulting Apollo in the Delphic oracles, see Tor 2017: 266, n. 113.

[27] Cf. Laks and Most 2016: 152–159 on Heraclitus' fragments D29–D45.

[28] Plut. *Mor.* 1118c. Heracl. DK22 B101 = D36 Laks-Most was quoted as well by the Roman emperor and philosopher Julian in his oration "To the uneducated Cynics" (362 CE), cf. Iulian. *Orat.* 6, 5, 32 (185a). On this Heraclitus' fragment, see Guthrie 1962: 416–419.

[29] See Curd 2015 and Tor 2017: 155–308.

Further in Parmenides' poem the noun δίζησις ("inquiry/investigation", derived from the verb δίζησθαι) occurs three times.[30]

> *I shall say... which are the only ways of inquiry* (ὁδοὶ μοῦναι διζήσιός) *for thinking* (νοῆσαι)*: the one which is... and the other which is not... And I am telling you that this (the second) is an utterly inscrutable/unknown path* (παναπευθέα ἔμμεν ἀταρπόν, DK28 B2 = D6 Laks-Most)

Archaic Homeric language is used here to pose deeply epistemological questions on what kind of knowledge can be acquired at all. The adjective παναπευθής (*"utterly inscrutable/unknown/not inquired into"*) is a hapax in Greek. It consists of the first part παν- meaning "totally/utterly" and of the more common adjective ἀπευθής meaning "unknown" derived from the verb πυνθάνομαι "to hear/to learn/to inquire".

In another fragment Parmenides formulates his famously obscure "being is/exists and nothing is not", and then adds:

> *For such is the first way of inquiry* (ἀφ' ὁδοῦ ταύτης διζήσιος)*, from which (I keep) you (away), but then also from this one, about which mortals whilst knowing nothing* (εἰδότες οὐδὲν) *create* (πλάττονται)*, two-headed...* (DK28 B6 = D7 Laks-Most)

In the same fragment but some verses further down, Parmenides criticizes the lack of capacity to judge (κρίνειν), using the passive Homeric adjective ἄκριτος meaning "undistinguishable, confused" actively about people as "being non-critical" (ἄκριτα φῦλα).

The same use of δίζησις ("inquiry/investigation") occurs once again:

> *But you do keep your thought/mind away from this path of inquiry* (τῆσδ' ἀφ' ὁδοῦ διζήσιος) *and do not let much-experiencing habit* (ἔθος πολύπειρον) *force you down onto this path* (DK28 B7 = D8 Laks-Most)

The various kinds or directions of inquiry, which presuppose different destinations, and the question of limits to human capacity to take responsibility for a true destination clearly belong to a strongly established

[30] Mogyoródi 2006: 126.

tradition of epistemological awareness, shared by Parmenides and his audience/readers.

For the concept of inquiry, Heraclitus employs a further Homeric (Ionian) lemma, ἵστωρ "the one who learned/ interested in acquisitive inquiry" (Curd 2015). In Homer and in early epics the word means generally "the one who knows"; Herodotus in the mid and second half of the 5th c. BCE would use the word ἱστορία for "inquiry, written account of inquiries".[31] Heraclitus' fragment says:

> *For men who love wisdom* (φιλοσόφους ἄνδρας) *should be inquirers* (ἵστορας) *into very many things* (εὖ μάλα πολλῶν, DK22 B35 (= D40 Laks-Most)

This interest in research and epistemology is shown by Heraclitus' use of the other verb which is used by Xenophanes meaning "to invent", referring to the acquisition of knowledge. The old Homeric verb ἐξευρίσκειν 'to find out' with the prefix ἐκ-, highlighting the "digging-out-process", is very similar to the verb with the prefix ἐπί- (ἐφευρίσκειν) from the fragment mentioned above. It occurs in Heraclitus in an epistemological context. In fragment DK22 B18 (= D37 Laks-Most) we read: "If one does not expect the unexpected, he won't find it out (οὐκ ἐξευρήσει), because it is not discoverable and not accessible (ἀνεξερεύνητον ἐὸν καὶ ἄπορον)". The object of the verb "to discover" and the adjective "discoverable" is knowledge itself, and the source of knowledge is being investigated.[32]

Heraclitus returns to the nature and acquisition of knowledge elsewhere, though his approach is metaphorical: "Those searching for gold (οἱ διζήμενοι), dig up much earth and discover (εὑρίσκουσιν) little"

[31] *Il.* 18, 501, 23, 886; Hes. *Op.* 792; *Hymn. Hom.* 32, 2; Hdt. 2, 99, 118, 119; 7, 96. Cf. Eur. fr. 910 TrGF as well on somebody who is happy/wealthy as a result of learning (μάθησις) from inquiry (ἱστορία).

[32] On the similar use of the verb "to discover" in the future tense, see Parmenides' verse "for without this being, in which (thinking) is spoken, you will not discover (οὐ... εὑρήσεις) thinking (τὸ νοεῖν, DK28 B8, 35–36 = D8, 40–41 Laks-Most). On the limits of human inquiry see two further much-discussed fragments by Heraclitus DK22 B45 (= D98 Laks-Most): "He would not find out (οὐκ ἂν ἐξεύροιο) the limits of the soul as he goes, who travels every road" (note the epistemological use of the verb "to discover"), and DK22 B40 (= D20 Laks-Most): "Much learning (πολυμαθίη) does not teach understanding (νόον)". See Betegh 2009 and Curd 2015.

(DK22 B22 = D39 Laks-Most)). Both verbs of "searching" (δίζησθαι) and "discovery" (εὑρίσκειν or ἐξευρίσκειν), employed in Archaic epics in the sense "X found Y in the house/ outside etc.", are paired here again as found in Xenophanes.

It is thus evident that at this early stage the interest in acquiring knowledge and questions on the origins, nature and limits of knowledge are present in the language. Nouns such as δίζησις ("inquiry/investigation") and ἵστωρ "the one who is interested in acquisitive inquiry" in Parmenides and Heraclitus signal a certain turn in epistemological thought. Verbs such as "to know" (εἰδέναι, ἐπίστασθαι, γιγνώσκειν), "to examine" (ἐξετάζειν), "to search" (μῶσθαι and ζητεῖν), "to reveal, illumine, discover" (ὑποδείκνυσθαι, δεικνύειν and ἐφευρίσκειν) though used in poetic context in Archilochus, Theognis and Xenophanes, and the verbs "to search" (δίζησθαι) in Parmenides' poem and in Heraclitus' prose fragments and "to discover" (εὑρίσκειν/ἐξευρίσκειν) in Heraclitus have already by this time acquired the essential meaning that will predetermine their future use as technical terms in epistemological and gnoseological scholarship.

2.2 "Discovery" as a Principle of Creative Process: Archaic Texts

Some earlier sporadic poetic fragments reveal the use of the verb "to invent / to discover" referring to (self-referential) intellectual activities. The melody of the song can be analyzed according to the co-occurrence of this particular verb "to invent" with nouns denoting the creative poetic product, such as 'song', 'poem', 'work', 'word', 'verse', 'melody' etc. The co-occurrence is in fact attested in Alcman, the earliest representative of Greek lyrics, a seventh century BCE choral lyric poet from Sparta who qualified his own activity as a composer by means of *sphragis* or a "personal seal" (fr. 39 PMGF):

> *These words and melody Alcman invented* (ϝέπη τάδε καὶ μέλος Ἀλκμὰν εὗρε) *by perceiving the tongued cry of partridges*

This is perhaps the oldest attestation of the verb "to invent" used self-referentially. The 1st person is lacking however: Alcman emphasizes authorship by citing his own name in the 3rd person. Both words and music are the objects of "finding out" by the author. The 'invention' of Alcman is in essence mimetic, he senses birds' voices and invents his songs in imitation of them. In this Alcman is not innovative but follows a standard pattern of Archaic poetry: poetry imitating birdsong.[33]

Stesichorus of Himera (ca. 630 – 555 BCE) slightly altered the model of poetic invention in both verse and music, when he stated in the opening of his *Oresteia* (fr. 212 PMGF = fr. 173 Davies-Finglass), meant for a civic festival (the Doric δαμώματα the scholiasts explained as 'compositions for public performance'):

> *they/we should sing these songs* (δαμώματα) *of beautifully-haired graces, discovering Phrygian melody* (Φρύγιον μέλος ἐξευρόντας) *delicately at the arrival of spring*

Those who sing and invent the melody here are the chorus members who perform the poem.[34] The object of the verb "to invent/discover" is the product Stesichorus is inventing, lyric song/music (μέλος) (as in Alcman above). 'Phrygian' probably suggests one of the modes of music (along with Dorian, Lydian and so on) established due to the spread of the seven-stringed lyre in the 7th c. BCE and which Sophocles in the 5th c. BCE was credited with having introduced into tragedy.[35] The Phrygian mode was considered appropriate for peaceful mental states such as praying, instructing, admonishing (Pl. *Resp.* 399a-b). It was associated with dithyramb (a form of choral performance in honor of Dionysus) and had orgiastic uses denoting "Dionysiac frenzy", being exciting and emotional

---

[33] See Calame 1983: 480–483. Cf. Alcman fr. 40 PMGF "I know (οἶδα) the melodies of all birds".

[34] Willi 2008: 80–81; Carey 2015: 52–53.

[35] Aristoxen. fr. 79. Davies and Finglass 2014: 495. West 1981: 125 n. 73; West 1992: 177–178, 180–181. Alcman mentions "Phrygian melody" as well, cf. Alcm. fr. 126 PMGF.

(Arist. *Pol.* 1342b2–12). Stesichorus here claims to have invented (ἐξευρόντας) such a melody in a soft and delicate way. The poetic text here is perhaps understood to be a divine gift received from the graces, while the melody has to be invented by the poet and his performers. It might be understood, however, that the Muses traditionally provide the 'divine' support, and the poet should invent himself his own 'human' constituent for this performance.

In the second book of the same *Oresteia* Stesichorus probably claimed that the Achaean hero Palamedes had invented (cf. εὑρηκέναι and εὑρετής) the alphabet (fr. 213 PMGF = fr. 175a and 175b Davies-Finglass). Here is a case when we have an evident attestation of interest in the concept of discovery, although we cannot be sure which word exactly was employed by Stesichorus, as both attestations (εὑρηκέναι and εὑρετής) belong to a later scholion paraphrasing Stesichorus.[36]

The "discovery"- process and the location of this process in language is thus crucial in the 7th and 6th c. BCE for the earliest development of concepts of (self)-recognition, analysis, learning and perception.[37]. In both Alcman and Stesichorus the act of exploring and identifying is referred to a specific poetic product, to "these words and melody" (ϝέπη τάδε καὶ μέλος) in Alcman, and to "Phrygian melody" (Φρύγιον μέλος) in Stesichorus. Again, although in a poetic and performative context, epistemological authentication, attested at this early stage of written literature, signals and pre-signals both an important shift taking place in society at large, and the terminological and conceptual background for Classical and Hellenistic scholarly thought.

[36] On Palamedes-myth and Stesichorus' use of it, see Davies and Finglass 2014: 498–500.

[37] Kleingünther calls this process "Entwicklung des Persönlichkeitsgefühls", cf. Kleingünther 1933: 21–24. 24; see also Desclos 2019.

3 "Discovery" as a Principle of Creative Process: Late Archaic Texts

The self-reflexive process of exploring and analysing in addition to an awareness of this process is thus attested in Greek as early as the 6th c. BCE. Attestations increase thereafter, and already for the first half of the 5th c. BCE more passages from both surviving texts and fragments should be considered.

The lyric poet Pindar (ca. 520-446 BCE), whose activity can be situated in the first half of the 5th c. BCE, is to a great extent engaged in epistemological considerations as well as analysis of the role and function of the poet in the society.[38] These include an awareness of poetic self-identification and self-consciousness, and a level of critical judgment with regard to one's own and others' poetry. The relationship between the poet, his audience and his work is reflected upon in Pindar's poetry. Neither a scholar nor a philosopher, Pindar nonetheless contributed greatly to the establishment of critical ideas on the literary text, and to the vocabulary of hermeneutic criticism.

The co-occurrence of the verb "to discover/find out" with nouns denoting poetic production applied to poet's self-referential claim to emphasize rational critical thinking is crucial in Pindar. Thus in his 1st *Olympian* (476 BCE) he states with regard to himself, writing in the 1st person singular (Pindar is perhaps the first attested author who used the verb "to invent" referring to his own creative act in the 1st person): "discovered/invented a road of words to his aid" (ἐπίκουρον εὑρὼν ὁδὸν λόγων, *Ol.* 1, 110). At the very end of his 4th *Pythian* (462 BCE) the poet "discovered/invented a fountain of immortal verses" (εὗρε παγὰν ἀμβροσίων ἐπέων, *Pyth.* 4, 299). In his 6th *Nemean* (460 s BCE?) he says that "the predecessors discovered/invented there/in such things the (poetic) highway (ὁδὸν ἀμαξιτὸν εὗρον); I myself follow alone, making it my concern (ἕπομαι δὲ καὶ αὐτὸς ἔχων μελέταν)" (*Nem.* 6, 53–54). In his 9th

[38] On Pindar's poetics in the historical and cultural context of Archaic Greek poetics as well as on the creation of literary structures in Pindar's work, see Maslov 2015.

Olympian Pindar creates an epithet out of the same verb: "may I be creative referring to verses" (εὑρησιεπής, *Ol.* 9, 80).

The tragic playwright Aeschylus (c. 523-c. 456 BCE) was Pindar's contemporary in Athens and thus participated in then current intellectual trends and discourses. He used the same syntagma in his *Eumenides,* where Athena asks whether the citizens intend to find (εὑρίσκειν) the road of good speech γλώσσης ἀγαθῆς ὁδὸν (Aech. *Eum.* 989).

The same verb emphasized by the prefix ἐκ- meaning "out of, from" in ἐξευρίσκειν, discussed above in Heraclitus' epistemological fragment is used in the 1st person in Pindar's 1st *Pythian* (470 BCE).

Pind. *Pyth.* 1, 60:

> ...*let us invent a friendly hymn* (φίλιον ἐξεύρωμεν ὕμνον) *to Aetna's king*

The object of "finding out" is here Pindar's work titled with the clear generic designation "hymn". From verse 58 on he proceeds to praise Deinomenes, the son of the king of Syracuse Hieron, and this follows on from Pindar's praise of Hieron himself for Hieron's victory in the chariot race and his military glory. The transition from one part of the ode to another, from one praise to another, is highlighted by the self-referential verb of innovation "to invent/discover". The poet has a task to create a new song for a new king. The adhortative use of the subjunctive is on the one hand typical for both choral and monodic song where 1st person singular and plural are interchangeable.[39] At the same time it suggests some common action of the poet and the listener/reader, the reader thus being involved in the creative act.

Xenophanes' verb ἐφευρίσκειν ('to find out'), discussed above and employed probably in an epistemological context, occurs in Pindar where a similar use refers also to mental creativity. Thus in Xenophanes people "discover" what is better because the gods did not show them everything (DK 21 B18 = D53 Laks-Most). In Pindar's *Pythian* 4 (462 BCE) Apollo granted the plain of Lybia to settlers of the future city of Cyrene. These Cyreneans (primarily the king of Cyrene Arcesilaus and his opponent Damophilus) are addressed in the 2nd person plural and called "you who

---

[39] See Lefkowitz 1991 in general and particularly p. 58.

discovered right-counselling craft/wisdom" (ὔμμι... ὀρθόβουλον μῆτιν ἐφευρομένοις, Pind. *Pyth.* 4, 258–262).[40]

Pindar's *Pythian* 12 (490 BCE) is composed for Midas of Akragas, the winner in the flute-playing competition. Midas won the victory in the art/skill, says Pindar, which Pallas Athena invented (τέχνᾳ, τάν ποτε Παλλὰς ἐφεῦρε) by imitation of the lament of Gorgons (Pind. *Pyth.* 12, 6–7). The syntagma "to invent an art" occurs for the first time in Greek here in Pindar and foresees the epistemological usage of this syntagma in Classical Greek later when the arts/skills (τέχναι) start flourishing and the issue of inventing them becomes important (cf. Hipp. *Prisc. med.* 3, 7; *Virg. morb.* 1; Plat. *Phdr.* 273c, *Lach.* 186c, *Minos* 314b). This recalls the famous monologue by Prometheus from the tragedy "Prometheus Bound" ascribed to Aeschylus in antiquity, but composed in all probability in the 430 s BCE by an unknown playwright ([Aesch.] *Prom. Vinct.* 436–506). Prometheus lists there the arts/skills he invented for mortals, and the verb "to discover" referring to Prometheus, is respectively prominent (v. 460 ἐξηῦρον "I invented", vv. 467–468 οὔτις ἄλλος ἀντ' ἐμοῦ... ηὗρε "nobody else but I invented", v. 469 ἐξευρὼν "having invented", vv. 502–503 τίς φήσειεν ἂν πάροιθεν ἐξευρεῖν ἐμοῦ; "who could say to have invented them before me?"). The monologue ends with an emphatic summary: πᾶσαι τέχναι βροτοῖσιν ἐκ Προμηθέως "all skills came to mortals from Prometheus" (v. 506).[41] Thus Pindar's use of the verb "to find out" referring to Athena and flute-playing signals the importance of this co-occurrence.

Aeschylus' tragedy *Palamedes* has survived in fragments. Here a character (in all probability Palamedes himself) claims to introduce the Greeks to the use of number.

Aesch. fr. 181a TrGF

[40] On the introtextual connection of Cyreneans' "finding out the craft" in v. 262 and Pindar's own "finding out a fountain of immortal verses" below in v. 299, in the metapoetic context, see Segal 1986: 160–161.

[41] Jouanna 1990: 39 n. 3. On the striking frequency of the epistemological vocabulary "to search" and "to invent" in Hippocratic treatise "On ancient medicine" (late 5th cent.-early 4th cent. BCE), see Jouanna 1990: 38–40. Cf. Kleingünther 1933: 66–90.

First of all I invented (εὕρηκ') *the most clever number, eminent among devices/methods*

The claim recalls Prometheus' list of his own inventions. It is probable that the author of the tragedy *Prometheus Bound* was inspired by Aeschylus' *Palamedes* (Sommerstein 2000: 121–122). As has been mentioned above, Stesichorus ascribed to Palamedes the discovery of the alphabet. Palamedes, like Prometheus, was credited with important inventions such as writing, military organisation, weights, measures, astronomy and arithmetic. Here the epistemological context is crucial: the verb "to invent/discover" is referred to as a method/skill/device (σόφισμα), number (ἀριθμός) being one of these.

One last example belonging to the period under discussion should be added. Pindar's contemporary lyric poet Bacchylides of Ceos (ca. 516-451 BCE) used the same verb in a paean (of which only a fragment survives).

Bacch. *Paean.* fr. 5 Snell-Maehler

Everyone gets skills (σοφός) *due to another, in the old days and now. For it is not easy to discover/invent the gates of words that have not been said* (ἀρρήτων ἐπέων πύλας ἐξευρεῖν)

The same innovative approach is significant here: the verb "to invent" refers not just to poetic work, but to the innovative power of poetry. The self-reflexive challenge of a poet is "poetry that has not been composed before".

The lyric examples discussed above reveal that the verbs of discovery co-occur with nouns denoting self-referential lyric poetry, consisting always of words and sometimes of music. Aeschylus, who was active at the same time as Pindar and Bacchylides, also used "to discover" referring to speech. In Alcman the object of discovery are these verses/words and song (ϝέπη τάδε καὶ μέλος), further in Pindar and in Aeschylus a road of words (ὁδὸν λόγων), a fountain of immortal verses/words (παγὰν ἀμβροσίων ἐπέων), the (poetic) highway (ὁδὸν ἀμαξιτὸν), a road of good speech (γλώσσης ἀγαθῆς ὁδὸν) all occur together with the verb "to discover/invent". In Bacchylides the object of discovering is the innovative product: the gates of (poetic) words that have not been uttered (ἀρρήτων ἐπέων πύλας).

As follows from the above arguments, epistemological vocabulary increases during the first half of the 5th c. BCE. This happens due to an increasing number of written texts and thus of material to provide information, and to an increasing variety of generic and authorial forms and patterns. The awareness of these forms and patterns is reflected in surviving texts (here Pindar, Bacchylides and Aeschylus) usually with the use of specific vocabulary and repeated syntactic formulas and structural models, and thus can be evaluated. In order to analyse different (fragmentary) texts where the same co-occurrence (the verb "to discover" plus a noun denoting intellectual/creative achievement) takes place, linguistic methods and tools prove their extensive value to the interpretation of these texts.

To conclude with a remark on grammar. All examples in this section contain the verb "to discover", and here the grammatical categories of tense, aspect and mood are important.[42] In Alcman, Stesichorus, Pindar and Bacchylides the aspect is the unmarked aorist (literally 'undefined') tense which is consistently employed (the aorist εὗρε in Alcman and ἐξευρόντας being aorist participle in Greek in Stesichorus, further εὑρόντι, εὑρὼν, εὗρε/ἐφεῦρε, εὗρον and ἐξεύρωμεν in Pindar, the aorist infinitive ἐξευρεῖν in Bacchylides). The aorist usually implies a past event in the indicative, but it does not assert pastness, it expresses the simple occurrence of an action, and can be used of present or future events. The aorist is a standard tense for telling a story, an (entire) action being considered in this case as a single undivided event and not as a continuous event.

Xenophanes, Heraclitus and Aeschylus, however, employ the verb in the present indicative tense (ἐφευρίσκουσιν, εὑρίσκουσιν and εὑρίσκειν) while Heraclitus also in future indicative (οὐκ ἐξευρήσει). The imperfective aspect is used for situations conceived as existing continuously or repetitively as time flows (in Xenophanes "they find out in process of time while searching", ζητοῦντες ἐφευρίσκουσιν the present indicative form

[42] See Moran 2016 as a useful survey for differentiating verbal categories and the 'standard' Goodwin 1875: 7–37.

emphasized by the participle present continuous denoting aspect[43]; in Hercalitus people who search keep on finding (εὑρίσκουσιν) little; in Aeschylus Athena asks whether the citizens intend to find continuously (εὑρίσκειν) the road of good speech). The same continuousness and repetitiveness focused on the future is emphasized through the imperfective aspect and the future tempus form in Heraclitus' claim that one will not find out (οὐκ ἐξευρήσει) the "not discoverable".

In Aeschylus, however, the verb is used in the 1st person singular perfect indicative (εὕρηκα), which represents an act as accomplished at the moment of speaking. Like the aorist tense, the perfect tense denotes perfective aspect; but unlike aorist tense, it describes a state persisting in the present and relates the time of this state to a prior point in time when the state did not exist. The character in Aeschylus (Palamedes?) thus highlights the accomplished and persisting fact of having discovered an intellectual device.

The grammatical form in this case not merely helps explain the meaning of the verb, but contributes to and enriches the interpretation of the claim of the continuous or repetitive process versus an accomplished act of searching and discovering. The aspectual criterion is important in order to differentiate a single occurrence (aorist) from a process and ongoing act (present or future), perfective from imperfective, to disclose a situation from the perspective of the internal temporal structure.

4 Linguistic Analysis and Literary Interpretation: A Case Study

This section will analyse three of Pindar's verses along the lines of the epistemological thought discussed above, with Pindar evaluating his own poetic achievements. The use of recent linguistic insights should enrich our understanding of the text.

---

[43] On the importance of the tense opposition of the present tense "they find out" (ἐφευρίσκουσιν) versus the aorist "they did show" (ὑπέδειξαν), in Xenophanes DK21 B = 8 D53 Laks-Most, s. Schäfer 1996: 123.

In the beginning of his *Olympian* 3 (476 BCE) the poet claims that the Muse has assisted him in his endeavors to compose this new ode in Dorian metre (or musical mode?). The use of the same verb "to invent" is crucial here.

Pind. *Ol.* 3, 4–6

Μοῖσα δ' οὕτω ποι παρέστα μοι νεοσίγαλον εὑρόντι τρόπον
Δωρίῳ φωνὰν ἐναρμόξαι πεδίλῳ
ἀγλαόκωμον...

And thus the Muse stood at some indeterminate place beside me, who invented a newly shining way to fit the sound of glorious celebration to the Dorian sandal...

These three verses represent an example of Pindar's self-evaluation, as they are full of self-referential pointers and allusions to contemporary intellectual streams on both levels of syntax and vocabulary.

The abrupt introduction of the Muse (Μοῖσα) as a poetic authority and source places the poem in the tradition of Archaic poetry with similar invocations appearing from the very beginning of the extant poetic texts.[44] Pindar is aware of the ambiguity of the poet's role: on the one hand it is the Muses who provide material for his poetry, on the other hand he "discovers" poetic devices himself. This poet's role as interpreter of the Muse is claimed by Pindar himself in a fragment

μαντεύεο, Μοῖσα, προφατεύσω δ' ἐγώ

Be a seer, Muse, whilst I will be a prophet (Pind. fr. 150 Snell)

poses him as "a theoretician (on the nature of poetry) as well".[45] The Archaic invocation to the Muse combined with the emphatic 1st person

[44] On poet's relationship with his Muses in Homer, Hesiod, and Homeric hymns and further on in lyric poetry, see Harriott 1969: 10–51 and Campbell 1983: 252–287.

[45] Ledbetter 2003: 62, n. 1. On both vocabulary and content recalling the poetics of Pindar's contemporary Empedocles, see Willi 2008: 235–236. Cf. Empedocles DK31 B3, 3–5 and DK31 B131. See Willi 2008: 249–251 on the Indo-European concept of prophet, poet, and healer. On the poet as interpreter in Pindar, see Ledbetter 2003: 62–68. Pind. *Ol.* 2, 83–88: *I have many swift arrows under my arm in their quiver, that speak to those who understand* (φωνάεντα συνετοῖσιν), *but for the whole they lack interpreters* (ἐς δὲ τὸ πὰν ἑρμανέων χατίζει). The term ἑρμηνεύς for "interpreter" is perhaps innovative in Pindar's time, as it does not occur before the 5th c. BCE.

singular personal pronoun and verb characterize the reciprocity of Pindar's poetic principle.

The creative poetic act is thus claimed to be a combination of the divine inspiration and his own capacities – again an important epistemological self-referential consideration on the nature of creativity which will later come to be one of the leading principles of literary criticism. Thus the Roman poet Horace in his *Ars of Poetry* (ca. 19 BCE) formulates it famously: "It is questioned whether a praiseworthy poem results from nature (*natura*) or skill (*arte*). I do not see any study (*studium*) without a rich vein, nor what rough talent (*ingenium*) brings. Thus, each seeks the aid of the other and swears a friendly pact" (Hor. *Ars Poet.* 408–411).

Further, the particle δέ here, often difficult to translate into modern languages, is important. It has been generally accepted that it "marks no more than a new step, a moment in time at which a new piece of information is activated in his [the narrator's] consciousness. The particle δέ is the most widely used linguistic *boundary marker* between foci of consciousness. And as an observable syntactic cue for such cognitive breaks in our text it is an important element for the study of how consciousness is turned into speech." (Bakker 1997: 63 [author's own italics]). The particle δέ occurs in Pindar, but not frequently, and it is generally used to mark significant boundaries in the discourse.[46]

Thus the combination of the first two words is supposed to create a certain tension in the audience's mind as a combination of a traditional and an Archaic invocation with a new turn in the narrative containing "a new piece of information".

One further word worth noticing is the deictic adverb οὕτω ("thus-there").[47] It belongs to the three-partite group of demonstrative pronouns in Greek, "this-here" (ὅδε), "that-there (with you)" (οὗτος), and "that" (ἐκεῖνος). They all are elements that channel the flow of information between the author and the reader/audience. While the first pronoun

[46] Bonifazi, A., A. Drummen, and M. de Kreij. 2016: https://chs.harvard.edu/CHS/article/display/6205

[47] See the thorough analysis of deictics in Greek in Bakker 2010: 152–161. On Emile Benveniste and Roman Jakobson who emphasized the function of deictics to connect language with discourse, see Godzich 1986: xvi.

(ὅδε) is seen as the deictic of the 1st person, used in a close proximity to the speaker and marking the arrival and onset of new information, and the third pronoun (ἐκεῖνος) is used for what is physically remote and absent for both speaker and his audience, the second pronoun (οὖτος) and the adverb οὕτω here derived from it, can be aligned with the recipient, in the sense that the recipient has better access to the item pointed at than the speaker himself. It is thought to refer back in the text and marks the point where information introduced becomes a basis, from which the discourse can start. In Pindar's verse the thought seems to be: "this is how, you know, recipient, what I mean, the Muse stood beside me". Another important feature of οὖτω here is that in contrast with ἐκεῖνος which frequently refers to concepts and ideas that already exist in the speaker's and audience' mind before the moment of utterance, both οὖτος and ὅδε represent newer and perceptually outstanding items in the context of speech. Both οὖτος and ὅδε tend to occur at moments of discontinuity, either transiting to explicitly announced novel information or referring to a previous topic as a step to a new one.

All this is significant for interpreting the verse. The reference to the Muse as a fact well known to the audience serves to emphasize the transition to the next set of information, the author's discovery of a new device for combining voice with rhythm.

παρέστα supports the divine/natural aspect of poetic creation: the Muse "stood beside" and thus protected the author while he invented a novel turn.

The Pindaric self-referential poet-composer and choral performer are both implied in the personal deixis of the personal pronoun μοι in the Dative singular ("me, myself"), while the poet claims originality for his song: "I invented a newly shining way" (νεοσίγαλον εὑρόντι τρόπον), "new" (νεο-) being significant in νεοσίγαλον.

Further, the much-discussed compound nature and "synthetic effect" of Pindaric performance is significant. Pindar posts a strong personal authorial presence working in the traditional choral genre. As has recently been noted, "in employing this notion, Pindar again emphasizes the effort of fusing, or fitting together, disparate elements" (Maslov 2015: 254). The sound of glorious celebration (φωνὰν ἀγλαόκωμον) refers to the fictionalized generic status of the choral performance. It is the voice/sound of the chorus which has to be tuned to a corresponding harmony (ἐναρμόξαι: ἐναρμόζω and ἐναρμόττω, "to fit or fix in", "to adapt", cf. Pind. *I*. 1.16; τι εἴς τι Pl. *Lg.* 819c). The main message of the passage is thus a personal authorial manifesto with an emphatic role for the authorial "ego" within the framework of the generic tradition in which he works.

The Dorian sandal (Δωρίῳ πεδίλῳ), however, remains unclear. This might be a reference to the poem's Doric dactylo-epitritic metre, although some scholars interpret it as a musical mode (Prauscello 2012: 77–79). Similarly in Aristophanes' comedy *Knights* (424 BCE) the chorus mentions the same construction: "he would fit his lyre only to the Dorian mode" (τὴν Δωριστὶ μόνην ἂν ἁρμόττεσθαι θαμὰ τὴν λύραν, Ar. *Eq.* 989–990). The Aristophanic "to fit one's lyre only to the Dorian mode" looks close to the Pindaric "to fit the sound of glorious celebration to the Dorian sandal" (Δωρίῳ φωνὰν ἐναρμόξαι πεδίλῳ ἀγλαόκωμον). Dorian was considered one of the most important of the melodic patterns (*harmoniai* or *tonoi*) in which Greek music was composed (along with others such as Lydian, Ionian and Phrygian, mentioned above in the fragments of Stesichorus and Alcman).[48] It was named after the Dorian Greeks and attributed to Doric societies and was associated with martial qualities. Applied to a whole octave, the Dorian octave species was built upon two tetrachords (four-note segments) separated by a whole tone.

The 'sandal' recalls the later metaphor of "foot" (πούς) as a unit of length, and finally the basic repetitive rhythmic unit in prosody as well, which has a long history. The Greek term πούς became the Latin term *pes*, which in turn became the English term 'foot'. The 'foot' forms part of a line of verse in most Indo-European traditions of poetry, including English accentual-syllabic verse and the quantitative meter of classical ancient Greek and Latin poetry. The 'foot' is composed of syllables, and is usually two, three, or four syllables in length. The most common feet in English are the iamb, trochee, dactyl, and anapest, while the Greeks recognized three basic types of feet as well, the iambic, the dactylic and the paeonic.[49] The term must have been fixed by the end of the 5th c.

[48] See Sommerstein 1981: 196–197 and West 1981.

[49] The iambic had the ratio of arsis to thesis 1:2, the dactylic had 2:2 and the paeonic had 3:2.

BCE along with other terms for metrics. The 4th cent. BCE Peripatetic Aristoxenus of Tarentum wrote a (now lost) treatise on rhythmics and metrics, where he seems to have specified that 'feet' (πόδες) must be classified according to the number of beats they contain (two, three, four or more).[50]

In Aristophanes' comedy *Frogs* (performed in 405 BCE) the character Aeschylus, dancing while singing, asks Euripides or Dionysus (both are present on stage) "do you see this foot?" (τὸν πόδα τοῦτον; v. 1323). This might represent the incorporation of a developing term for the stage. "Foot" preserves its direct meaning, emphasized by the deictic pronoun "this" and perhaps the actor's gesture pointing to his foot. At the same time it refers "to a physical movement accompanying a certain sequence of syllables" (Dover 1993: 356).[51] The emphatic (repeated four times in two verses) deictic use of the verb of visual perception combined with the personal deixis "you" and the local deixis "this" has a peculiar importance here, as attention is drawn to something physically happening while the metric structure of the verse is actually intended.

The 'Dorian sandal' (πέδιλον) in Pindar thus might symbolize this fusion of choral dancing (movement) and reciting (metrics) and presupposes the future development of the technical prosodic terminology. A footwear metonymy used for a generic designation is common. Thus Aristophanes rendered from buskins (κόθορνοι, *cothurni*) worn by the tragic actors a metonymy of tragedy in the person of Dionysus (Ar. *Ra.* 47, 557). Other characters recognize Dionysus from his buskins, emphatically pointing to them on stage: "you did not expect I would recognize you again, because you had your buskins on" (Ar. *Ra.* 556–557). The slipper (*soccus*) or sandal (*solea*) and buskin were in the post-classical period to become respectively the two ancient metonymies of comedy (slippers and sandals) and tragedy (buskins).[52] Horace's *Art of poetry* (ca. 19 BCE?) can illustrate well the development of the term and the image, the metaphor of foot and the metonymies of tragic and comic footwear. Two genres, tragedy and comedy, are juxtaposed, with two established

[50] Aristox. *Elem. rhythm.* 2, 16–18. See Pearson 1990: 10–13, 59–60.

[51] For the 4th cent. BCE vocabulary cf. for example Pl. *Rep.* 400a, 400c.

[52] The emblems do not mean the exclusiveness of these types of shoes, nor are the terms for each type clear. Tragic actors could be depicted barefoot as well. See the discussion in Stone 1977: 232–253 with bibliography.

(Vergil, Horace, Ovid) (Brink 1971: 168–169) symbols with an iconic status, the socks or slippers of comedy and the buskins or high boots of tragedy, both made for the iambic "foot":

Hor. *AP* 79–82:

> *Madness armed Archilochus with its own iambus; this foot took slippers and big buskins* (hunc socci cepere pedem grandesque coturni), *because it was suitable for dialogue (changing speeches), winning the shouts of the mob and born for a life of action*

The whole spectrum of technical terminology is present here in Horace, but for our understanding of Pindaric verses on the 'Dorian sandal' the metaphor of 'shoe' for the metric 'foot' is important, standing for genre or style or, perhaps, metre. More importantly, Pindar's use of the metaphor of 'sandal' foresees and thus marks a stage in the creation of a certain image in the vocabulary: a 'shoe' for a metric 'foot'. A linguistic perspective on the phenomenon of metaphor includes its use in different types of public discourse, as also in everyday language where it can lose its novelty through frequent use and become conventionalized (or 'dead' such as 'the foot of the mountain'). To quote Mark Johnson, "metaphor is not a mere linguistic entity, but rather a process, by which we experience the world" (Johnson 1981: 15).

Thus the other way to interpret the innovation (νεοσίγαλον τρόπον) in Pindar's *Olympian* 3 can be "to fit the sound of glorious celebration" to the Dorian metric model.

The central verse containing the verb "to invent/discover" (εὑρόντι), discussed above, is significant. The Dative singular of the first person of Pindar's poetic "I" ("me who discovered") highlights the central role of the authorial claim to find out, juxtaposed to the Muse standing beside and supporting him. The double nature of the creative process, a skill supported by talent, is emphasized.

This close-reading-approach to three Pindaric verses, with a special attention to individual items, to the syntax, to the order in which the words unfold content, as well as to formal structures, emphasizes the necessity and inevitability of the combination of linguistic and hermeneutic approaches in interpreting epistemological texts.

5 Conclusion

To sum up, a fruitful interaction between linguistic and hermeneutic analysis is inevitable for any interpretation of Ancient literary texts. As has been argued above, within the context-dependent categories of 'perspective' or 'focalization' the creation of specific perspectives is signaled by linguistic indicators (such as deictics).[53] A linguistic and cognitive perspective on metaphor, accepted now generally as a ubiquitous phenomenon in human discourse, provides a new interpretation of Greek epistemological and poetological statements. The formulaic pattern created by an author in the process of scheduling and arranging words and music must be adapted to the structural characteristics of a language.[54]

Further, the fields of corpus and computational linguistics address fundamental goals of philological research and challenge us to rethink philological structure. For an analysis of the passages presented in this paper a number of research programs were used such as the TLG (*Thesaurus Linguae Graecae*, a large digital corpus of Ancient and Medieval Greek literary texts)[55] and PHI Latin (*Classical Latin Texts*, a digital corpus of Latin literary texts of approximately 350 authors).[56] In order to provide novel research, philologists have to rethink their traditional methods or find new forms for their application with the active use of developing linguistic and computational tools.[57]

In the programmatic volume of 2007 edited by Fritz Hermanns and Werner Holly with the speaking title *Linguistische Hermeneutik*, the necessity of textual understanding on the basis of cognition is theoretically and empirically followed up by a plea for an understanding-oriented

[53] König and Pfister 2017: 136: "It is in particular with the categories of 'perspective' or 'focalization' that a fruitful interaction between literary and linguistic analysis is possible and even called for".

[54] On the literary-linguistic analysis focused on metre and rhythm, see König and Pfister 2017: 15–39.

[55] TLG® research activities combine the traditional methodologies of philological analysis with computational technologies (http://stephanus.tlg.uci.edu/tlg.php, the University of California, Irvine).

[56] PHI Latin is a resource by The Packard Humanities Institute, Los Altos, Santa Clarita, California (https://latin.packhum.org/browse).

[57] On the use of quantitative data in the interpretation of classical texts, see Chronopoulos-Maier-Novokhatko 2017 and Chronopoulos-Maier-Novokhatko 2020.

linguistics. Understanding is the basic condition of any linguistic behavior and linguistics should regard itself as a science of language understanding ("Sprachverstehenswissenschaft"), while linguistic hermeneutics constitutes a chance for linguistics to take its place alongside other sciences of understanding.[58] Linguistics has its basis in communicative – that is necessarily interpretative and hermeneutic – experience. Linguistic hermeneutics is thus meant as an interactive game between openness and definiteness of meaning, a continuous process which proves particularly efficient in its contribution to the understanding of ancient texts.

References

Allan, Rutger J. 2020. Narrative immersion: Some linguistic and narratological aspects. In *Experience, narrative, and criticism in ancient Greece. Under the spell of stories*, ed. J. Grethlein, L. Huitink, A. Tagliabue, 15–35. Oxford: Oxford University Press.

Allan, Rutger, and Michel Buijs, eds. 2007. *The language of literature. Linguistic approaches to classical texts*. Leiden/Boston: Brill.

Bakker, Egbert. 1997. *Poetry in speech: Orality and Homeric discourse*. Ithaca/London: Cornell University Press.

———. 2010. Pragmatics: Speech and text. In *A companion to the ancient Greek language*, ed. by E.J. Bakker, 151–167. Malden/Oxford: Blackwell.

Bamman, David, and Gregory Crane. 2010. Corpus linguistics, treebanks and the reinvention of philology. In *Proceedings of INFORMATIK 2010: Service science*, Leipzig, vol. 2 of *Lecture notes in informatics*, ed. K.-P. Fähnrich and B. Franczyk, 542–551. Leipzig, Germany.

Betegh, Gábor. 2009. The limits of the soul: Heraclitus B45 DK. Its text and interpretation. In *Nuevos ensayos sobre Heráclito*, ed. E. Hülsz, 391–414. Mexico City: Universidad nacional autónoma de México.

Biere, Bernd. 2007. Linguistische Hermeneutik und hermeneutische Linguistik. In *Linguistische Hermeneutik. Theorie und Praxis des Verstehens und Interpretierens*, ed. F. Hermanns and W. Holly, 7–21. Tübingen: Niemeyer.

[58] Hermanns and Holly 2007: 2 with a vast bibliography on the subject. See also Biere 2007 and Fix 2010: 34.

Bonifazi, Anna, Annemieke Drummen, and Mark de Kreij. 2016. *Particles in ancient Greek discourse: Five volumes exploring particle use across genres,* Hellenic studies series 74. Washington, DC: Center for Hellenic Studies. https://chs.harvard.edu/CHS/article/display/6205

Bossi, Francesco. 1990. *Studi su Archiloco*. 2nd ed. Bari: Adriatica.

Bowra, Cecil. 1940. The fox and the hedgehog. *The Classical Quarterly* 34 (1/2): 26–29.

Brink, Charles. 1971. *Horace on poetry. The 'Ars Poetica'*. Cambridge: Cambridge University Press.

Brinker, Klaus; Cölfen, Hermann; Pappert, Steffen: Linguistische Textanalyse: eine Einführung in Grundbegriffe und Methoden 8., neu bearb. und erw. Aufl., Berlin: Erich Schmidt Verlag, 2014 (Grundlagen der Germanistik; 29)

Calame, Claude. 1983. *Alcman, Introduction, texte critique, témoignages, traduction et commentaire*. Roma: Edizioni dell' Ateneo.

Calame, Claude. 2019. Poèmes "Présocratiques" et formes de poésie didactique, quelle pragmatique? in M.-L. Desclos (ed.), La Poésie archaïque comme discours de savoir, Paris, 53-72.

Campbell, David. 1983. *The golden lyre: The themes of the Greek lyric poets*. London: Duckworth.

Carey, Chris. 2015. Stesichorus and the epic cycle. In *Stesichorus in context*, ed. P.J. Finglass and A. Kelly, 45–62. Cambridge: Cambridge University Press.

Chang, Kevin. 2015. Philology or linguistics? In *World philology*, ed. S. Pollock, B.A. Elman, and K.K. Chang, 311–331. Cambridge, MA: Harvard University Press.

Chronopoulos, Stylianos, Maier, Felix K., und Novokhatko, Anna. 2017. Quantitative Daten und hermeneuztische Verfahren in den „digital classics". In: Schweiker M., Hass J., Novokhatko A., Halbleib R. (eds) Messen und Verstehen in der Wissenschaft. J.B. Metzler, Wiesbaden, 57–68

——— (Hrsg.). 2020. Digitale Altertumswissenschaften:: Thesen und Debatten zu Methoden und Anwendungen, Heidelberg: Propylaeum (Digital Classics Books, Band 4).

Curd, Patricia. 2015. Thinking, supposing, and physis in Parmenides. *Études platoniciennes Online* 12. http://journals.openedition.org/etudesplatoniciennes/741

Davies, Malcolm, and Patrick Finglass. 2014. *Stesichorus: The poems*, intr., ed., comm. Cambridge: Cambridge University Press.

De Beaugrande, Robert, and Wolfgang Dressler. 1981. *Introduction to text linguistics*. London: Routledge.
De Man, Paul. 1986. *The resistance to theory*. Minneapolis: University of Minnesota Press.
Desclos, Marie-Laurence. 2019, Le tissage des savoirs et de la langue dans la poésie archaïque, in M.-L. Desclos (ed.), La Poésie archaïque comme discours de savoir, Paris: Classiques Garnier, 19–51.
Detienne, Marcel, and Jean-Pierre Vernant. 1978. *Les ruses de l'intelligence. La mètis des Grecs*. Paris: Champs Essois.
Dover, Kenneth. 1993. *Aristophanes' frogs*. Clarendon: Oxford University Press.
Fix, Ulla. 2010. Literary studies and linguistics. The "LiLi" project in a contemporary linguistics perspective. *Journal of Literary Theory* 4/1: 19–40.
Ford, Andrew. 1985. The seal of Theognis: The politics of authorship in archaic Greece. In *Theognis of Megara: Poetry and the polis*, ed. T.J. Figueira and G. Nagy, 82–95. Baltimore/London: Johns Hopkins University Press.
Garrett, Matthew. 2018. *The Cambridge companion to narrative theory*. Cambridge/New York: Cambridge University Press.
Gerson, Lloyd. 2009. *Ancient epistemology*. Cambridge: Cambridge University Press.
Godzich, Wlad. 1986. The tiger on the paper mat. In *The resistance to theory*, ed. P. De Man, ix–xviii. Minneapolis: University of Minnesota Press.
Goodwin, William. 1875. *Syntax of the moods & tenses of the Greek verb*. Boston: Ginn and Company.
Guthrie, William. 1962. *A history of Greek philosophy. Vol. 1: The earlier Presocratics and the Pythagoreans*. Cambridge: Cambridge University Press.
Hamilton, John T. 2018. Philology of the Flesh. Chicago: Chicago Univ. Press.
Harriott, Rosemary. 1969. *Poetry and criticism before Plato*. London: Methuen.
Heitsch, Ernst. 1994. *Xenophanes und die Anfänge kritischen Denkens*, Akademie der Wissenschaften und der Literatur. Stuttgart: F. Steiner.
Hermanns, Fritz, and Werner Holly, eds. 2007. *Linguistische Hermeneutik. Theorie und Praxis des Verstehens und Interpretierens*. Tübingen: Walter de Gruyter.
Hogg, Richard. 1994. Linguistics, philology, chickens and eggs. In *English historical linguistics: Papers from the 7th international conference on English historical linguistics*, ed. F. Fernández, M. Fuster and J. José Calvo, Valencia, 22–26 September 1992, Amsterdam; 3–16.
Janich, Nina (Hg.) 2008. Textlinguistik. 15 Einführungen. Narr Verlag, Tübingen.

Johnson, Mark, ed. 1981. *Philosophical perspectives on metaphor*. Minneapolis: University of Minnesota Press.
Jouanna, Jacques. 1990. *Hippocrate 2,1. De l'Ancienne médecine; texte établi et trad*. Paris: Les Belles Lettres.
Kleingünther, Adolf. 1933. *ΠΡΩΤΟΣ ΕΥΡΕΤΗΣ. Untersuchungen zur Geschichte einer Fragestellung*. Leipzig: Dieterich.
Koerner, Konrad. 1997. Linguistics vs philology: Self-definition of a field or rhetorical stance? *Language Sciences* 19 (2): 167–175.
König, Christoph. 2009. *Das Potential europäischer Philologien: Geschichte, Leistung, Funktion*, Philologien. Theorie – Praxis – Geschichte. Göttingen: Wallstein Verlag.
König, Ekkehard, and Manfred Pfister. 2017. *Literary analysis and linguistics*. Berlin: Erich Schmidt Verlag.
Laks, André, and Glenn Most. 2016. *Early Greek philosophy. Vol. 3: Early Ionian thinkers*, Loeb classics library. Cambridge, MA/London: Harvard University Press.
Lampe, Geoffrey. 1961. *A patristic Greek lexicon*. Oxford: Oxford University Press.
Lanata, Giuliana. 1963. *Poetica pre-platonica*. Firenze: La Nuova Italia.
Ledbetter, Grace. 2003. *Poetics before Plato: Interpretation and authority in early Greek theories of poetry*. Princeton: Princeton University Press.
Lefkowitz, Mary. 1991. *First-person fictions: Pindar's poetic "I"*. New York: Oxford University Press.
Lesher, James. 1991. Xenophanes on inquiry and discovery: An alternative to the "hymn to progress" reading of fr. 18. *Ancient Philosophy* 11: 229–248.
———. 1992. *Xenophanes of Colophon: Fragments*. Toronto: University of Toronto Press.
———. 2013. A systematic Xenophanes? in: Early Greek Philosophy: the Presocratics and the emergence of reason, ed. by. J. McCoy, CUA, Washington D.C.: 77–90.
Liveley, Genevieve. 2019. Narratology. Oxford Univ. Press.
Maslov, Boris. 2015. *Pindar and the emergence of literature*. Cambridge: Cambridge University Press.
Mogyoródi, Emese. 2006. Xenophanes' epistemology and Parmenides' quest for knowledge. In *La costruzione del discorso filosofico nell' età dei Presocratici*, ed. M.M. Sassi, 123–160. Pisa: Edizioni Della Normale.
Momma, Haruko. 2013. From Philology to English Studies. Language and Culture in the Nineteenth Century. Cambridge: Cambridge University Press.

Moran, Jerome. 2016. Tense, time, aspect and the ancient Greek verb. *The Journal of Classics Teaching* 17 (34): 58–61.

Nagy, Gregory. 1989. Early Greek views of poets and poetry. In *Classical criticism*, ed. G. Kennedy, vol. 1, 1–77. Cambridge: Cambridge University Press.

Pearson, L. 1990. *Aristoxenus, Elementa rhythmica. The fragment of book II and the additional evidence for Aristoxenean rhythmic theory*. New York: Oxford University Press.

Pollock, Sherman, Benjamin Elman, and Kevin Chang. 2015. *World philology*. Cambridge, MA: Harvard University Press.

Prauscello, Lucia. 2012. Epinician sounds: Pindar and musical innovation. In *Reading the victory ode*, ed. P. Agócs, C. Carey, and R. Rawles, 58–82. Cambridge: Cambridge University Press.

Saussure, Ferdinand. 1922 [1916]. *Cours de linguistique générale*. Paris/Lausanne: Payot.

Schäfer, Christian. 1996. *Xenophanes von Kolophon: ein Vorsokratiker zwischen Mythos und Philosophie*. Stuttgart/Leipzig: Springer.

Segal, Charles. 1986. *Pindar's mythmaking: The fourth Pythian ode*. Princeton: Princeton University Press.

Sommerstein, Alan, ed. 1981. *The comedies of Aristophanes. Knights*. Liverpool, UK: Liverpool University Press.

———. 2000. The prologue of Aeschylus' *Palamedes*. *RhM* 143: 118–127.

Stone, Laura. 1977. *Costume in Aristophanic comedy*. Dissertation, University of North Carolina, Chapel Hill.

Swift, Laura, 2019. Archilochus: The poems. Introduction, text, translation, and commentary. Oxford Univ. Press.

Thomas, Richard. 1990. Past and future in classical philology. *Comparative Literature Studies* 27 (1): 66–74.

Tor, Shaul. 2017. *Mortal and divine in early Greek epistemology. A study of Hesiod, Xenophanes and Parmenides*. Cambridge: Cambridge University Press.

Van Groningen, Bernhard. 1966. *Theognis, le premier livre, édité avec un commentaire*. Amsterdam: N.V. Noord-Hollandsche Uitgevers Maatschappij.

West, Martin. 1981. The singing of Homer and the modes of early Greek music. *Journal of Hellenic Studies* 101: 113–129.

———. 1992. *Ancient Greek music*. New York: Oxford University Press.

Willi, Andreas. 2008. *Sikelismos: Sprache, Literatur und Gesellschaft im griechischen Sizilien (8.-5. Jh. v. Chr.)*. Basel: Schwabe Verlag.

Woodbury, Leonard. 1991. Poetry and publication: Theognis 769–772. In *Collected writings*, ed. L.E. Woodbury, C.G. Brown, R.L. Fowler, E.I. Robbins, and P.M. Wallace Matheson, 483–490. Atlanta: Scholars Press.

Ziolkowski, Jan. 1990a. *On philology*. University Park: The Pennsylvania State University Press.

———. 1990b. Introduction, in: Ziolkowski 1990a, 1–12.

13

On the History of Models in American Linguistics

Jacqueline Léon

1 Introduction

The debate on the use of models has recently known a revival of interest in the history of sciences, considering that the term 'model' is now massively used in linguistics as well as in other human sciences, with considerable variations of usages and meanings (see, e.g., Armatte and Dalmedico 2004; Blanckaert et al. 2016; Varenne 2018).

In linguistics, the emergence of the term and the notion of model is a recent phenomenon dating back to the 1940s–1950s. It concerns

A first version of this chapter was given as a plenary talk at the Colloquium of the Henry Sweet Society (University of Edinburgh, 5–7 September 2019).

I am grateful for the helpful comments of Ryan Nefdt on an earlier version of this paper. Obviously, remaining errors of facts and interpretations in the text are entirely my own.

J. Léon (✉)
Laboratoire d'Histoire des Théories Linguistiques (CNRS, Université de Paris, Université Sorbonne Nouvelle), Paris, France
e-mail: jacqueline.leon@u-paris.fr

R. M. Nefdt et al. (eds.), *The Philosophy and Science of Language*,
https://doi.org/10.1007/978-3-030-55438-5_13

American linguistics and is contemporary with its mathematization.[1] It should be said that the apparition of the term 'model' cannot be confused with the formalization and mathematization of linguistics since some linguists made an early use of it in a non-formal sense, as we will see.[2] However, from the 1950s, the notion of model was stabilized when the formalization and mathematization of language led to the rise of generative grammars and, in the 1960s to the development of computational linguistics.

The process of formalization and mathematization of linguistics is very well documented by Tomalin (2006). At the beginning of the twentieth century, Hilbert's proof theory had a profound impact on linguistic research, to begin with Bloomfield's axiomatic method as early as 1926.[3]

In the 1930s, the works of the Polish school of logic and the Vienna Circle, especially Carnap's, were very influential on Goodman, Quine and Bar-Hillel, who, in turn, influenced Harris and Chomsky. In 1953, Bar-Hillel's article on recursion is an important milestone. He argues that recursive definitions can be used not only in mathematics but also in empirical sciences, namely linguistics. Chomsky's notion of generative grammar was also greatly influenced by Post's work on recursively enumerable sets (see also De Mol 2006, Pullum 2011). Thus in the 1950s, the use of the axiomatic-deductive method, recursive definitions and logical categories were relatively well spread in linguistic research, at least among some linguists.

The rise of the term and notion of 'model' in linguistics also owes to Model theory, first applied in physics and later used in other sciences. Model Theory was introduced in linguistics by the works of Tarski, Carnap and Quine who played a crucial role, notably for the development of the logical model of syntax.[4] These works were well known to

[1] In European linguistics, first occurrences of the term 'model' appeared in the late 1960s in the wake of the spread of Information theory (Malmberg, 1968). Neither Saussure, Trubetzkoy, Hjelmslev, Jespersen, Bally, Meillet, Tesnière, Martinet and Firth used 'model' in another sense than 'copy' or 'analogy'.

[2] See Nefdt (2019: 1672) on the difference between formalization and mathematisation.

[3] Although, as Seuren (2009: 104) notes, Bloomfield's method in sets of postulates (1926) is not axiomatic in the technical sense.

[4] Although the term Model theory, at the intersection of logic and mathematics, was coined by Tarski only in 1954, the basis of Model theory dated back to 1915 with the emergence of metal-

some American linguists, and, from the 1940s onward, the use of the term 'model' in linguistics had become connected with logical mathematics.

In this chapter, I will first address the early uses of the term 'model' in American linguistics and examine how 'model' took a mathematical turn gradually and contributed to the rise of generative grammars. Three linguists played a crucial role in this modeling turn: Zellig S. Harris (1909–1992), Charles F. Hockett (1916–2000) and Noam Chomsky (b. 1928). All three of them were aware of the work of Tarski, Carnap, Goodman, Quine and Post.[5] Harris and Chomsky had close personal relationships with Yehoshua Bar-Hillel.

As everyone knows, these authors were aware of each other's works, most notably Harris was Chomsky's PhD supervisor. Harris and Hockett exchanged letters in 1950–51.[6] They wrote reviews on one another's work, so that the term 'model' became an underlying issue for controversies leading them to settle theoretical scores as well as to stabilize the notion of model. I will examine to what extent the use of different meanings of 'model' was a controversial issue, and how it evolved into the emergence of a common approach to models and modeling in linguistics in the 1960s.

It is beyond the scope of this chapter to give a definition of the notion of model and to account for the copious literature on the topic. However it is necessary to provide some benchmarks to identify how the linguists used the term 'model' in the pioneering period of the mathematization of linguistics. The uses of 'model' are various, and the early ones pertains to ordinary language: model-as-draft, model-as-exemplar, model-as-copy, model-as-analogy, model-as-paradigm, axiomatic-deductive model, model-as-diagram, model-as-representation. With the process of

ogic, and Tarski himself began to discuss related questions as soon as 1926 (Vaught "Model Theory", 159).

[5]Tomalin, *Linguistics*, 103–104 recalls that Harris was well aware of the sources of the mathematical description of language as indicated in his last book *Theory of Language and Information* published in 1991. Even if he may not have read all these works in the 1940–50s, Harris (1991, 145) mentions the Polish School of logic (Lukasiewicz, Lesniewski and Ajdukiewicz), the Intuitionist mathematicians as Brouwer, the constructionist techniques of Post and Gödel, Quine and Goodman, the Turing machine and automata theory.

[6]On all these points see Barsky, *Zellig Harris*.

mathematization, the sense of 'model' had gradually joined the meaning given to it by historians of science. For them (Israel 1996, Sinaceur 1999, Armatte and Dalmedico 2004), the origin of model is the mathematical modeling of physics after the 1920s which is distinct from other forms of mathematization such as formalization and axiomatisation. Modeling consists in finding the mathematical expressions which represent a physical process schematically and analogically.

The mathematical model acts as an intermediary instance and a representation for empirical validation of a theory.[7] Other forms than mathematical models exist which do not correspond to this definition. In the case of engineering sciences, operational research, for example, models are imperfect and reducing but they are effective tools for analyzing reality. In some cases, they supply when observation is impossible. (Sinaceur 1999) evokes the original meaning of 'model', namely that of mock-up draft on which it is easier to make calculations and measurements than the original. She emphasizes the status of models as artificial objects, their role as intermediaries, their heuristic and representational function. They assume the role of intermediary between an enigmatic situation and the questions we addressed to solve the enigma and understand the situation. They also assume a function of theoretical or technical knowledge.[8] For Israel (1996), one of the essential components of models is analogy: a model imitates the phenomenon studied.

Considering models in current linguistics, especially grammars, Nefdt (2019) distinguishes three grades of mathematical involvement in linguistics arguing that the first one, where the mathematics are a modeling tool, is more adapted to linguistics. He considers that grammars are closer to mathematical models than to theories because "the mathematics involved in grammar construction is merely a helpful aid and not directly

---

[7] "La modélisation mathématique de la physique [...] est assez proche de la conception d'un modèle qui « représente' un réel capturé à la fois par une théorie et par une observation quantifiée, en bref d'un modèle qui joue le rôle d'instance intermédiaire de validation empirique d'une théorie" (Armatte et al., "Models", 244).

[8] "Tout matériel qu'il puisse être, un modèle n'est pas un objet réel, mais un objet artificiel qui appartient au registre de l'invention. C'est un intermédiaire entre une situation qui nous paraît énigmatique et les questions que nous posons pour tâcher de réduire l'énigme et comprendre la situation. Les modèles assument donc une fonction de connaissance théorique ou technique." (Sinaceur, "Modèle", 649).

structurally committing to the target system" (p. 1685). His notion of model meets that defined by historians of science: grammars are abstract objects like mathematical models; models are representations that allow to empirically account for complexity. Models and grammars are theoretical intermediaries.

On the methodological level, this study on the emergence of the notion from the apparition of the term 'model' widely relies on the use of the database CTLF (Corpus des Textes Linguistiques Fondamentaux) overseen by Bernard Colombat and Arnaud Pelfrêne, comprising digitized versions of texts by Bloomfield, Boas, Harris, Hockett and Chomsky, among others.

2 Early Uses of 'Model' in American Linguistics

2.1 Bloomfield's Models as Analogies and Grammatical Paradigms

As early as *Language* (1933), Bloomfield uses the term 'model' when he deals with analogies in grammatical paradigms. In the quotation below, 'model' refers to grammatical paradigms:

The model set *(sow: sows)* in this diagram represents a series of models (e.g. *bough: boughs, heifer: heifers, stone: stones,* etc.), which, in our instance, includes all the regular noun-paradigms in the language (Bloomfield 1933, 405).

That use was taken up by his pupils and disciples. See for example Wells (1947) when defining immediate constituents. He uses 'model' in a theoretical way:

> The leading idea of the theory of ICs here developed is to analyse each sequence, so far as possible, into parts which are expansions; these parts will be the constituents of the sequence [...]If A is an expansion of B, B is a model of A. (Wells 1947, 83)

Wells gives the example of *oldish* as an expansion of the model *old.*

2.2 Bloomfield's Models and Sapir's Patterns, Patterning and Configurations

In parallel with 'model' in Bloomfield, the terms 'pattern', 'patterning' and 'configuration' had already appeared in American linguistic terminology, found in work by Sapir.

In the 1940s, following Sapir, many linguists used the term 'pattern' to designate what Bloomfield named 'models', such as Pike, who uses 'patterns' to name Bloomfield's analogies, i.e. grammatical paradigms, in his article on 'Taxemes and immediate constituents':

> Analogies (or patterns) are used to build up new utterances, as in the sample (276) *dog: dogs, radio: radios, pickle: pickles* for in analogic change. (Pike 1943, 80)

2.3 Sapir Revisited by Harris: Process Models

It is important to note that neither Boas nor Sapir used the term 'model'. But Sapir's patterns and configurations led Harris to discuss new forms of linguistic models. Harris's stance concerning 'model' is particularly interesting as it evolves from an unstabilized use to a use inspired by model theory and certain forms of mathematisation.

Just a word about Sapir's patterns. Among various commentaries on Sapir's work, let us mention Pottier and Dreyfus (1970, 98) who, while assimilating the terms 'pattern', 'patterning' and 'configuration' to 'structure', insists on the interchangeable use Sapir made of them and on the variations of meanings they underwent over time.

If we follow Fortis (2015, 155), Sapir refers to several grammatical entities when he uses the term pattern. And these entities varied over the years from 1921 to 1931: classification of nouns in singular and plural; system of vocalic alternance in English (*goose-geese, sing-sang-sung*), pronominal verbs in French etc.

In spite of this vagueness of the notion of 'pattern', Harris gave it a significant role in American linguistics of the 1930s. In his review of Newman's Yokuts grammar, Harris (1944), uses the term 'model' with two different meanings, one of which concerns the relevance of Sapir's patterns as models.

(i) model = a copy, something one imitates or represents.
(ii) model = deductive model, inherited from logical mathematics.

type (i) 'In several places, the model is that of a living organism' (Harris 1944, 208).

type (ii) 'model' qualifies the deductive method of scientific description, used by Sapir to deal with configuration and pattern. In this text, Harris defines 'deductive method' without referring to the axiomatic-deductive method inherited from Hilbert. He simply defines it as working from general to particular and where the general relation controls the particular relation. Particular events are described in terms of general systemic relations.

Harris does not say that Sapir's patterns are models, but that Sapir, dealing with patterns and configuration, used a deductive model, that is a specific deductive method.

It is the method used by Newman in his Yokuts grammar from which Harris extracts examples. Sapir's patterns and configurations, as used by Newman, make it possible to describe particular events in terms of general systemic relations. With the example below, extracted from Newman's grammar, Harris points out that the use of a protective form, here the use of a vowel to separate two consonants in a morpheme, constitutes a deduced and general form protecting a general pattern:[9]

Both Newman's method and the alternative methods indicated above are essentially similar in that they describe particular events or relations in terms of general systemic relations. This was indeed the great contribution of Sapir's talking about configuration and pattern. [...] Newman can say that in *pilaw* (*in the road)* there has been added to the morpheme *pil- (road)* and *-w* (*in)* a protective vowel [*a*] in order to preserve the general pattern of not more than one consonant finally (§ 22:3). Today, the tendency is to use as model the deductive system of scientific description,

[9] Note that in *Structural Linguistics*, Harris (290) rephrases the example extracted from Newman's grammar in terms of model. The example that time deals with two consecutive vowels that should be separated by a protective consonant.

so that we might say that no final clusters occur and deduce from that that when the morpheme *pil-* occurs before the morpheme *-w* it has a phonemic form ending in a vowel. A protective form protects a general pattern; a deduced form is a special case of a general statement, which statement would not have been true if the actually occurring forms had not fitted in with it. (Harris 1944, 199).

In the same text, along with model, appears the idea of 'style' concerning methods of research. Sapir's method is considered a heuristic device that each researcher is likely to appropriate. 'Style' and 'model' are equivalent.

In *Structural Linguistics* (1951a), Harris uses the term 'model' in yet another sense, that of 'representation' this time. Model is a diagram. One can observe that the notion is not yet stabilized as the term refers to different entities:

> The most detailed diagram or model may state the occurrence of each actual morpheme sequence. (Harris 1951a, 351)

> by the construction of geometric models (diagrams). (Harris 1951a, 373)

The same year, in 1951, Harris publishes a long review of the *Selected Writings of Edward Sapir* in *Language*, where he develops his idea of process model concerning Sapir's method. However, he said that patterning of data is Sapir's greatest contribution to linguistics, "patterning" meaning linguistic context, that is distribution:

> Sapir's greatest contribution to linguistics, and the feature most characteristic of his linguistic work, was not the process model but the patterning of data. [...] he pointed out that what is linguistically significant is not what sounds are observed in a given language but under what linguistic circumstances (i.e. in what distribution) those sounds occur. (Harris 1951b, 292)

In the second extract, Harris characterizes Sapir's patterning, not only as distribution but as how distribution can be identified from the context of use of language. Now, Harris takes use into consideration:

> Sapir's patterning is an observable (distributional) fact which he can discover in his data and from which he can draw those methodological and psychological considerations which he cannot observe directly, such as function and relevance, or perception and individual participation. He can the more readily do this because his patterning is established not directly on distributional classification but on an analysis in depth of the way in which the various elements are used in the language. (Harris 1951b, 294)

Still according to Harris, Sapir's approach is postulational, that is derived from axioms, and dynamic. Patterns are the results of processes within descriptive linguistics, that is the result of a process model, comprising postulates:

> Sapir puts the essential statements of modern linguistics in postulational or definitional form: [...] Phonemes are presented not as a classification of phonetic events or types, but as the result of a process of selection [...] Sapir thus sees the elements of linguistics and the relations among them as being the results of processes in language. [...] *Process or Distribution*. Sapir, however, also used this model of an 'entity as a result of process' within descriptive linguistics proper. (Harris 1951b, 289–290)

For Harris, the process model opens the way to a more subtle descriptive analysis:

> It has the greater advantage of opening the way to a more subtle descriptive analysis – something always dear to Sapir's heart – by giving a special secondary status to some parts of the descriptive structure. (Harris 1951b, 290)

From the point of view of the history of science, the process model constitutes a stage for separating descriptive method from historical analysis and from the older psychologizing of grammatical forms. Second, it makes it possible to distinguish morphological from historical analysis, contrary to what 'older' grammars did.

> [...] it [the process model] also occupies a determinate position from the point of view of the history of science. It seems to constitute a stage in the separation of descriptive method both from historical analysis and from

> the older psychologising of grammatical forms. [...] Finally, the older grammars frequently failed to distinguish morphological from phonological considerations [...] The formulations in terms of process give expression to all this while at the same time separating descriptive linguistics from the rest. (Harris 1951b, 291)

Here, Harris points out works anterior to those of Bloomfield. As for Bloomfield's works, he says, they constitute an intermediary phase:

> [Bloomfield] 'presents phonemes no longer as the result of process but as direct classification, whereas the morphology is still largely described in terms of process.' (Harris 1951b note 8, 291)

Finally, from the point of view of mathematisation, Sapir's views on meaning comprise a formal validity of the same order as logic and mathematics:

> All these investigations involving meaning, when carried out with the kind of approach that Sapir used, have validity and utility. The formal analysis of language is an empirical discovery of the same kinds of relations and combinations which are devised in logic and mathematics; and their empirical discovery in language is of value because languages contain (or suggest) more complicated types of combination than people have invented for logic. (Harris 1951b, 301)

To sum up that section on early uses of 'model', we saw how 'model' evolved from Bloomfield's grammatical paradigms and Sapir's patterning to Harris's diagrammatic representations. Harris played a significant role in pointing out the advantages of a process model that made distributional analysis and descriptive linguistics more dynamic. He sees in Sapir's process model an important stage in the history of linguistics: first by separating descriptive analysis from both historical method and psychological analysis of grammatical forms; second by separating morphology from phonology.

3 Markov Models: From Information Theory to Hockett's General Headquarter

Shannon's use of Markov processes is one of the main ways in which mathematical models have been introduced into linguistics. In order to define stochastic processes, Shannon & Weaver use the term 'model' only once in their 1949 book *The mathematical Theory of Communication.* 'Model' refers to a mathematical model defining stochastic processes:

> A physical system, or a mathematical model of a system which produces such a sequence of symbols governed by a set of probabilities, is known as a stochastic process. (Shannon and Weaver 1949, 5)

It is the first time that Markov processes are used outside of mathematics and that the term 'model' is used to name Markov processes (Léon 2015). Recall that the aim of the Russian mathematician, Andrej Andreevič Markov (1856–1922), attempted to search probabilities constancies by studying the succession of vowels and consonants in a text, Pushkin's Eugene Onegin. His method, called Markov's chains, is a finite state automata where transitions from one state to another are ruled by probabilities.

Shannon, within the framework of Information theory, developed a probabilistic model of sequences of letters and words in English, based on Markov chains.[10] With this unique occurrence, Markov's processes, hitherto a mathematical entity, felt outside of the proper framework of mathematics. They endorsed heuristic functions, likely to describe, represent and explain systems from other sciences, the theory of communication in this case.[11] This use of model is close to one of the definitions given by historians of science for mathematical physics (see introduction above):

[10] For example, if a text has 100 occurrences of 'th', with 60 occurrences of 'the', 25 occurrences of 'thi', 10 occurrences of 'tha', and 5 occurrences of 'tho', Markov model predicts that the next letter which follows the digram 'th' is 'e' with a probability of 3/5; 'i' with a probability of 1/4; 'a' with probability of 1/10, and 'o' with probability of 1/20. In other words, Markov model *can be used to approximate natural languages such as English. In our example, it is second-order approximation: the symbol probabilities are dependent upon the previous two symbols.*

[11] As Segal (*Zéro*, 308) points out, in Europe 'information theory' refers to both theory of communication and cybernetics while in the USA, it is cybernetics that encompasses the other two.

'Modeling' consists of using mathematical expressions which represent a physical process schematically and analogically. One can consider that in Shannon and Weaver's book, 'model' refers to the communication system conceived of as a physical system providing sequences of symbols ruled by probabilities, which can be represented by a mathematical entity, i.e. stochastic process.

With the general success of Information theory – sometimes an excessive success[12] – Markov's model became very popular among American linguists, as a way in which mathematical models could be applied to linguistics.

Hockett was one of the first linguists who introduced Information theory into linguistics.

He wrote two articles in the early 1950s on that subject before using Information theory more largely in his work. Hockett (1952) uses the term 'model' for the first time by applying Information theory to linguistics as well as some of its concepts, such as redundancy, entropy and semantic noise.[13] Actually, model corresponds to the diagram of communication defined by Shannon: a transmitter sends a coded message through a canal to a receiver, so that his conception of model corresponds to the classical notion of application of mathematics to physics, here the theory of communication (or Information theory). With the quantification of semantic noise, he studies the discrepancy between the hearer and the speaker who nevertheless succeed in understanding each other.

Hockett (1953) is a substantial review of Shannon and Weaver's book where he warns against excessive applications of Information theory to linguistics, in particular against the polysemic uses that linguists made of some of its notions, such as redundancy. He uses the term 'model' only to refer to his 1952 article and qualifies his own use of Information theory in this article as a 'highly oversimplified model'. What matters to him is

[12] Shannon "Bandwagon", 3, criticizes all the scientists eager 'to jump in the bandwagon' at all costs.
[13] Quantification of semantic noise is 'the probability of misunderstanding any given message' (Hockett "Semantic noise", 260).

the compatibility between linguistics and mathematics. He suggests modifying Shannon's mathematical machine in order to account for certain linguistic phenomena.

4 Debates on Model as a Theoretical Notion

From Hockett's article "Two Models of Grammatical Description" (1954), the notion of 'model' started to be discussed as a theoretical notion.[14] It is probably the first time that 'model' appears in the title of a linguistics article.

Recall that in this article, Hockett discusses the respective advantages of three models of grammars. The first model is 'Item and Process', worked out by Boas and Sapir and discussed by Harris under the name of 'Process Models' – indeed, Hockett quotes Harris's article (1944) (Hockett 1954, 210); the second one is 'Item and Arrangement', drafted by Bloomfield and developed by Harris, Bloch, Wells, etc. This model is more formalized but without any recourse to logical mathematics. Hockett also mentions WP (Word and Paradigm) traditionally used for the description of ancient languages.[15] In this article, models are grammars, defined as frameworks for the description of languages. He names them 'archetypical frames of reference' and gives a long definition of what a grammatical model is: it is a generative grammar, generating all the utterances of a language, and not only the utterances that have been observed, which have to be validated, by native speakers:

A grammatical description must be a guidebook for the analysis of material in the language […] one must be able to generate any number of utterances in the language, above and beyond those observed in advance by the analyst—new utterances most, if not all, of which will pass the test of casual acceptance by a native speaker (Hockett 1954, 232).

---

[14] As we saw, 1954 is also the year of publication of Tarski' first article using the term "Model theory".

[15] See Robins "In defence of WP".

In his text, Hockett puts forward criteria of validation for the model: it has to be general and can be applied to every language, it has to be specific for the language, inclusive to cover all the observed data; prescriptive as it must generate all the utterances of a language; productive as it has to generate an indefinite number of utterances; and finally, highly effective as it should use as little machinery as possibly.

Discussion of models in Hockett's article includes a whole part on the articulation between mathematics and linguistics, called 'mathematical interlude'. More than probabilities and stochastic processes, he addresses algebraic operations and their dynamic character, in particular addition and its properties of commutation and association, as well as the possibility of computation offered by mathematics. Computing 'underscores the importance of the dynamic or generative nature of operations'. He underlines the difference between linguists and mathematicians: 'the mathematician derives his notions by abstraction from language, whereas we are deriving, not language itself, but a way of handling language,[16] from mathematics' (Hockett 1954, note 14, 229).

That requirement of compatibility between both disciplines is reasserted later in his 1967 book *Language, Mathematics and Linguistics*. He claims that one can make the analysis of language more accurate thanks to mathematisation, but only if one can find a mathematical model corresponding to linguistic facts already known. If a mathematical model does not match empirical facts, it is the model which is wrong, not the facts.

In addition to his 1952 communication model of semantic noises and his 1954 models of grammar, Hockett put forward a third type of model in his *Manual of phonology* in 1955. He addresses the relation between the phonological pattern of a language and the rest of the linguistic system, an issue on which linguists of the time disagree. He 'presents a mechanical and mathematical model of a human being regarded as a talking animal', (Hockett 1955.3) allowing him to analyse how information is carried from the speaker to the hearer in a conversational situation. This model, called the *Grammatic Headquarters* (GHQ), has several sources: information theory and cybernetics, and articulatory phonetics.

---

[16] Stressed by Hockett himself.

It is a cybernetic machine, a robot made of control charts and electrical wires:

> We shall be presenting a picture which looks vaguely like the block-diagrams ('control-flow charts') used by electrical engineers, and shall be assigning various names to the units portrayed in the diagrams ... given such a device, and the solution to the problem of neuron-to-wire linkage, one could produce a talking dog. (Hockett 1955, 3–4)

The mode of presentation of this model is 'analogy': 'it is a type of "as if" mode'. The model assigns units that may not exist, since current knowledge in physiology cannot ascertain their existence. However, Hockett decides to act 'as if' these units exist somewhere in the central nervous system, allowing humans to talk and understand.

Grammatic Headquarters (GHQ) is a model both of the speaker and the hearer as it sends and receives the flow of morphemes treated by transducers. Hockett assumes the stochastic nature of grammar, i.e. the sequence of linguistic units emitted by GHQ is determined by Markov's processes. In short, sequences with very low probabilities are interpreted as grammatically impossible while sequences with high probabilities are interpreted as grammatically possible, with a possibility of grading between the two. GHQ is a finite-state grammar, with a neurophysiological component representing the speaker and hearer behavior.

To sum up Hockett's views on models, let us stress the richness of his reflection on the relation between linguistics and mathematics, such as the compatibility between algebraic operations and immediate constituents, and the stochastic properties of morpheme chains. Models range from simple frameworks of grammatical description inherited from early American structuralist descriptive linguistics to the representation of human brain activity through a generative machine mobilizing reasoning by analogy, inherited from cybernetics and first connectionism. He claims that the model is a generative grammar which is a recursively enumerable formal device based on discrete mathematics but also holds onto a continuous mathematical understanding for his GHQ. (It should be added that, in his 1968 *State of the Art*, Hockett rails against the Chomskyan formal grammar concept and recursive grammar formalisms generally).[17]

5 Chomsky: Toward Computational Models

Two years later, in 1956, Chomsky wrote his famous paper 'Three models for the description of language', an article which takes over the term 'model' in his title. One could imagine, from a title mentioning '3 models', that Chomsky would have criticized Hockett's '2 models'. Actually, that was not exactly the case.

In this article, Chomsky compares formal grammars in their capability to generate English sentences. He sets up a hierarchy of 3 classes showing that neither Markov's processes or finite-state grammars (type 0 grammars) nor context-sensitive grammars (type 1 grammars) can generate every sentence of English. Only phrase-grammars (context-free grammars, type 2), with a transformational component, are able to do this. In this text, as well as in *Syntactic Structures* (1957a) where the same arguments are put forward, the term 'model' is in collocation essentially with terms from Information theory, such as 'statistical model', 'probabilistic model', 'communication theoretic model', 'simple order of approximation model', 'Markov processes models', 'the finite state Markov process model', 'the finite state process model'. Chomsky only mentions Hockett once in a note in *Syntactic Structures* (1957a, 20) when he criticizes Markov's chains and finite-state grammars. The criticism is more developed in his review of Hockett's *Manual of Phonology* also published in 1957. The 'model' he criticized is that of GHQ *Grammatic Headquarters*, and in no way the grammatical models (IA, IP and WP). One of the main arguments of his criticism, as far as Markov models are concerned, is the links of GHQ with information theory: the communication diagram, codes, source of information, finite-state machine (which will produce infinitely many non-grammatical sequences). Besides, Chomsky

[17] Thanks to Ryan Nefdt for this remark.

stresses that an infinite amount of equipment would be required to build Hockett's hardware GHQ.

In addition, Chomsky criticizes Hockett to be concerned with use while grammar should focus on the structure of language. That is why, he says, 'statistical models turn out to be of little relevance to grammar':

> Grammar, after all, is concerned with the structure of language, not the uses to which language is put. It is thus concerned with the set of sentences that belong to the language and the formal properties of this set. The fact that a certain sentence is frequent or infrequent does not seem to have any obvious relevance to its concerns. (Chomsky 1957b, 224)

Nonetheless, it should be mentioned that Chomsky credited Hockett with some innovation in generative grammar. He acknowledges (1957a, 49–50) that Hockett (as well as Hjelmslev actually) has set up external conditions of adequacy on grammars: (i) an external condition: 'the sentences generated will have to be acceptable to the native speaker' (ii) a generality requirement, according to which phoneme and phrase are defined independently of any particular language.

Chomsky resumed his criticism of Hockett in 1963, when he introduced the competence/performance opposition. In his article 'Finitary Models of Language Users', written with the psychologist George A. Miller, he re-examined Hockett's proposals. In this text, Chomsky and Miller put forward models for users of language, that is models for performance. Considering that contrary to Hockett, a single model should be developed both for speakers and hearers,[18] they compare the advantages and shortcomings of stochastic vs. algebraic models. In particular, the use of statistical and stochastic methods would require the computerized treatment of an enormous number of texts to give it an empirical basis, this treatment being impossible with 1960s computers.[19]

[18] See Chomsky's *Aspects*, 9 "When we say that a sentence has a certain derivation with respect to a particular generative grammar, we say nothing about how the speaker or hearer might proceed, in some practical or efficient way, to construct such a derivation. These questions belong to the theory of language use – the theory of performance."

[19] We know that this kind of treatment became feasible in the 1990s with the development of computers and software.

Chomsky and Miller acknowledge the interest of Markov model for treating phonemes, letters, syllables etc. but not for treating syntax, an argument that Chomsky had put forward many times since 1956.[20]

Turning now to Chomsky's stance concerning Hockett's generative model. Several historians of generative grammar have addressed this point and shown that Chomsky's criticisms against Hockett were eminently strategic. Pullum and Scholz (2007) note that Chomsky completely disregarded a crucial aspect of GHQ model, that is that Hockett was an advocate of both generative grammars and their neuropsychological reality (Pullum and Scholz 2007, 711). Besides, Pullum (2011) shows that *Syntactic Structures* contains no mathematical proof that English is beyond the power of finite state description while Chomsky's main criticism of Hockett rests on his use of finite-state grammars, that is Markov's processes.

According to Radick (2016), criticizing GHQ through the defects of Markov's chains was a way for Chomsky to discredit the whole work of Hockett, disregarding his innovations. By eliminating Hockett, one of the most respected and talented Neo-Bloomfieldians, Chomsky appeared to be the one true pioneer of generative grammar.

To sum up this point, Chomsky confiscated the term 'model' to assign it to his own conception of generative grammar. Hockett used the term with different meanings, among them: (i) to characterize grammars in the wake of Bloomfield, Sapir and Harris: that is process models with algebraic modeling, (ii) to devise GHQ as a generative grammar with a neuropsychological aspect. Chomsky, by ignoring its neuropsychological aspect and by limiting GHQ to a finite-state machine, trivialized Hockett's contribution to early generative grammars.

[20] After this incursion in performance, Chomsky definitively lost interest in Markov models, bringing most linguists, at least American linguists to give up statistical and probabilistic methods for a long time.

6 From 1959 on, the Generalization of the Use of Models

The generalization of the use of models by American linguists was acted by a symposium 'Operational Models in Synchronic Linguistics' organized within the Meetings of the American Anthropological Association in 1958, and published in 1959. Among the different contributions, the most interesting and substantial is by far Harris's text called 'The Transformational Model of Language Structure'. In this text, Harris makes the notion of model in linguistics clearly explicit and distinguishes three levels of models, according to the type of mathematization, type of results and criteria of validation:

(i) in a non-technical sense, a model is 'a particular style of grammar making', 'any framework in respect to which language is described'. Harris had used the term 'style of grammar' since 1944 in his text on Newman's Yokuts grammar (see above) to indicate a method of research, actually 'the deductive style of talking'. In 'Distributional Structure' (1954), both terms 'style and 'model' are equivalent. He distinguishes item style or combinatorial style from process style (where one can recognize Hockett's item and arrangement and item and process models). In 1959, 'style' refers to a non-technical sense of model. It refers to the linguist's way of making grammar while 'model' refers to the language structure.

(ii) A second type of model, in a technical sense this time, is an uninterpreted system of marks, which becomes a theory of the structure of something, when it is interpreted or when particular values are given to its parameters, so that several structures can have the same model.

(iii) The transformational model is mathematically different.

In 'model 1', in a non-technical sense, Harris gives the following examples of styles of grammars:

> the grammatical statements of Panini, the traditional grammars (mostly of Indo European languages) which are still the staple of language teaching, the list-and-occurrence grammars of Boas and his students, the element-

and-process grammars of Sapir[21] (and Stanley Newman and others), and the subclass-and-occurrence grammars of Bloomfield. (Harris 1959, 27)

Model 1 and model 2 say that the sentence is composed out of certain smaller stretches. The basic relations, then, are composition and classification. The mathematical equipment is concatenation algebra for composition and some form of Boolean algebra for classification.

The result of the combinatorial method is a hierarchy of inclusive structures (immediate constituents). All these styles may be called combinatorial linguistics.

As to transformational models: here we ask about each sentence not 'How is it composed out of smaller stretches?' but 'From what other sentence or sentences can it be said to have been transformed?'. The mathematical structures used here are an equivalence relation and an operation (transformation). The result of the transformational method is families of equivalent sentences or sentential structures.

The transformational model is able to characterize types of constructions which cannot be characterized in the combinatorial model: for example, some types of ambiguities, such as 'the two constructions represented by *Flying planes can be dangerous.* Here *Flying planes* comes from two transformational sources, from two different kernels (*Planes fly; Someone flies planes*)' (Harris 1959, 29).

From then on, the notion of model seemed to be well established in linguistics. Some milestones can be identified:

In his 1962 article, *'Explanatory models in Linguistic',* Chomsky uses the term model to define a theory of linguistic structure, a theory of linguistic universals, a theory of the linguistic intuition of a native speaker:

> What we seek, then, is a formalised grammar that specifies the correct structural descriptions with a fairly small number of general principles of sentence formation and that is embedded within a theory of linguistic structure that provides a justification [below an evaluation procedure] for the choice of this grammar over alternatives. Such a grammar could

[21] i.e. Hockett's 'item and process' model.

properly be called an explanatory model, a theory of the linguistic intuition of a native speaker. (Chomsky 1962, 533)

The mathematical basis for this model is Emil Post's production systems (1944), first introduced in Chomsky's article of 1959 'On certain formal properties of grammars'. For some historians of logic and linguistics (De Mol 2006; Pullum 2011), production systems and their capacities to generate any recursively enumerable set of strings is quite original to Chomsky's views on generative grammar.

In 'The Logical Basis of Linguistic Theory' (1964), Chomsky puts forward two objectives for linguistic theory: (a) build a perceptual model to assign a structural description to the utterance; (b) build a learning model, by making a generative grammar on the basis of primary linguistic data and by using 'faculté de langage'.

In addition, an important path leading to the generalization of the use of 'model' in linguistics was the path of computer science. Chomsky's 1956 article on the hierarchy of grammars played a significant role in the formalization and computerisation of grammars and the use of the term 'model' in computational linguistics. Actually, from the early 1960s, 'model' was used to name any computerized system of syntactic analysis (also named 'parsers'). One of the first articles on automatic syntactic analysis "A Model and an Hypothesis for Language Structure", was written in 1960 by Victor Yngve (Yngve 1960), then the director of the MIT translation center.[22] It should be mentioned that context-free grammars, after being rejected for the benefit of Chomsky's transformational grammars, had been acknowledged as particularly efficient by computer scientists for programming languages (Chomsky and Schützenberger 1963). In his article 'On the Equivalence of Models of Language used in the Fields of Mechanical Translation and Information Retrieval', Maurice Gross (1966) shows that every model of language used for machine translation and for information retrieval is equivalent to a context-free grammar. One can also mention 'dependency models' and 'adjunct tree models'. It is even more the case now when every system of natural

[22] It should be noted that Chomsky, even if he never worked on Machine translation directly, was recruited in 1955 in Yngve's Research Laboratory of Electronics at MIT, like Harris's several other pupils.

language processing is called a model. For instance, *Hidden Markov Models –HMM,* the descendants of Markov's models which saw a revival of interest in the 1990s with the technological development of computers and the appearance of large corpora and later big data.

7 Conclusion

The term 'model' in linguistics has a complex history and is vastly polysemous now. We saw that the notion of model led to the emergence of different forms of mathematization of linguistics, algebra, information theory, statistics and probabilities in addition to logical mathematics stemming from proof theory. We also saw that through this notion the reflection on the relations between linguistics and mathematics could be explained.

Further research will involve several issues concerning this relation: does model imply an analogy of functioning between linguistics and mathematics? Does it imply the use of common metalanguage of common objects and to what extent? Does it mean resemblance or compatibility between notions belonging to both disciplines?

References

Armatte, Michel, and Amy Dahan Dalmedico. 2004. Models and modelling, 1950–2000: New practices, new implications. *Revue d'Histoire des Sciences* 57 (2): 243–303.

Bar-Hillel, Yehoshua. 1953a. On recursive definitions in empirical science. *Proceedings of the XIth International Congress of Philosophy* 5: 160–165.

———. 1953b. A quasi-arithmetic notation for syntactic description. *Language* 29: 47–58.

Barsky, Robert. 2011. *Zellig Harris. From American linguistics to socialist Zionism.* Cambridge, MA: MIT Press.

Blanckaert, Claude, Jacqueline Léon, and Didier Samain, eds. 2016. *Modélisations et sciences humaines. Figurer, interpréter, simuler.* Paris: L'Harmattan.

Bloomfield, Leonard. 1926. A set of postulates for the science of language. *Language* 2: 153–164.

———. 1933. *Language*. New York: H. Holt & Company.

Chomsky, Noam. 1956. Three models for the description of language. *IRE (Institute of Radio Engineers) Transactions on Information Theory* 3: 113–124.

———. 1957a. *Syntactic structures*. London: Mouton.

———. 1957b. Review of Hockett, manual of phonology. *International Journal of American Linguistics* 23 (3): 223–234.

———. 1959. On certain formal properties of grammars. *Information and Control* 2: 137–167.

———. 1962. Explanatory models in linguistics. In *Logic, Methodology and Philosophy of Science*, ed. Ernest Nagel, Patrick Suppes, and Alfred Tarski, 528–550. Stanford: Stanford University Press.

———. 1964. The logical basis of linguistic theory. In *Proceedings of the 9th international congress of linguists, 1962*, ed. Horace Lunt, 914–1008. The Hague: Mouton.

———. 1965. *Aspects of the theory of syntax*. Cambridge, MA: MIT Press.

Chomsky, Noam, and Marcel-Paul Schützenberger. 1963. The algebraic theory of context-free languages. In *Computer programming and formal systems*, ed. P. Braffort and D. Hirschberg, 118–161. Amsterdam: North Holland.

De Mol, Liesbeth. 2006. Closing the circle: An analysis of Emil Post's early work. *The Bulletin of Symbolic Logic* 2 (12): 267–289.

Fortis, Jean-Michel. 2015. Sapir et le sentiment de la forme. *Histoire Epistémologie Langage* 37 (2): 153–174.

Gross, Maurice. 1966. On the equivalence of models of language used in the fields of mechanical translation and information retrieval. NATO Advanced Study Institute on Automatic Translation of Languages, Venice 15–31 July 1962, 123–138. Pergamon Press

Harris, Zellig. 1944. Yokuts structure and Newman's grammar. *International Journal of American Linguistics* 10 (4): 196–211.

———. 1951a. *Methods in structural linguistics*. Chicago: University of Chicago Press.

———. 1951b. Review of *Selected Writings of Edward Sapir in Language, Culture, and Personality* by David G. Mandelbaum. *Language* 27 (3): 288–333.

———. 1954. Distributional structure. *Word* 10 (2–3): 146–162.

———. 1959. The transformational model of language structure. *Anthropological Linguistics* 1 (1): 27–29.

———. 1962. String analysis of sentence structure. In *Papers on Formal Linguistics 1*. The Hague: Mouton.
———. 1991. *A theory of language and information: A mathematical approach*. Oxford/New York: Clarendon Press.
Hockett, Charles. 1952. An approach to the quantification of semantic noise. *Philosophy of Science* 19: 257–260.
———. 1953. Review of Shannon Cl. E. et Weaver W. *The mathematical theory of communication* 1949, Urbana Ill. Language 29 (1): 69–93.
———. 1954. Two models of grammatical description. *Word* 10 (2–3): 210–234.
———. 1955. A manual of phonology. In *Memoir 11 of international journal of American linguistics*. Baltimore: Waverley Press.
———. 1968. *The state of the art*. The Hague: Mouton.
Israel, Georgio. *La mathématisation du réel*, 1996, Paris, Le Seuil.
Léon, Jacqueline. 2015. *Histoire de l'automatisation des sciences du langage*. Lyon: ENS-Editions.
Malmberg, Bertil. 1968. *Manual of phonetics*. Amsterdam: North-Holland.
Miller, George, and Noam Chomsky. 1963. Finitary models of language users. In *Handbook of mathematical psychology*, ed. Robert D. Luce, Robert R. Bush, and Eugene Galanter, vol. II, 419–491. New York: Wiley, 1963.
Nefdt, Ryan. 2019. Infinity and the foundations of linguistics. *Synthese* 196 (2019): 1671–1711.
Pike, Kenneth L. 1943. Taxemes and immediate constituents. *Language* 19 (2): 65–82.
Post, Emil. 1944. Recursively enumerable sets of positive integers and their decision problems. *Bulletin of the American Mathematical Society* 50 (1944): 284–316.
Pottier, Bernard, and Simone Dreyfus. 1970. A propos de Sapir. *L'Homme* 10 (1): 94–99.
Pullum, Geoffrey K. 2011. On the mathematical foundations of syntactic structures. *Journal of Logic, Language and Information* 20: 277–296.
Pullum Geoffrey K and Barbara Scholz. 2007. Review article: "Tracking the origins of transformational grammar, Marcus Tomalin, 2006, *Linguistics and the formal Sciences*; The origins of generative grammar". *Journal of Linguistics* 43–3: 701–723.
Radick, Gregory. 2016. The unmaking of a modern synthesis: Noam Chomsky, Charles Hockett, and the politics of behaviorism, 1955–1965. *Isis* 107: 49–73.
Robins, Robert. 1959. In Defence of WP. *Transactions of the Philological Society* 58 (1): 116–144.

Segal, Jérôme. 2003. Le Zéro et le Un. Histoire de la notion scientifique d'information au 20e siècle.
Seuren, Pieter. 2009. Concerning the roots of transformational generative grammar. *Historiographia Linguistica* 36 (1): 97–115.
Shannon, Claude. 1956. The bandwagon. *Institute of Radio Engineers, Transactions on Information Theory* 2: 3.
Shannon, Claude E., and Warren Weaver. 1949. *The mathematical theory of communication*. Champaign: University of Illinois.
Sinaceur, Hourya. 1999. Modèle. In *Dictionnaire d'histoire et philosophie des sciences*, ed. D. Lecourt, 649–651. Paris: PUF.
Tomalin, Marcus. 2006. Linguistics and the formal sciences. Cambridge: Cambridge University Press.
Vaught, Robert L. 1974. Model theory before 1945. In *Proceedings of the Tarski symposium*, ed. L. Henkin et al., 153–172. Providence: American Mathematical Society.
Varenne, Franck. 2018. Models to Simulations.London: Routledge.
Wells, Rulon. 1947. Immediate constituents. *Language* 23: 81–117.
Yngve, Victor. 1960. A model and an hypothesis for language structure. *Proceedings of the American Philosophical Society* 104 (5): 444–466.

14

Poetics and the *life of language*: Disciplinary Transfer in the Guise of Metaphor

Carita Klippi

Rien n'embellit tant le discours que le bon usage des *métaphôres;* mais il faut pour cela quelles soient justes et naturelles, qu'elles soient sensibles au comun des lecteurs; et que, dans le discours relevé, elles soient nobles et décentes. Leurs défauts sont de manquer de justesse, d'être tirées d'Arts et de Sciences peu conus; d'être forcées et recherchées, d'être trop multipliées et entassées; d'être basses et rampantes dans le style élevé.
(Jean-François Féraud (1787–1788), *Le Dictionaire critique de la langue française*)

C. Klippi (✉)
Tampere University, Tampere, Finland

Laboratoire d'Histoire des Théories Linguistiques (CNRS, Université de Paris, Université Sorbonne Nouvelle), Paris, France
e-mail: carita.klippi@tuni.fi

R. M. Nefdt et al. (eds.), *The Philosophy and Science of Language*,
https://doi.org/10.1007/978-3-030-55438-5_14

1 Introduction

From the *Rhetoric* of Aristotle to the first stylistic and pragmatic studies of different text types and genres at the beginning of the twentieth century, it has been maintained that scientific language does not lend itself to the use of figurative images. As the aim of scientific language is to show the objective facet of things—to describe reality and to demonstrate the truth—the choice of words and the turns of phrase should be as neutral as possible (Bally 1909, 117–119; Bally 1921, 148). In contrast to ordinary and literary language, scientific language in its ideal axiomatic sense seeks to convey thought in its purest form, and therefore it should be devoid of any ambiguities, approximations, confusions and ill-defined contents (Molino 1979, 84). This perspective is in line with the realist tradition of science, which postulates an unequivocal correspondence between the truth of a scientific theory and reality. Besides, on the back of the positivist hierarchy of sciences, it has been claimed that the farther a discipline departs from the formality of mathematics, the more it is likely to have recourse to figures of speech (Molino 1979, 85). At the same time, however, it has been recognized that human thought processes are largely metaphorical (Lakoff and Johnson 1980, 194), and for this reason, scientific language is also loaded with figurative terms and heuristic images, with mathematics being no exception to this. The constructivist philosophy of science argues that between a scientific theory and reality there is always a human element, and therefore reality cannot be apprehended directly as such; it is instead conceptualized through human experience and manmade artifacts, including language use. In this realm of thought, scientific phenomena exist in accordance with linguistic products, terminologies, discourses, texts, and metaphors. Language use is thus an inherent part of scientific paradigms and conceptual tools, up to the point of rising above reality, and even being likely to construct a 'reported' reality that in turn may become an object of analysis (Kuhn 1962; Berger and Luckmann 1966; Boyd 1993 [1979]).

The aim of this chapter is to illustrate how 'literature' is used to talk about science. Special attention is paid to the role of metaphor in the making of linguistics in the latter half of the nineteenth century, when it had become quite customary to talk about the *life of language* in the context of linguistic thought. Linguistic thought is revisited[1] here with reference to modern metaphor studies, which have constantly highlighted that metaphor is one of the most important and omnipresent tools in our conceptualisation of reality. What lies behind the 'poetics' of the life of language—poetics in the sense borrowed from literary theory as a technique to convey meaning? What was the legitimacy of the literal meaning of the life of language? Was it rather a manner of speaking for want of a better term, or was it only one among many attempts to capture the uncapturable nature of language?

The metaphor of the life of language can be approached from three different methodological perspectives. It can be examined in the light of cognitive linguistics with a special focus on its conceptual level. It can also be studied within the framework of pragmatics and discourse analysis, which emphasizes its context of production and the intentions and objectives of the particular authors who use it (Charteris-Black 2004; Zanotto and Palma 2008). It can also be studied in the realm of corpus linguistics (see e.g. Deignan 2005) in order to make a quantitative account of the occurrences of the noun phrase in a large dataset—in this case, in the German, French, and English linguistic texts of the nineteenth century. Be it systematic and rigorous in providing an overall picture of the presence of the metaphor, this approach nevertheless has the weakness of taking into account a rather narrow lexical and syntactic context, and therefore it alone might appear to be ill-adapted to give access to the multifaceted historical meaning of this metaphor.[2] The study of the life of language is limited here to the cognitive and discursive point of view, with the focus on the scientific discourse of the nineteenth century as a relevant context to interpret the role of the metaphor in the making of linguistics. This perspective is also in line with the idea that

---

[1] See Klippi (2010).

[2] In digital humanities, the text mining would provide a valuable and systematic device to check the extent and frequency of naturalistic concepts in linguistic texts, but at its best it would provide only a point of departure for a more thorough analysis using the 'old-fashioned' methods of close reading and understanding. Even if an individual researcher cannot reach the efficacy of computers, he is nevertheless able to accomplish the tasks that a machine could hardly ever do: to grasp and solve scientific problems, to make use of creativity, empathy and larger contextual knowledge, to cease the nuances of communication and to interpret implicit meanings behind the words.

metaphor is an interdisciplinary phenomenon also from a historical perspective, rather than being restricted to the realm of poetry, literature, and rhetoric. The conceptual mappings of the life of language are deconstructed here through the notation derived from the classic work of Lakoff and Johnson, *Metaphors We Live By* (1980), and their validity is reflected against the background of naturalistic culture. Criticism of the life of language ran concurrently with its success, but its critics appeared to be beneficial to the whole science of language insofar as they obliged linguists to find different solutions to explicate thoroughly the nature of language. Therefore, it might not be an exaggeration to claim that linguistics as a modern science arises from the dynamics of the literal and metaphoric interpretation of the life of language.

2 Positioning with Respect to Metaphor

Traditionally, metaphor has been seen as a figurative means to produce meaning that is added to the basic, literal meaning of language, and in this role, it has been considered simply as an aesthetic ornamentation of reality (Aristotle 1997; Ricœur 1975, 32). The French lexicographic tradition has defined metaphor as 'a figure of rhetoric: a kind of abbreviated comparison, by which one conveys the proper meaning of a word to a figurative sense' (*Dictionnaire de l'Académie* 1832–1835, *s.v. métaphore*).[3] Metaphor has received different epithets according to its success and woes: it has been described either as beautiful, happy, and appropriate, or as bold, exaggerated, and forced (*Dictionnaire de l'Académie* 1832–1835, *s.v. métaphore*). The critics of the lame, over-elaborated, and inaccurate use of metaphors have reported from literature examples of the lack of a relationship between the objects of comparison, thereby demonstrating in a concrete manner that occasionally even the best authors are sloppy in their language use (Féraud 1787–1788, *s.v. métaphôre*). Historical lexicography has also retained from the scholars of antiquity one of the fundamental features of metaphor—its close connection to analogy. Analogy establishes a structural relationship between things from one domain of

[3] All translations from the original French are by the author.

knowledge to another. In a similar manner, a metaphor requires the recognition of an analogical connection between these domains, but it also implies a purposeful ignorance of the radical differences between them. If we say, to use an Aristotelian example, that a man is a lion, we underline his bravery and courage on the basis of popular imagery, without actually assessing whether lions are literally more courageous than other predators or animals in general. Metaphor is thus a subtype of analogy, but the relationship is asymmetrical insofar as 'all metaphors are analogies, but not all analogies are metaphors' (Itkonen 2005, 41). By combining discourses of different sources, lexicography highlighted that a metaphor is a figure of speech by which the proper meaning of a word is transferred to another meaning, which suits it only by virtue of a 'comparison made in spirit'[4]—and not at the level of words—thus paving the way for future cognitive studies on metaphor.

In the process of metaphorisation, there is a transposition from one object to another, usually with a leap from the known to the unknown or from the concrete to the abstract, the former allowing new ways of representing the latter. This also explains why metaphor has become an interdisciplinary object of study instead of being reduced to literature alone (see Ortony 1993 [1979]). Modern cognitive theory suggests that metaphor plays an important role as a conceptual instrument in all aspects of our everyday life, and the use of metaphor extends to all areas in which people need to order their surroundings and analyse reality (Lakoff and Johnson 1980). This theory rests upon the latent assumption that human intelligence cannot depend on a ready-made, innate, *a priori* biologico-cognitive structure, but is the result of the empirical interaction of humans with their environment (see Auroux 1998, 7, 297). People sort the observed facts by means of tools of various natures, including

[4] La métaphore est une figure par laquelle on transporte, pour ainsi dire, la signification propre d'un nom à une autre signification qui ne lui convient qu'en vertu d'une comparaison qui est dans l'esprit. Un mot pris dans un sens métaphorique, perd sa signification propre, et en prend une nouvèle qui ne se présente à l'esprit que par la comparaison que l'on fait entre le sens propre de ce mot, et ce qu'on lui compare (Du Marsais 1730, 125).

Metaphor is a figure by which the proper meaning of a name is conveyed, so to speak, to another meaning that suits it only by virtue of a comparison which is in the mind. A word taken in a metaphorical sense loses its own meaning, and it takes a new one that comes to mind only by the comparison that is made between the proper meaning of this word and what is compared to it.

different elements of the environment and the human body, which in turn become symbols and signs. Our conceptual system is thereby entrenched in the way we interact with our living space, and language contains traces of this activity in the form of metaphors without us even being conscious of it (Lakoff and Johnson 1980, 194). For instance, in some languages the grammaticalisation of lexical words referring to the parts of body (*head*, *breast*, *flank*) demonstrates the process of the abstraction of the human body into a grammatical device. The same abstraction shows in the historical systems of measurement (*digit*, *finger*, *nail*, *span*, *cubit*, *fathom*, *foot*), which profits from the human body as a tool to quantify the environment. However, instead of appearing always objective or universal, reality is relative to a given space and time, and consequently, the physical, cultural, and social experience is reflected in the formation of metaphors (Lakoff and Johnson 1980). Thus, there are on one hand universal metaphors tied to the human condition, so to speak, and on the other hand, cultural metaphors that pick up certain aspects of reality considered important in a particular context (Kövecses 2005). In the case of the latter, each language has metaphors of its own, consecrated by usage and possessing no equivalent in other languages (Du Marsais 1730, 142). Given the significance of metaphor in the structuring of our conceptual system, words do not adjust to reality—and conversely reality does not adjust to words—in a simple and straightforward manner. In having the potential to change our conceptual system, metaphor influences our way of perceiving and acting in the world (Lakoff and Johnson 1980, 145–146).

2.1 Metaphor in Science

Metaphor opens up dynamic processes on reality, and these processes are based on the 'transgression of acquired boundaries' (Ricœur 1975, 298). By consolidating abstract thought, metaphor is also used—despite the requirements of the genre—in scientific explanation insofar as transgressing boundaries between different domains results in a better understanding of the object of knowledge. The history of science demonstrates the possibility of using several metaphors to conceptualize an abstract system using a more concrete one, witnessing also a quest for a model that best corresponds to the truth (see Klippi 2010, §2). This may be illustrated by examples borrowed from physics, whose conceptual value is to grasp the behavior of electrons. The traditional metaphor to depict the relationship of electrons with respect to the nucleus of an atom was to compare electrons to the planets orbiting the sun. This metaphor was replaced by a new one in the electromagnetic model of Nils Bohr, the same relationship being rendered by means of the alternating position of billiard or ping pong balls (Kuhn 1993 [1979], 538). More recently in density functional theory, the relative position of electrons has been compared to a cloud where its constituting drops are divided differently in terms of probability in the different areas of the cloud: in its inner parts and at the edges.

These examples reveal that one metaphor may be replaced by another with the advance of science. On the other hand, they show that a metaphor allows us to examine the object of knowledge only from one perspective at the same time, thus leaving aside other ways of approaching the object. They also bring out the epistemological division between the realist and the constructivist traditions of science (see Boyd 1993 [1979]). A scientist of the realist persuasion holds that scientific practice aims at discovering truth. His relationship to what is real is univocal: p and ¬p are mutually exclusive. At the same time, however, being aware of the connection of truth and temporality, he must accept uncertainty to a certain extent, because in the course of time, the given theory may become invalidated by a more plausible one. By contrast, a scientist of a constructivist bent would claim that there is no one single theory (or by extension one single metaphor) to explain the same data; he would rather maintain that the data are created by different theories (or metaphors)—in other words, facts are overdetermined by theory (or to use Saussurean terms, it is the point of view that creates the object). The object is thus constructed within the epistemological, conceptual, and methodological framework of a given theory, and as a consequence, brute facts do not exist independently of theory. For this reason too, the criterion of falsifiability is not valid in this realm of science as it is in the realist context, because all theories are basically equivalent to the constructivist (see Popper 2002 [1934], §4). The overdetermination is, however, also present in the more realist-oriented approaches, for instance, in grammatical theories whose mutual emulation is based on the classical ideal of providing a sound and complete—and thus, the best—description of language. In this sense, a linguist may claim (without being a constructivist) that the spatiotemporal level cannot empirically falsify a theory—only an analysis carried out by one of the theories can falsify the analysis of the other (see Itkonen 1978).

As a general rule, the natural sciences follow the realist pattern, whereas the human sciences, being by necessity more tolerant with respect to simultaneously existing perspectives, cleave to the idea of epistemological relativism (Kuhn 1962). The duck-rabbit illusion (Fig. 14.1) by Joseph Jastrow (1899)—along with its simplified version by Wittgenstein (1968 [1953])—has been exploited both by the philosophy of science and metaphor studies to illustrate, on one hand, the difference between the realist and constructivist traditions, and on the other hand, literal and metaphorical meaning (Kuhn 1962; Davidson 1978; Klippi 2010, 33).

A realist seeks to permeate the truth behind the lines and contours that compose the drawing to find the interpretation that best corresponds to reality. However, the truth is likely to alter. Paradigmatic changes in the course of the history of (natural) science provide new, more accurate conceptions of reality: what in the former theory was seen as a duck—or a gull to be more accurate with the realist intention of the artist—becomes a rabbit in the revised theory. For a constructivist, such things as lines and contours do not exist as themselves; there are only different points of view, which create different versions of reality that exist in parallel. The image may show a duck or a rabbit, depending on the angle from which it is viewed. According to the philosophical tradition, metaphor either

Fig. 14.1 Duck-rabbit (Jastrow 1899)

structures reality on the basis of the literal meaning or, as a symbolic instrument, provides different versions of reality.

Despite the contingency of the truth of a metaphor, metaphorisation itself is a recurring phenomenon in science. Richard Boyd (1993 [1979], 482, 486–7) has pointed out that as a linguistic machinery shared by an entire scientific community, metaphor is an integral part of the development and articulation of scientific theories. In the absence of a corresponding literal paraphrase, metaphor is not just a didactic or heuristic tool, but rather a link between scientific language and reality, allowing epistemic access to a particular phenomenon (Kuhn 1993 [1979], 538; Boyd 1993 [1979], 486–487, 490).

2.2 Metaphor in the History of Linguistics

Language as an atheoretical concept might be said to belong to a democratic space where everyone as an ordinary language user may hold an opinion on its nature. However, when one attempts to provide a clear definition of language as a theoretical term, one may risk running into an autonymous tautology, LANGUAGE IS LANGUAGE, which does not say anything more about the object than the name itself. On the contrary, a brief overview of the classics in the history of linguistics reveals immediately that the basic endeavor to grasp the object of linguistic thought has been to bend it around different metaphors (Klippi 2010, 10–11). Plato's Cratylus (388c) suggested that language, sentences, and words are *instruments* 'capable of teaching and distinguishing reality, like the shuttle to unravel the threads'. In the same fashion, Wittgenstein (1968 [1953], I, 11) maintained that the function of words can be compared to the function of *tools*, such as a hammer, saw, chisel, and so on. The basis of this metaphor lies in the fact that all tools, including language, have formal and functional features, the function of the tool being determined in the conception of its form. Language has also been perceived as an *institution*, a *custom*, and a *usage*—that is, a set of practices and rules shared by individuals who together constitute a community. The existence of an institution can be measured in the way these practices and rules channel the behavior of the members of a community. This position has been defended by, among others, William Dwight Whitney (1827–1894), who considered that 'every language as an institution' (1877 [1875]: 231) imposes its rules on language users from above. Language has also been regarded as an *organ* in the same way as the heart or lungs. Indeed, Noam Chomsky (1965, 1980) has asserted that universal grammar is a mental organ whose anatomy and physiology are determined by biological necessity. This list could be extended to *code*, *social fact*, and so forth.

These metaphors encountered throughout the history of linguistic thought are sufficient enough to underline the omnipresence of metaphor in scientific explanation, but at the same time they also serve to show the open character of the interpretation of language as an object of knowledge. Due to their different internal structures, the aforementioned metaphors highlight that language can be grasped from a multiplicity of perspectives, suggesting thereby the absence of a fixed, inerrant definition of language. Metaphorisation is a way to provide substance to an abstract object, that is, to create a discrete category and to reify the object of knowledge, but even at its best, metaphor is a means to transfer the evasive nature of language only through externality and otherness: LANGUAGE IS SOMETHING ELSE. However, whatever the metaphor, its use aims at creating a conceptual apparatus capable of explaining language in a relevant and coherent manner within a given system of thought. While establishing a link between two concepts, it emphasizes some details and removes other details, either by a filtering or shielding effect (Black 1962; Ricœur 1975, 114).

Ivor Armstrong Richards defined metaphor as a 'transaction between contexts'. In *The Philosophy of Rhetoric* (1936), he stated that the relevance of a metaphor could be measured by means of the concepts of *tenor*, *vehicle*, and *ground*. *Tenor* is the object of comparison, *vehicle* refers to the source domain of the figurative image, and *ground* constitutes the link between the two objects—that is, the proper meaning of the metaphor. LANGUAGE, in this case, is thus the tenor or target domain requiring further clarification through the use of different vehicles or metaphorical expressions, and the convergence of their qualities and characteristics serve to legitimate their connection, allowing one domain to be understood in terms of another.

In *Metaphors We Live By*, Lakoff and Johnson started the cognitive era of metaphor studies. By profiting from the studies of their predecessors, they renewed the terminology and adopted an abstract annotation, TARGET IS SOURCE, using small capitals as a means to refer to the underlying conceptual deep structure of the metaphor, whereas the properly linguistic level was regarded as its surface structure (Deignan 2005, 14; Vereza 2008, 164). In this model, the target constitutes a problem, while the source is supposed to provide the solution. The aforementioned metaphors encountered throughout the history of linguistic thought can thus be represented schematically as follows:

LANGUAGE IS A TOOL.
LANGUAGE IS AN INSTITUTION.
LANGUAGE IS AN ORGAN.

As in analogy, in the process of metaphorisation, there is a transposition of similar and contiguous elements from one domain to another based on a structural relationship, not a literal identity. The legitimacy of a metaphor depends on its capacity to establish an accurate system of mappings between the target and the source: what is valid for the source domain should also be valid for the target domain (Gentner 1989, 201; Sullivan 2013). In other words, an appropriate metaphor has the potential to raise features that were not understood to exist before, and therefore the metaphor is subject to revealing some essential aspect of the object of knowledge. The Fig. 14.2 groups together some basic terminology that has been used in metaphor studies to take into account the interaction between the two systems of mapping:

However, as Max Black (1954, 289) emphasized in his famous article on metaphor, the system of mappings (or the *system of associated commonplaces* in his terminology) 'may include half-truths or down-right mistakes'. Representing a point of view in the constructivist sense, metaphor may lead to erroneous ideas or false perceptions that might not be accepted straightforwardly by all scholars and schools of thought. This is also the reason why there have been a number of metaphors to transfer the supposedly true nature of language, which, one by one, have been deemed unsatisfactory over the course of history.

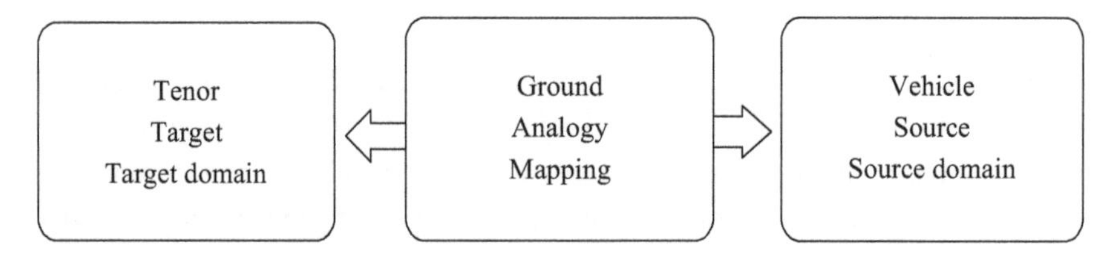

Fig. 14.2 Target is source

3 The Life of Language as a Meme

As an emerging discipline, in the 19th century linguistics did not exist in a vacuum. On the contrary, it was inseparable from the more general scientific *zeitgeist*. Rather than being considered as an intangible cloud, the spirit of time should be seen as a concept imbued with historicity in the sense of condensing circulating imaginaries and discourses of various origins. Therefore, it is closely intertwined with the Foucauldian notion of *épistémè,* which determines for a given space and time what it is possible to know and talk about (Foucault 1969). The spirit of the time also enables the 'contagion of ideas'—that is, an involuntary epidemic-like transmission of ideas within and across different communities of practice, including scientific fields (Sperber 1996).

Linguists adopted a particular style of science from natural history, whose semiotic code spread over a number of other domains over the course of the century (Schlanger 1971; Blanckaert 2004). In fact, no one could avoid the terminological or ideological infection of naturalism, whatever the field of study, even if it was not always quite clear from where it took its origins (Morpurgo Davies 1987, 94, 102). In the absence of comprehensive studies on this issue, it has been maintained that the history of the humanities has not taken properly into account the extent, coherence, and international fortune of the naturalistic discourse, which played a major role in the emergence of the humanities' rationale (Blanckaert 2004, 10). If, in this respect, things have henceforth changed in the history of the numerous disciplines involved in studying human actions and artifacts (sociology, geography, architecture, etc.), it might be claimed that linguists have 'always' been sensitive to the historical connection between biology and linguistics (see e.g. Normand 1976; Tort

1980; Maher 1983; Hoenigswald and Wiener 1987; Desmet 1996; Alter 1999; Klippi 2007, 2010).

The linguistic thought of the long nineteenth century can thus be constructed through and around the metaphor of the life of language as a part of the recurring naturalistic theme of scientific discourse. It dates back to German Romanticism and coincides with the emergence of linguistic comparativism, but it was also closely intertwined with the more general discussions on the place and autonomy of new scholarly fields in the classification of the sciences. Language was even considered to be the fourth kingdom of nature, in parallel with the animal, vegetable, and mineral kingdoms, which was the eighteenth-century chemists' metaphorical manner of classifying natural beings (see Du Marsais 1730, 133; Desmet 1996, 65). The interplay between natural history and linguistics was reinforced in the 1870s with a critical undertone, but despite opposing views, the metaphor was still discussed in linguistic texts as late as in the 1920s, the most famous of them certainly being the work of Edward Sapir (1921) and Otto Jespersen (1922). In fact, from the 1870s on, many authors adopted the naturalistic metaphor in the titles of their work, including *The Life and Growth of Language* (1875) by William Dwight Whitney (translated into French in 1877 with the title *La vie du langage*), *Biographies of Words* (1887) by Friedrich Max Müller, *La vie des mots étudiée dans leurs significations* (1887) by Arsène Darmesteter, and *La vie du langage* (1910) by Albert Dauzat. These titles are convincing enough to give substance to the hypothesis of the existence of a 'linguistics of the life of language'. Due to its all-pervasive character, this metaphor may be depicted in terms of poetics as an extended, conceited, or sustained metaphor that is repeated coherently throughout the linguistic work of the century, with some variations arising from the same semantic field. It can also be regarded as an 'inherited concept' in the sense that it is not forged by the historian, but is validated in a given spatiotemporal context by different scholars who convey a plurality of meanings to it in their scientific narratives (see Koselleck 1990 [1979]). To use a more modern term, the life of language is a meme or a cultural gene (Dawkins 1976; Blackmore 1999) that is replicated in the network of linguists with some modifications, thereby shaping the mindset of the whole community of practice, independent of the linguists' scholarly divergences.

3.1 The Life of Language: Literal or Metaphoric Naturalism?

The *life of language* in its heuristics is more ambiguous compared to more straightforward metaphoric expressions, insofar as one must first deconstruct the noun phrase in order to determine which element is the *tenor* or target and which is the *vehicle* or source, and, thereafter, reconstruct the potential underlying sentence(s):

LANGUAGE IS A LIVING BEING.
LANGUAGE HAS A LIFE.
LANGUAGE LIVES.

If the target constitutes the problem and the source provides the solution, it is necessary to figure out in this particular case what the source—that is, LIFE—means as a component of the conceptual metaphor. LIFE belongs to universal metaphors insofar as it does not arise exclusively from some specific cultural conceptualisations, but is a physical aspect of any living being (see Lakoff and Turner 1989; Kövecses 2005). Instead of engaging in any profound philosophical discussion on the concept of 'life', linguists first relied merely on its everyday conception, taking its existence at face value. Therefore, the outcome of the mappings creating similarities between language and a living being seems rather trivial. Life consists of a beginning (birth) and an end (death), and between the two extreme points there are several stages of life corresponding literally to childhood, adulthood, and old age. Projected onto language, these mappings with all their absurd entailments were vehemently criticized for several decades, especially by the French linguists Victor Henry (1896), Michel Bréal (1897), Ferdinand de Saussure (CLG/E 1974), Charles Bally (1965 [1913]), and Joseph Vendryes (1968 [1921]).

> Tout en reconnaissant ce que cette métaphore a d'inexact et d'ambigu, on peut cependant l'utiliser à titre d'hypothèse pour orienter la recherche ou rendre plus commode l'exposé didactique. Mais les données sur lesquelles nous avons pu opérer jusqu'ici n'étaient que des abstractions, créées par l'esprit du linguiste; et c'est presque un abus de parler de vie du langage

pour désigner ce qui est justement dépourvu de vie, les sons, les formes grammaticales et les mots (Vendryes 1968 [1921], 260–261).

While recognizing how this metaphor is inaccurate and ambiguous, it can nevertheless be used as a hypothesis to guide the research or to make the didactic presentation more convenient. But the data upon which we have been able to operate until now were only abstractions, created by the mind of the linguist; and it's almost an abuse of talking about the life of language to refer to what is so devoid of life, sounds, grammatical forms, and words.

It thus became necessary to determine the exact value of the source domain (LIFE), because every time *language-as-a-living-being* was used, it advanced some aspects at the expense of others. Instead of being an innocent figure of speech to decorate the style, it imposed on the target certain features that had pregnant epistemological consequences on linguistics as a science.

The indeterminacy of language-as-a-living-being may be illustrated by a more scientific attempt to provide a definition for the concept of 'life'. Different editions of the *Dictionnaire de l'Académie française* (*DA*) reflect the prevailing narrative of the time. The conception of 'life' extended from 'the union of the soul with the body' (*DA* 1694[1], *s.v. vie*) to 'the state of living beings which lasts as long as they have in them sensations and movement' (*DA* 1762[4], 1798[5], 1832–1835[6]). The *Grand dictionnaire universel du XIXe siècle* by Pierre Larousse, published in 1866–1877 (and 1888), asserted rightfully that from a common sense point of view, life is easy to recognize, whereas from a scholarly point of view, it is the most difficult concept to define:

Il n'est personne qui ne distingue une matière animée d'une matière inerte, un corps vivant d'un corps brut; il n'est personne qui ne sache apercevoir la vie quand elle se présente; rien de plus facile à reconnaître, mais rien de plus difficile à définir; rien de plus manifeste ni de plus inexplicable. (*Larousse* 1876, tome 15, *s.v.* vie)

There is no one who would not distinguish an animate matter from an inert matter, a living body from a raw body; there is no one who would not discern life when it appears to him; but nothing is more difficult to define; nothing more obvious or more inexplicable.

Although the dictionary's objective was to give an account of the concepts, discoveries, and inventions of a wide range of fields (from history and geography to the natural and moral sciences), it provided the most accurate definition of 'life' for the purposes of the present context. Consistent with the contemporaneous achievements in natural history, 'life' was defined as the 'state of organised beings, manifested by the functioning of their organs, contributing, through assimilation, to the development or preservation of the subject and his proper state':

> Etat des êtres organisés qui se manifeste par le fonctionnement de leurs organes, concourant, par l'assimilation, au developpement ou à la conservation du sujet et de son état propre (Larousse 1876, tome 15, *s.v. vie*).

In its literal, physiological, and biological sense, 'life' was thus a property of a living organism considered as an individual whose functions aim at its growth and survival.

As was observed above, the conceptual metaphor of LIFE existed in various fields of scholarly work, and 'linguistic research has in this respect been in full accordance with tendencies observed in many other branches of scientific work during the last hundred years' (Jespersen 1922, 7). To apply features of life to objects that do not possess a 'separate existence in the same way as a dog or a beech has' (Jespersen 1922, 7) can be explained in the first place by the following proportional analogy, which allows comprehension of the overlapping mappings between living beings and entities that in a rigorous sense do not possess life:

$$\frac{\text{anatomy}}{\text{physiology}} = \frac{\text{form}}{\text{function.}}$$

A BUILDING IS A LIVING BEING. (Architecture)
SOCIETY IS A LIVING BEING. (Sociology)
A LANDSCAPE IS A LIVING BEING. (Geography)

The relationship of the anatomy of an animate being to its physiology is the same as the relationship of the form of an inanimate entity to its function. For this reason, the mapping of language-as-a-living-being also overlaps with the metaphor of *language-as-a-tool*: the function of an organ or a tool depends on the constitution of that organ or tool, as it was claimed in the early biology, and vice versa, an organ or a tool has been designed in the most suitable way for its function. In this sense, a building can be considered a living being, but the meaning of the metaphor can be extended to cover other fundamental features of life: a building is conceived, a building gets old and has to be taken care of, a building can change its function, etc. The interpretation of life in terms of linear development that tends toward progress also explains why the cycle of life could be applied from animate matter to inanimate matter, to abstract artifacts, such as religions, civilisations, and sciences, whose final end is to achieve intellectual maturity (see Comte 1996).

Even if the *life of language* as a meme rose from and spread in the *zeitgeist* of the nineteenth century, with other lexical items from the same semantic field (*organ*, *organism* and so forth), a decisive turning-point could nevertheless be attributed to it. This was also the most identifiable event from which its use started to become controversial among linguists, as it was supposed to have a determining influence on the future direction of the linguistic science.

3.2 Causal Theory of Reference: August Schleicher

Although the metaphor of the *life of language* is merely a mental representation, its discursive origin can be illustrated by the causal theory of reference, which has been primarily formulated to deal with proper names (and thereafter with names of a natural kind). In the causal theory of reference, an individual (i.e. a referent) receives upon baptism a name that is first used by the relatives of the individual, which then diffuses into wider usage. In the case of a famous person, more and more people can refer to the person by this name, even if—apart from the name—they hardly have any other information about the referent. According to this theory, the name belongs to our social heritage and its users form a chain from the baptism on, while each link in the chain makes at least an implicit reference to the original act of name-giving (Donnellan 1972; Kripke 1980) (Fig. 14.3).

Similarly, every new occurrence of the metaphor of the *life of language* is anchored implicitly in its literal use, with every further user having in mind the ideas August Schleicher (1821–1868) developed in *Die Darwinische Theorie und die Sprachwissenschaft* (Fig. 14.3):

> Die Sprachen sind Naturorganismen, die, ohne vom Willem des Menschen bestimmbar zu sein, entstunden, nach bestimmten Gesetzen wuchsen und sich entwickelten und wiederum altern und absterben; auch ihnen ist jene Reihe von Erscheinungen eigen, die man unter dem Namen 'Leben' zu verstehen pflegt. Die Glottik, die Wissenschaft der Sprache, ist demnach eine Naturwissenschaft; ihre Methode ist im Ganzen und Allgemeinen dieselbe, wie die der übrigen Naturwissenschaften. (Schleicher 1863, 7)
>
> Languages are organisms of nature; they have never been directed by the will of man; they rose, and developed themselves according to definite laws; they grew old, and died out. They, too, are subject to that series of phenomena which we embrace under the name of 'life'. The science of language is consequently a natural science; its method is generally altogether the same as that of any other natural science. (Schleicher 1869 [1863], 20–21)

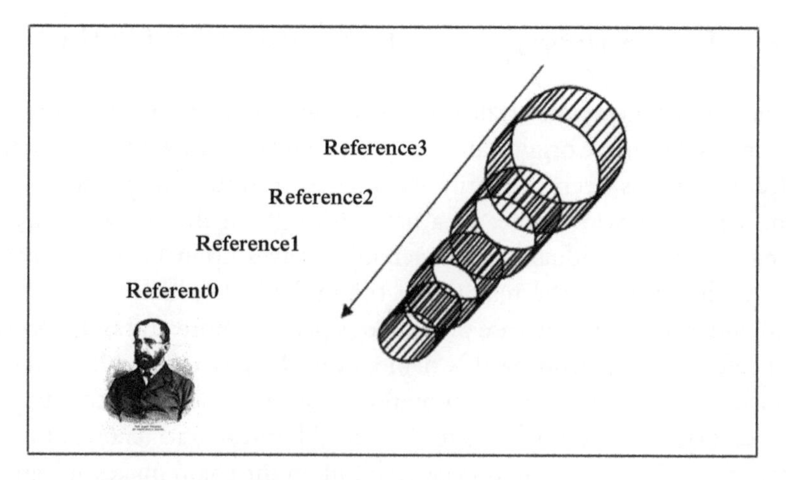

Fig. 14.3 Application of the causal theory of reference to the literal baptism of the life of language. (Klippi 2010, 45)

If the first comparatists had taken the term 'organism' merely in the sense of a system[5] (Morpurgo Davies 1987; Wells 1987), Schleicher considered that the mappings between language and organism, and, following *On the Origin of Species* (1859) by Charles Darwin (1809–1882), between language and species, were supposed to be taken literally:

LANGUAGE IS AN ORGANISM,
that is, an individual that grows and develops in evolving stages.
LANGUAGE IS A SPECIES,
that is, an entity likely to undergo transformation over time (in order to become a different entity).

The organicist conception of language was in harmony with the definition of 'organism' by comparative anatomy, which studies the forms and functions of a static being, in addition to the anatomical regularities that different organisms hold with respect to each other:

> Tout être organisé forme un ensemble, un système unique et clos, dont les parties se correspondent mutuellement, et concourent à la même action définitive par une réaction réciproque. Aucune de ces parties ne peut changer sans que les autres changent aussi; et par conséquent chacune d'elles, prise séparément, indique et donne toutes les autres. (Cuvier 1825, 95)
>
> Every organized being forms a whole, a unique and closed system, the parts of which correspond to each other, and contribute to the same definitive action by a reciprocal reaction. None of these parts can change without the others changing too; and therefore, each of them, taken separately, indicates and gives all the others.

In the same way, it is possible to study and compare the individuality of particular languages—provided they are immune to any structural variation—and thereby show their mutual relationships through a

[5] A comparison of the lexicographic definitions of the dictionaries of the period reveals that the terms 'machine', 'organism', and 'system' were regarded as synonyms due to their common features. They all compose a unitary whole, a differentiated complex that serves as a support for their constituent parts, and these parts are conceived and arranged in order to ensure the proper functioning of the whole.

taxonomic classification. The identification of language with a natural organism does not represent a specific issue as long as the linguist agrees to describe the structure of a 'linguistic organism', but as soon as the ideas of evolutionary logic and historical dynamics are introduced as integral parts of life, the use of the term 'species' becomes more appropriate. In the nineteenth century natural history, there were three main definitions of the concept of species:

1. Reproductive conception of species: the natural grouping of individuals who possess a continuous fertility as opposed to infertile hybrids (Georges-Louis Leclerc, Comte de Buffon).
2. Organic conception of species: the natural grouping of individuals endowed with mutual resemblance, based on a comparative analysis (Georges Cuvier).
3. Relative conception of species: the artificial grouping of individuals who have a tendency to vary within certain limits (Jean-Baptiste Pierre Antoine de Monet, Chevalier de Lamarck; Charles Darwin).

Individuals who share a similar appearance and behavior and are capable of reproducing belong to the same species, but even more importantly, individuals are supposed to transform over the course of generations in order to become a new, distinct species (Lamarck 1809, 106). At the time of the publication of *On the Origin of Species* (1859), the terms 'organism' and 'species' were roughly used as synonyms in the linguistic context, but in actual practice, language was studied either as a static anatomical entity or as a supra-individual category prone to modification. Even if the idea of the transformation of species had started to take root already in the latter half of the eighteenth century, it is not quite clear whether linguists were able to cease and implement the subtleties of Darwin's theory. Darwin's evolutionary model was based on the idea that each individual bears the characteristics of its species, but that each individual differs a little from its neighbor, and it is this difference that contains in it the germ of evolution. Linguists emphasized the evolutionary history of languages and recognized variation in the level of idiolects, but at the same time, they were tempted to associate evolution either with progress from simple beings toward complex beings (incorporated already in the ancient Great Chain of Being) or with the lifecycle of an individual organism (birth, growth, maturity, death).

Schleicher's pamphlet was translated into French in 1868 (*La théorie de Darwin et la science du langage*) with a foreword by Michel Bréal (Tort 1980). A whole school of thought was founded around Schleicher's naturalistic conceptions in France, as Piet Desmet (1996) has scrupulously demonstrated. With Abel Hovelacque as their spokesman, the members of this school contributed to developing and diffusing the literal meaning of the *life of language* in the footsteps of their German paragon:

> In fact, languages, like all living beings, are born, grow, grow old and die. They first go through the embryonic stage, reach their full growth and finally surrender in front of a regressive metamorphosis. Indeed, this perception of the life of language [...] distinguishes the modern linguistics from the past speculations. (Hovelacque 1877 [1876], 9)

However, despite their prolific output, the tenants of this school remained quite marginal with respect to academia because of the disputable implications of the literal interpretation of the life of language. The 'mainstream' linguists in France took a critical stance toward the widespread use and mappings of the life of language, which merely corresponded to a complex synthesis of different conceptions of natural history, from comparative anatomy and transformationism to embryology and anthropology. In addition, the fundamental unit of classification became a major issue. The concept of language encountered the same problems as the concept of species, because it could be considered as either a natural or an artificial category, depending on the epistemological posture of each linguist. Contrary to the common belief regarding the ontology of these categories, in France the idea prospered that species as well as language are abstract units not found in nature. In nature, there are only individuals, and the potential transformations occur only at the local level of individuals.

Schleicher's pamphlet was translated into English in 1869 (*Darwinism tested by the Science of Language*) and reviewed by (Friedrich) Max Müller (1823–1900) the following year in the newly founded journal *Nature*. Müller had already developed similar ideas in his *Lectures on the Science of*

Language at the Royal Institution of Great Britain, especially on the question whether the science of language should be treated as a natural or a historical science. As far as the *life of language* is concerned, he estimated that we must not let ourselves be fooled by metaphors:

> Ever since Horace it has been usual to compare the growth of languages with the growth of trees. But comparisons are treacherous things. What do we know of the real causes of the growth of a tree, and what can we gain by comparing things which we do not quite understand with things which we understand even less? [...] If we must compare language with a tree, there is one point which may be illustrated by this comparison, and this is that neither language nor the tree can exist or grow by itself. Without the soil, without air and light, the tree could not live; it could not even be conceived to live. It is the same with language. Language cannot exist by itself; it requires a soil on which to grow, and that soil is the human soul. To speak of language as a thing by itself, as living a life of its own, as growing to maturity, producing offspring, and dying away, is sheer mythology; and though we cannot help using metaphorical expressions, we should always be on our guard, when engaged in inquiries like the present, against being carried away by the very words which we are using. (Müller 1861, 51)

Even if Müller expressed his distrust of the literal interpretation of the naturalistic metaphor, which involved the independence of language from human will, he nevertheless subscribed to the idea that the science of language was a natural science by its methods. Like Schleicher, Müller declared his positivism openly and considered that modern linguistic science should be based on the objective observation of facts as opposed to having a prior theoretical framework or a rational system as its point of departure. Consequently, any general conclusions were the result of inductive reasoning based on a careful observation of a representative sample of particular and individual facts. However, a simple classification based on the detailed description of facts could not guarantee the scientificity of a theory, because a scientific theory in the full sense of the term consisted of the interaction between empirical verification and falsification.

As a critical response to his colleagues' position, William Dwight Whitney published *The Life and Growth of Language* (1875). He denounced one by one the mappings of the literal interpretation of the 'life and growth of language' and proposed an alternative reading in order to show what 'life' really meant when associated with language:

> Life, here as elsewhere, appears to involve growth and change as an essential element; and the remarkable analogies which exist between the birth and the growth and decay and extinction of a language and those of an organized being, or of a species, have been often enough noticed and dwelt upon: some have even inferred from them that language is an organism, and leads an organic life, governed by laws with which men cannot interfere. (Whitney 1875, 34)

For Whitney, 'life' is thus the problem requiring a solution, but instead of looking for the answer in the natural sciences, a linguist should merely take recourse to history which, operating with the meanings of the same source domain, makes the whole concept of life quite superfluous. The indeterminacy of the concept of life with respect to language may be highlighted by more figurative mappings. The target concept of life is explained as growth, development, evolution, progress, change, variation, and movement. Only in this symbolic sense can the metaphor of life be extended from a living being to other objects of knowledge involving change and variation.

LIFE (OF LANGUAGE) IS TRANSFORMATION.
LIFE (OF LANGUAGE) IS CHANGE.
LIFE (OF LANGUAGE) IS VARIATION.
LIFE (OF LANGUAGE) IS MOVEMENT.

Taking up the wording of Müller, Ferdinand de Saussure (CLG/E 1974, N 10) attacked the use of metaphors that his time had adopted. He complained that in its quest to identify the nature of language, linguistics was filled with 'mythological beings' and 'all kinds of analogies between heaven and earth', whose meaning depended on 'the infinite game of epithets' attributed to them. Saussure was well aware of the fact that the 'epithets'—that is, the conceptual mappings of the metaphor—determine the fundamental meaning of the linguistic metaphor. On the other hand, however, his lessons left a lasting impression on his students because of the numerous metaphorical devices he deployed when explaining language, thereby showing that even the most rigorous scientific mind cannot help using figurative language, be it only for heuristic purposes:

> 'La langue est le vaisseau en mer, non plus en chantier'
> (Language is a ship at sea, not a ship under construction) (Capt-Artaud 1994, 143)

This metaphor, reminding us of the more famous Humboldtian metaphor (*ergon—energeia*), is one of the leftovers that was not integrated into the *Cours de linguistique générale* by the editors. It is, however, an illustration of the numerous means a linguist may have at his disposal when trying to approach the slippery ontology of language, which is intuited as something evermoving. It is not difficult to establish overlapping mappings between this particular metaphor and life—like a ship, a living language moves with the natural motion of waves. The question of the master of the vessel opens up the way to new metaphors.

3.3 Scientific Implications of the Life of Language

Even if some of the conceptual mappings of 'life' seemed to overlap in the theories of Schleicher and Whitney, their implications for linguistics as a science did not. Lucien Adam (1833–1918) –an atypical representative of the naturalistic school in France, according to Piet Desmet (1996)—aimed to raise linguists' awareness of 'the opposite solutions offered by great masters to the most important problems of the science of language' (Adam 1882, 53). He formulated three explicit assertions that are the key to understanding the whole complex of the place of linguistics in the classification of sciences in this context of production (Adam 1882, 53; Klippi 2010, 177):

1. "Linguistics is a natural science" (August Schleicher, Max Müller, Abel Hovelacque, Julien Vinson).
2. "Linguistics is a historical science" (William D. Whitney).
3. "Linguistics is a historical science by its object, a natural science by its method" (Friedrich Max Müller).

The object of linguistics seemed to be both natural and historical, and in this respect, linguistics was as controversial as other new fields of knowledge whose object shared these characteristics. The integration of the idea of transformation, variation, change, and movement in the commonly received classification of sciences reversed the usual order of things and led to interdisciplinary discussions on the meaning of 'natural' and 'historical' as possible epithets of science. Ever since Aristotle, it had been considered that 'there is no science of individual, particular, contingent and variable' (Cournot 1975 [1851], 360). Accordingly, one of the main distinctive features of science was the absence of history, which, by definition, involved particular facts and events and was left at the mercy of chance and contingency, whereas the real sciences in the positivist spirit sought to reveal the immutable laws and constants of nature without taking into account any historical data (Cournot (1975 [1851], 365). In addition to positive, empirical sciences (physics and chemistry), there were negative, abstract sciences (mathematics, logic, and even grammar[6]) that reasoned in terms of theorems and syllogisms. Despite their close connection to human activity, these sciences did not involve any historical element. However, even the positive philosophy of science was obliged to yield its principles before a new type of facts:

> Il faut distinguer, par rapport à tous les ordres de phénomènes, deux genres de sciences naturelles: les unes abstraites, générales, ont pour objet la découverte des lois qui régissent les diverses classes de phénomènes, en considérant tous les cas qu'on peut concevoir; les autres concrètes, particulières, descriptives et qu'on désigne quelquefois sous le nom de sciences naturelles proprement dites, consistent dans l'application de ces lois à l'histoire effective des différents êtres existants (Comte 1996, 94).

[6] Dans les langues, la structure grammaticale est l'objet d'une théorie vraiment scientifique; à part quelques irrégularités qu'il faut imputer au caprice de l'oreille et de l'usage, le raisonnement, l'analogie rendent compte des lois et des formes syntaxiques; tandis que la composition matérielle des mots et les liens de parenté des idiomes ne peuvent en général s'expliquer que par des précédents historiques. (Cournot 1975 [1851], 314)

In languages, the grammatical structure is the object of a truly scientific theory; apart from some irregularities which must be imputed to the caprice of the ear and of use, reasoning and analogy account for laws and syntactic forms; while the material composition of words and the kinship ties of idioms can in general only be explained by historical precedents.

> It is necessary to distinguish, with respect to all the orders of phenomena, two kinds of natural sciences: the abstract and general have as their object the discovery of the laws which govern the different classes of phenomena in all conceivable cases; and the concrete, particular and descriptive, sometimes referred to as natural sciences in a restricted sense, whose function is to apply these laws to the actual history of different existing beings.

New scientific fields (cosmology, geology, paleontology, theory of evolution, anthropology, and the genealogical classification of languages) emerged in the nineteenth century, challenging the dominant positivist attitude. These sciences may be termed *onto-historical*, because 'temporality is an essential part' and 'an identification tool of their object' (Auroux 1998, 145–157).

The importance of temporality and historicality in explaining language can be highlighted by some sparse contemporaneous testimonies. According to Hermann Paul (1846–1921), linguistics equals the historical study of language. Michel Bréal (1832–1915) considered that synchronic norms seem arbitrary and illogical without an explanation and justification by history (Bréal 1864–1865, 45). Looking back over the history of linguistics in the nineteenth century, Albert Dauzat (1906, 73) held that the idea of evolution had revolutionized all the sciences of observation and changed the old empirical and descriptive grammar into a linguistic science. In a lighter tone, Ferdinand de Saussure claimed that diachrony is more fun than synchrony, and, in a more serious manner, 'there are very few objects available to our knowledge that escape historical succession': 'everything in this world has a historical side' (CLG/E 1974, 23, 26). Despite the positivist credo, the nineteenth century was not considered the century of history in vain, because all the knowledge ultimately had an onto-historical basis.

The readings of the *life of language* are thus rooted in a more general epistemological discussion that developed throughout the century (for instance, in the writings of Auguste Comte (1798–1857), William Whewell (1794–1866), and Antoine Augustin Cournot (1801–1877)) and gravitated around the controversial issue of the place and status of linguistics with respect to the rest of the sciences, and as a consequence, the nature of its object. The literal reading of the *life of language* by Schleicher shows that linguistics, like several other fields of study, had to cope with an inferiority complex in relation to the exact natural sciences, and in a similar way, it aimed at reaching the standards of natural laws. The same is true regarding natural history as a 'natural science'. It has been pointed out that the theory of evolution was not based on any universal natural law before the discovery of the role of genes in heredity, but was rather a historical assertion about particular and unique events (Popper 1989 [1957], § IV). For this reason, natural history (or biology) was not a science in the same way as physics or chemistry, but rather a 'hypothesis' based on abduction or a 'metaphysical research program' (Popper 1989 [1957], § IV).

4 Conclusion

The fundamental difference between the literal and metaphorical interpretations of the *life of language* is that they are not defined as equidistant from four different disciplinary types: formal, normative, historical, and nomological. A discipline receives particular epistemological properties depending on the mode of knowledge validation. The content of knowledge is conveyed by four types of propositions that correspond to their disciplinary type (see Auroux 1998, 138–139):

1. Formal sciences: theorem or correct proposition within a given system ($a^2 + b^2 = c^2$);
2. Normative sciences: normative rule (*Nje plivatz na pol, Älä sylje lattialle, Spotta ej på golfvet* 'Don't spit on the floor')[7]
3. Historical sciences: fact or non-universal assertion (*Urho Kaleva Kekkonen, the eighth president of Finland, was bald*);
4. Nomological sciences: law or universal assertion (*Water freezes at zero degrees Celsius*).

[7] Signboards with this text could be found in the public transports in Finland in three territorial languages at the beginning of the twentieth century, and still in the seventies in the two national languages.

However, this classification may not be permanently fixed: one and the same discipline may have several properties, and one discipline may change its validation type throughout its history. This is particularly the case with linguistics. The distance from different types of disciplines also appears to be specific to each reading of the *life of language*. The main axis of linguistics in the nineteenth century is between the historical and nomological sciences, with serious reservations from the 'normative' side that show a tendency to associate language with a human artifact used by real speakers. On the one hand, the literal interpretation of the *life of language* coincides with the onto-historical implications of natural history, which deals with facts or non-universal assertions, leading thereby to a greater distance from the [nomological] natural sciences, despite their ultimate desire to establish universal laws for language. On the other hand, the onto-historical interpretation of the *life of language*, which results from the dynamic nature of language (development and history), comes closer to the normative sciences in the sense that normative rules and institutions are also likely to change with time and space,[8] but unlike the object of other onto-historical sciences, the object of knowledge in linguistics is not always expected to lead an autonomous life with respect to its human support, albeit in a metaphorical way. However, the idea of autonomy, intrinsic in the comparative grammar, resurfaces in a new guise with a renewed focus on linguistics as a formal science: LANGUAGE IS CALCULUS.

The *life of language* shows thus that there is not necessarily an insurmountable gap between different types of sciences. On the contrary, in actual practice it is possible to 'crossbreed' the ontological, theoretical and methodological features, as Sylvain Auroux has pointed out:

> Il n'y a aucune raison pour que les objets réels, historiques et culturels, que sont les disciplines scientifiques, correspondent à des types idéaux. Il faut donc changer le point de vue classificatoire habituel. La seule méthode valable consiste à aborder la question des caractéristiques d'une discipline,

[8] For instance, in certain parts of the world *Spit on the floor* may have been considered as a normative rule, because according to the local vision of world concerning human health 'what has to come out, must come out', whereas such behavior would be seen in our modern Western world as vulgar, inappropriate, and even contrary to public health.

> non pas d'un point de vue classificatoire préalable, mais en établissant des propriétés attestées. Bien entendu, une discipline identifiée peut posséder différentes propriétés et en partager certaines avec d'autres disciplines. (Auroux 1998, 137–138)
>
> There is no reason why scientific disciplines as real, historical and cultural objects should correspond to some ideal types. We must therefore change the point of view of the usual classification. The only valid method is to approach the question of the characteristics of a discipline by establishing its attested properties, instead of a prior classificatory point of view. Of course, a particular discipline may have different properties and share some with other disciplines.

To consider science in the constructivist light as a social practice that takes a discursive form, the discourse over the literal and metaphorical meaning of the *life of language* raises the societal question of power: who exerts and possesses power over scholarly matters, and on what basis are scientific ideas accepted and adopted, favored or rejected within a community of scholars? Even if the *life of language* allowed multiple openings as a theoretical explanation, each linguist was ultimately convinced of the correspondence between their theory and reality. At the same time, however, as Michel Bréal (1897: 257) observed in an insightful manner, the classification of linguistics among natural sciences, with the use of the life of language, was a political act not only to attract the public opinion, but also to convince positivistically oriented fellow scientists and decision-makers of the legitimacy of linguistics as a science of its own right.

References

Adam, Lucien. 1882. *Les classifications, l'objet, la méthode, les conclusions de la linguistique*. Paris: Maisonneuve et Compagnie.

Alter, Stephen G. 1999. *Darwinism and the linguistic image: Language, race and natural theology in the nineteenth century*. Baltimore/London: Johns Hopkins University Press.

Aristotle [Aristoteles]. 1997. Retoriikka ja runousoppi. *Teokset IX*. Helsinki: Gaudeamus.

Auroux, Sylvain. 1998. *La raison, le langage et les normes*. Paris: Presses Universitaires de France.

Bally, Charles. 1909. *Traité de stylistique française I*. Paris: Librairie C. Klincksieck.

———. 1921. *Traité de stylistique française II*. Paris: Librairie C. Klincksieck.
———. 1965 [1913]. *Le langage et la vie*. Genève: Librairie Droz.
Berger, Peter, and Thomas Luckmann. 1966. *The social construction of reality. A Treatise in the Sociology of Knowledge*. New York: Doubleday.
Black, Max. 1954. Metaphor, Proceedings of the Aristotelian Society, 55, 273–294.
———. 1962. *Models and metaphors*. Ithaca: Cornell University Press.
Blackmore, Susan. 1999. *The meme machine*. Oxford: Oxford University Press.
Blanckaert, Claude. 2004. *La nature de la société. Organicisme et sciences sociales au XIXe siècle*. Paris, Harmattan.
Boyd, Richard. 1993 [1979]. Metaphor and theory change: What is 'metaphor' a metaphor for? In *Metaphor and thought*, ed. Andrew Ortony, 481–532. Cambridge: Cambridge University Press.
Bréal, Michel. 1864–1865. De la méthode comparative dans l'étude des langues. In *De la grammaire comparée à la sémantique. Textes de Michel Bréal publiés entre 1864 et 1898*, ed. Piet Desmet and Pierre Swiggers, 63–79. Louvain/Paris: Peeters.
———. 1897. *Essai de Sémantique*. Brionne: Gérard Monfort.
Capt-Artaud, Marie-Claude. 1994. *Petit traité de rhétorique saussurienne*. Genève: Librairie Droz.
Charles, Darwin. 1859. *On the origin of species*. Harmondsworth: Mentor.
Charteris-Black, Jonathan. 2004. *Corpus approaches to critical metaphor analysis*. New York: Palgrave Macmillan.
Chomsky, Noam. 1965. *Aspects of the theory of syntax*. Cambridge, MA: MIT Press.
———. 1980. *Rules and representations*. Oxford: Blackwell.
CLG/E 1974 = Saussure, Ferdinand de. 1974. *Cours de linguistique générale*. Wiesbaden: Otto Harrassowitz. Edition critique par Rudolf Engler.
Comte, Auguste. 1996. *Philosophie des sciences*. Paris: Gallimard.
Cournot, Antoine Augustin. 1975 [1851]. *Essai sur les fondements de nos connaissances et les caractères de la critique philosophique*. Paris: Vrin.
Cuvier, Georges. 1825. *Discours sur les révolutions de la surface du globe et sur les changemens qu'elles ont produits dans le règne animal*. Paris: G. Dufour et Ed. D'Ocagne.
Darmesteter, Arsène. 1979 [1887]. *La vie des mots étudiée dans leurs significations*. Paris: Librairie Delagrave.
Dauzat, Albert. 1906. *Essai de méthodologie linguistique*. Paris: Librairie Honoré Champion.
———. 1910. *La vie du langage*. Paris: Librairie Armand Colin.

Davidson, Donald. 1978. What metaphors mean? *Critical Inquiry* 5 (1): 31–47.
Dawkins, Richard. 1976. *The selfish gene*. Oxford: Oxford University Press.
Deignan, Alice. 2005. *Metaphor and corpus linguistics*. Amsterdam/Philadelphia: John Benjamins.
Desmet, Piet. 1996. *La linguistique naturaliste en France (1867–1922). Nature, origine et évolution du langage*. Leuven/Paris: Peeters.
Dictionnaire de l'Académie française. 1832–1835. Paris: Imprimerie et librairie de Firmin Didot Frères. Imprimeurs de l'Institut de France. [Sixth edition].
———. 1694. Paris: La Veuve de Jean Baptiste Coignard, Imprimeur ordinaire du Roy, & de l'Académie Françoise. Avec privilège de Sa Majesté. [First edition].
———. 1762. Paris: La Veuve de Bernard Brunet, Imprimeur de l'Académie Françoise. [Fourth edition].
———. 1798. Revu, corrigé et augmenté par l'Académie elle-même. Paris: J.J. Smits et Ce. [Fifth edition].
Donnellan, Keith S. 1972. Proper names and identifying descriptions. In *Semantics of Natural Language*, ed. Donald Davidson and Gilbert Harman, 356–379. Dordrecht: Reidel.
Du Marsais, César Chesneau. 1730. *Des Tropes ou Des diferens sens dans lesquels on peut prendre un mème mot dans une mème langue*. Paris: La Veuve de Jean-Batiste Brocas.
Féraud, Jean-François. 1787–1788. *Le Dictionaire critique de la langue française*. Marseille: Jean Mossy Pere et Fils. Avec approbation, et privilège du Roi.
Foucault, Michel. 1969. *L'archéologie du savoir*. Paris: Gallimard.
Gentner, Dedre. 1989. The mechanisms of analogical learning. In *Similarity and analogical reasoning*, edited by Stella Vosniadou and Andrew Ortony, 199–241. Cambridge: Cambridge University Press.
Henry, Victor. 1896. *Les antinomies linguistiques*. Paris: Didier Erudition.
Hoenigswald, Henry, and Linda F. Wiener. 1987. *Biological metaphor and cladistic classification: An interdisciplinary approach*. London: Frances Pinter.
Hovelacque, Abel. 1877 [1876]. *La linguistique*. Paris: C. Reinwald et Compagnie.
Itkonen, Esa. 1978. *Grammatical theory and Metascience*. Amsterdam: John Benjamins.
———. 2005. *Analogy as structure and process*. Amsterdam/Philadelphia: John Benjamins.
Jastrow, Joseph. 1899. The mind's eye. *Popular Science Monthly* 54: 299–312.
Jespersen, Otto. 1922. *Language: Its nature, development and origin*. London: George Allen & Unwin Ltd; New York: Henry Holt and Company.

Klippi, Carita. 2007. La première biolinguistique. *Histoire Epistémologie Langage. Les désordres du naturalisme* 29/II: 17–41.

———. 2010. *La vie du langage. La linguistique dynamique en France de 1864 à 1916*. Lyon: ENS Éditions.

Koselleck, Reinhart. 1990 [1979]. *Le futur passé. Contribution à la sémantique des temps historiques*. Paris: EHESS.

Kövecses, Zoltán. 2005. *Metaphor in culture: Universality and variation*. Cambridge/New York: Cambridge University Press.

Kripke, Saul. 1980. *Naming and necessity*. Oxford: Blackwell.

Kuhn, Thomas S. 1962. *The structure of scientific revolutions*. Chicago: Chicago University Press.

———. 1993 [1979]. Metaphor in science. In *Metaphor and thought*, ed. Andrew Ortony, 533–542. Cambridge: Cambridge University Press.

Lakoff, George, and Mark Johnson. 1980. *Metaphors we live by*. Chicago: The University of Chicago Press.

Lakoff, George, and Mark Turner. 1989. *More than cool reason*. Chicago/London: The University of Chicago Press.

Lamarck, Jean-Baptiste. 1809. *Philosophie zoologique I-II*. Paris: Dentu et L'Auteur.

Larousse, Pierre. 1866–1888. *Le grand dictionnaire universel du XIXe siècle*. Paris: Administration du Grand dictionnaire universel.

Maher, J. Peter. 1983. Introduction. In *Linguistics and Evolutionary Theory: Three Essays by August Schleicher, Ernst Haeckel, and Wilhelm Bleek*, ed. Konrad Koerner, xvii–xxxii. Amsterdam/Philadelphia: John Benjamins.

Molino, Jean. 1979. Métaphores, modèles et analogies dans les sciences. *Langages* 54: 83–102.

Morpurgo Davies, Anna. 1987. 'Organic' and 'organism' in Franz Bopp'. In *Biological metaphor and cladistic classification: An interdisciplinary approach*, ed. Henry Hoenigswald and Linda F. Wiener, 39–80. London: Frances Pinter.

Müller, Friedrich Max. 1861. *Lectures on the science of language*. London: Longman, Green, Longman, and Roberts.

Müller, Max. 1870. Darwinism tested by the science of language. *Nature* 1 (10): 256–259.

Müller, Friedrich Max. 1887. *Biographies of words and the home of the Aryas*. London: Longmans, Green and Co.

Normand, Claudine. 1976. *Métaphore et concept*. Bruxelles: Editions Complexe.

Ortony, Andrew. 1993 [1979]. *Metaphor and thought*. Cambridge: Cambridge University Press.

Popper, Karl. 2002 [1934]. The Logic of Scientific Discovery. London/ New York: Routledge.

Popper, Karl. 1989 [1957]. The Poverty of Historicism. London / New York: Ark Paperbacks.

Plato [Platon]. *Gorgias, Menon, Meneksenos, Euthydemos, Kratylos. Teokset 2.* Helsinki: WSOY.

Richards, Ivor Armstrong. 1936. *The Philosophy of Rhetoric.* Oxford: Oxford University Press.

Ricœur, Paul. 1975. *La métaphore vive.* Paris: Seuil.

Sapir, Edward. 1921. *Language. An introduction to the study of speech.* New York: Harcourt, Brace.

Schlanger, Judith. 1971. *Les métaphores de l'organisme.* Paris: Vrin.

Schleicher, August. 1863. *Die Darwinische Theorie und die Sprachwissenschaft.* Weimar: Hermann Böhlau.

———. 1869 [1863]. *Darwinism tested by the science of language.* Trans by. Alex V. W. Bikkers. London: John Camden Hotten.

Sperber, Dan. 1996. *La contagion des idées: théorie naturaliste de la culture.* Paris: O. Jacob.

Sullivan, Karen. 2013. *Frames and Constructions in Metaphoric Language.* Amsterdam/Philadelphia: John Benjamins.

Tort, Patrick. 1980. *Evolutionnisme et linguistique.* Paris: Vrin.

Vendryes, Joseph. 1968 [1921]. *Le langage. Introduction linguistique à l'histoire.* Paris: Albin Michel.

Vereza, Solange. 2008. Exploring metaphors in corpora: A study of 'war' in corpus generated data. In *Confronting metaphor in use: An applied linguistic approach,* ed. Mara Sophia Zanotto, Lynne Cameron, and Marilda do Couto Cavalcanti, 163–180. Amsterdam/Philadelphia: John Benjamins.

Wells, Rulon S. 1987. The life and growth of language: Metaphors in biology and linguistics. In *Biological metaphor and cladistic classification: An interdisciplinary approach,* ed. Henry Hoenigswald and Linda F. Wiener, 39–80. London: Frances Pinter.

Whitney, William Dwight. 1875. *Life and Growth of Language.* London: Henry S. King and Co.

———. 1877 [1875]. *La vie du langage.* Paris: Didier Erudition.

Wittgenstein, Ludwig. 1968 [1953]. *Philosophical Investigations.* Oxford: Basil Blackwell.

Zanotto, Mara Sophia and Palma, Vesaro Dieli. 2008. 'Opening Pandora's Box: Multiple readings of a metaphor' *Confronting metaphor in use: An applied linguistic approach* edited by Mara Sophia Zanotto, Lynne Cameron, and Marilda do Couto Cavalcanti, 11–43. Amsterdam/Philadelphia: John Benjamins.

15

Linguistics and Philosophy: Break Up Song

Kate Hazel Stanton

1 Introduction

Linguistics and philosophy have enjoyed a special relationship spanning over half a decade. From the mid-twentieth century to the present, there has been much fruitful cross-pollination; work in philosophy has contributed directly to advances in linguistics, and work in linguistics has been offered as evidence for philosophical conclusions. Journals and major conferences present both in tandem; sharing out space between advances in either discipline and anticipating audiences with overlapping interests. A standard introductory class in philosophy of language will now contain a little linguistic semantics (perhaps a few chapters of Heim and Kratzer (1998), which itself begins with Frege and Davidson) and an introductory linguistic semantics class will contain a little philosophical

K. H. Stanton (✉)
University of Pittsburgh, Pittsburgh, PA, USA

Pittsburgh Center for Philosophy of Science, Pittsburgh, PA, USA
e-mail: katehazelstanton@pitt.edu

R. M. Nefdt et al. (eds.), *The Philosophy and Science of Language*,
https://doi.org/10.1007/978-3-030-55438-5_15

rumination on the nature of meaning or meaning-related notions such as reference, modality and truth conditions. In the heyday of exchange, partisans of both camps waxed optimistic about the potential for a progressive strengthening of the interdisciplinary ties. A small selection:

- "linguistic theory incorporates answers to significant philosophical problems." (Katz 1965: 590)
- "linguistics and philosophy, like steak and barbecue sauce, have much to give each other" (Higginbotham 2002: 575).
- "[e]ven as the role of linguistics in the philosophy of language is constantly changing, one may recognize it as a constant fact that the relationship between the disciplines is greatly productive for both." (Moss 2012: 11)

The usual setting for such optimism is reflection on a litany of actual or promising potential successes in interdisciplinary coordination. Sarah Moss (2012: pp. 5–8) for example, details an impressive range of interdisciplinary work on the nature of conditionals both subjunctive and indicative, descriptions both definite and indefinite, alternative semantics, situation semantics and more. Lepore and Pelletier in their (2008) offer a detailed discussion of the philosophical heritage and justification of event semantics. Katz (1965) argues for the promise of a linguistic theory of logical form. But how optimistic should we be? This paper takes a snapshot of what underpins the successes of the special relationship, and argues that it is time for linguistics and philosophy to talk; key points of contact that formerly sustained easy exchange between the disciplines may be facing a path to atrophy. Landmark successes in interdisciplinary work were made possible by similarities both at the level of semantic theory and at the level of data collection and handling; focusing on semantic theory, I argue that these paths are beginning to diverge between the disciplines, leaving an unclear prognosis for continued close exchange.

Before discussing the interactions of linguistics and philosophy, however, I ought to spell out what I will mean by 'linguistics' and what I will mean by 'philosophy'. Linguistics comprises a very wide range of subfields and their interactions; some, like semantics, pragmatics, syntax and perhaps phonology, will be familiar at least in name to philosophical audiences. Others, such as socio-linguistics, computational linguistics, psycholinguistics, will be a little more alien (we will get to that later). Similarly, linguists will be familiar with certain areas of philosophy, such as philosophy of language, and perhaps metaphysics and epistemology, but less concerned with other areas, such as philosophy of sociology or biology. So to discuss interactions between the fields I must first identify the subparts with which I will be concerned.[1] The closest connection, and so the most natural site for observing the state of the special relationship, is between those subfields concerned with the recovery of speaker meaning: linguistic semantics, along with a little syntax and pragmatics and philosophy of language. Talk of 'semantics' and 'pragmatics' is shared currency between the disciplines; it is not rare for researchers in either field to represent their work as contributing to thematically unified bodies of inquiry distributed across departments, in something like the way that research programs in biochemistry are distributed across biology and chemistry. To make things simpler, hopefully without distorting the big picture, I will narrow the scope even further and restrict my discussion primarily to semantics, which I will take (somewhat controversially) to be the investigation of meaning determined by linguistic convention. This restriction to semantics will not be an exclusive one—here and there I will draw from work that integrates neighboring fields on either side of the division—but it will serve to anchor discussion.

2 Three Programs for Success

There are many ways to frame the reasons for success in bidirectional exchange between linguistics and philosophy, but in what follows, I will take my lead from a short dialogue between Jerrold Katz and Zeno Vendler. In his (1965) paper 'The Relevance of Linguistics to Philosophy', Katz makes a case for the philosophical significance of linguistic theory; linguistic theory, he claims 'specifies the universals of language – the principles of organization and interpretation that are invariant from one

[1] I should also note at the outset that I am concerned with delimiting fields and not individuals; I take it that individuals can contribute to either field at different times.

language to another' (1965: 590–1). The crossover seems natural: philosophy of language also deals in what is invariant across languages; while it might take cases from one or another language, it is not interested in cataloging the idiosyncrasies of the language in question (in producing what Katz calls 'linguistic descriptions'). And so both disciplines, insofar as they have had eyes on the bigger, language-invariant picture, stand to gain from dialogue. Zeno Vendler, in his comments on Katz' paper, draws attention to the interest of linguistic data for philosophy. He suggests that 'Conceptual investigations based upon the facts of language cannot but profit by a more systematic presentation and a deeper understanding of these facts.' (1965: 604). By this he means that the linguist's analysis of particular data, as containing e.g. ambiguity, polysemy, hidden structure, hidden constituents, can direct philosophers down the path of valid argumentation to insight concerning language-external conclusions.

Though Katz and Vendler aimed to show what linguistics can give philosophy, they identify points of contact that can be used to sort contributions from both directions. In what follows I will discuss semantic theory, though I will make one tweak to Katz' account. Since our focus is not on linguistics generally but on semantics in particular, I will talk about semantic theory: the search for principles of conventional interpretation and organization that are invariant from one language to another. I will review three metasemantic unifiers: research programs, that have directed much joint, fruitful semantic theory in the twentieth century. These are: the Fregean Program, Compositional Semantics, and Expanding the Fragment. I will suggest that mutual engagement in these research programs has sustained the relatively high bidirectional literacy that has characterized the special relationship; the ability of both sides to both consume and contribute to a range of core theoretical concepts and structures. I will then suggest that their grip on (at least) linguistics is beginning to fade, in light of a rise in statistical methods and models in linguistics that do not require them.

To introduce the three research programs it will be useful to start with sentences, as one of the early loci of widespread agreement, both in importance and interpretation, across linguistics and philosophy. Sentences denote propositions that determine sets of truth conditions. *Propositions* typically make their introduction with a list of characteristic properties: they are the objects of attitude verbs; the referents of that-clauses; types that share what is common between sentence tokens, even across translations (McGrath and Frank 2018). The content of Pepper's conviction that Lucky is happy, Lucky's belief that *Lucky khush hai*, and the that-clause of the sentences that I used to articulate those attitude states are one and the same proposition. The *truth conditions* of a proposition i.e. the conditions under which it is true, are typically stated by quoting proposition on the left hand side of a biconditional, which specifies that the enquoted sentence is true just in case the disquoted metalanguage translation on the right hand side holds. This tradition extends seminal work in logic on the interpretation of quantified first-order logic (Tarski 1935) to the interpretation of sentential meaning in natural language (c.f. Davidson (1967)). The truth conditions for the proposition that 'Lucky is happy' are expressed like this:

1. 'Lucky khush hai' is true iff Lucky is happy.

For many, the meaning of a sentence is identified with its truth conditions; grasp of those truth conditions is equated with grasp of sentence meaning (Lewis 1975). I will sidestep the range of discomforts and questions that propositions in general and the disquotational account of their truth conditions in particular typically occasion (that's not giving the meaning at all!).[2] (What is a proposition anyway?! (Quine 1960, King 2009, Schiffer 2003) Why truth conditions only and not other sorts of conditions, say, use or proof conditions? (Schroeder-Heister 2018, Horwich 2005, Potts 2005, 2007, Gutzmann 2015) Propositions have structure or no structure? (Cresswell 1985) Can they have centers? (Ninan 2010)). Instead, I want to focus on the program of specifying the way that the meaning of a sentence is built up from the meanings of its parts. I split this inquiry into two programs: the Fregean Program, and the program of compositionality.

[2] I do not include a particular citation for this sentiment because I take it to be the general reaction on learning that meaning is specified in this way.

2.1 The Program of Compositionality

When asked how the meaning of a sentence, or any complex expression, is built up from the meanings of its parts, linguists and philosophers will typically trot out (versions of) one answer:

> The meaning of a compound expression is a function of the meanings of its parts and of the way they are syntactically combined. (Partee 1984: 281)

This is the *principle of compositionality*, and while apparently intuitive, it is as stated in fact almost entirely uninformative. Its uninformativeness does not stem from its not actually saying what a meaning is—after all we asked a how-question, not a what-question. Rather it stems from gaps in the formulation. The principle says only that there must be some function (whichever that is) from part meaning (whatever that is) and syntactic ordering (whatever parts are) to complex expression meaning (whatever a complex expression should be). That's a lot of whatevers and if we leave them unconstrained that leaves an uncomfortable amount of room for the principle to endorse useless gibberish. Imagine that 'hello world' were syntactically chunked up as follows:

[1] [Hell][o wo][rld].

We can introduce a Humpty Dumpty-style interpretation rule to stipulate that 'Hell' means hello; [o wo] is a meaningless cry of despair, and [rld] means world. From this interpretation of parts and that syntax we will still generate a the desired complex meaning as a function of part meaning and syntax: [1] means: hello world (along with a bit of meaningless affect). But this reveals not much more about natural language than that it can be used as a code.

Say we commit to not doing anything unintuitive with syntactic ordering and interpretation of parts. This still leaves the principle almost entirely uninformative. Consider syntactic structure: does the following topicalized and non-topicalized have the same syntactic ordering?

[2] Pizza, I had for lunch.
[3] I had pizza for lunch.

The surface structure differs, but syntax for a long time held that this surface structure is the result of a superficial reorganization of one underlying form. This underlying form has gone by a range of names (LF, logical form, deep structure), as have the reorganization rules (transformations, *move*) and it's very existence has been called into question but the point here is merely that the syntactic ordering is not pretheoretically given. And now consider part meaning in the following:

[4] I good morninged him *sotto voce* but he cottoned on and good morninged me right back tit-for-tat.

Do idioms and dialect expressions have independently meaningful parts? To 'good morning' someone is a novel transitive verb, and so does it have meaning in the way that more established sentential neighbors do? Is its meaning in some sense 'built on the spot' from other known meanings, and if so do those other known meanings need to be factored in to our computation of whole meaning? Do we take apparent word boundaries at face value and count hyphenated expressions as having internal parts? Is it a *part* of [4] that it is a declarative utterance and not a question or command? Should tense count as its own part? Once more the point of these questions is to draw out that the principle is not a self-standing piece of semantic theory about natural language. It is a motivation and a guiding principle for to conducting further semantic theorizing, and it is one whose work has been conducted by both linguists and philosophers.

Even fixing the parthood and structure, a lot of possible funny business at the level of specifying the function still needs to be scored out. There is, for example a function that takes the meaning of 'hello' and the meaning of 'world', combines them in the order given by the syntax, and generates the compound meaning in two stages: first by giving an intermediate value that reverses the order of the constituents, then by reordering them properly and popping out the desired final value. Functions are cheap and most of them tell us nothing about the generation of natural language meaning; in order for a compositional meaning function to be useful for semantic theory, the kind of function that we will accept must be independently, empirically constrained. This general point has been widely acknowledged since early work from philosophy showed that compositional meaning functions can be constructed for non-compositional systems (Zadrozny 1994; cf. Dever (1999), Szabó 2000).

I have spent quite some time saying what the principle of compositionality does *not* say. This is worthwhile because part of its power as an interdisciplinary program comes from its being at once so intuitive as to appear almost necessary and so underspecified as to be almost contentless. The rest comes from being coupled with theories that fill in many of the blanks. These theories regularize part meaning and syntax; they place constraints on the way in which the function from these to whole meaning can be constructed; and they specify what the whole should be. If we believe that a proposition gives the meaning of a sentence and truth conditions express the content of a proposition then this is our target for the whole. We then need to represent part meaning in a way that allows us to combine parts to build up to disquotational truth conditions at the sentence level. If we believe that our grammar is generative, that is that it specifies the complex expressions of the language by recursive application of a finite stock of rules, then we need to be able to map from the output of this recursive grammar and part meaning to disquotational truth conditions.

This line on compositionality associates the program with a set of open research questions; those that concern how to deal with cases in which it is not clear how to characterize part meanings and syntax in such a way that we can move from them to disquotational truth conditions. These include idioms (Chae 2015); varieties of tricky cases in anaphora (King and Lewis 2016); presuppositions (Beaver 2001); integration of affect, gesture, prosody and more. The shape of these problems are also regular enough that there are high-level generalizations about how to resolve common tricky cases (Zimmerman 2012). The work under the scope of this program has been carried out both by both philosophers and by linguists.

2.2 The Fregean Program

I will now turn to the characterization of part meaning, which presents its own program. I believe that some of the most deeply entrenched framework assumptions of a field are to be found in its introductory texts, and in their highly influential *Introduction to Semantics in Generative Grammar*, Heim and Kratzer write:

> Frege, like Aristotle and his successors before him, was interested in the semantic composition of sentences. [...] he conjectured that semantic composition may always consist in the saturation of an unsaturated meaning component. But what are saturated and unsaturated meanings, and what is saturation? [...] Frege construed unsaturated meanings as functions. Unsaturated meanings, then, take arguments, and saturation consists in the application of a function to its arguments. (1989: 2–3)

We can tell the following story about (1) above, for example, 'Lucky' is a name denoting an individual (the one, in fact, sitting opposite from me right now). It can serve as the argument for the predicate 'is happy', which can be represented as a function that sorts entities into those that are happy, and those that are not.[3] Now applying 'Lucky' to 'is happy' gives us the truth conditions in (1) above: 'Lucky is happy' is true iff Lucky is happy. And we can evaluate these truth conditions to true or false depending on the state of the world: he appears to be smiling, so let's say he is.

'The Fregean program' is intended as a general moniker for the quest to identify function-argument structure in natural language; it is a useful, if perhaps somewhat misleading label. Frege himself was not attempting to identify a semantic parthood structure in natural language that would be true to linguistic parthood as revealed in competent speaker behavior.[4] Nor did he aim to build a representation that that might be informative with respect to some language-external domain, such as the state of the cognitive system at some level of abstraction, or sets of norms governing

---

[3] For simplicity in exegesis, I here ignore tense, aspect and so on.

[4] Many of his own insights have informed that project. Semanticists have widely adopted his tool of choice for working with natural language expressions, namely a first-order logic enriched with apparatus from set theory (we will address a further enrichment for function expression, the typed lambda calculus, in the next section). His analysis of quantificational expressions as relations between sets, for example, was widely taken up in later semantic theory (cf. Barwise and Cooper, 1981) and arguably paved the way for the Montagovian theory of generalized quantifiers. His interpretation of quantifiers as variable binders was also widely taken up.

patterns of linguistic behavior among speakers (cf. Yalcin 2018; Lewis 1975). Nor indeed did all semanticists searching for function-argument structure think of themselves as explicating and developing Frege's insight. A *caveat emptor* is thus in order: a great deal of additional baggage has been introduced by linguists and philosophers, in order that the quest might yield empirically or conceptually informative fruits.

For the systematic study of natural language the Fregean program has one great virtue: under its guidance it is legitimate to replace the diaphanous, opaque question: what are meanings and how do they mean, with a much more precise and tractable one: what kind of function-argument structure is a defensible representation of a given natural language expression? How do we characterize the functions? How do we characterize their arguments? How do we characterize their composition? Or equivalently, trading on the isomorphism between sets and their characteristic functions, how do we characterize the meanings of expressions and the meanings of their combinations as sets and operations on sets.

This project has driven, and continues to drive, much semantic theorizing. 'Giving a semantics' for an expression, across linguistics and philosophy usually means making a proposal for a function that is defensible in the sense: satisfies a range of theoretically motivated constraints. These include, but are not limited to: generates the desired truth conditions, or entailments; enters into the desired inference patterns; combines with the right types of expressions; helps capture patterns in meaning intuitions of competent speakers; captures evidence of dependence on context; can be integrated with our best syntactic, or psychological, or metaphysical or epistemological commitments. To see how meaning combination is projected into function argument structure in the Fregean program in a way that allows us to conduct general semantic theorizing, let's work with an example. Consider this well-known generalization over the meanings of privative adjectives (adjectives like 'fake', 'counterfeit' 'ersatz', 'phony', 'pseudo') from Richard Montague:

[2] $\forall Q_{<s,et>} \forall x_e [\text{ADJ}' \rightarrow \neg {}^{\vee}Q(x)]$.

This says that privative adjectives are, necessarily, functions that entail that for some property Q (say, being a gun), it is not the case that its argument has that property.[5] The meaning of a particular privative, 'fake' will be along the lines of:

[3] $\lambda Q_{<s,et>}\ \lambda x_e[\text{FAKE}' \rightarrow \neg{}^{\vee}Q(x)]$.
and fake gun:

[4] $\lambda x_e[\text{FAKE}' \rightarrow \neg{}^{\vee}\text{GUN}'(x)]$

This secures both that 'fake' can combine with a property-specifying common noun, and also cover the speaker intuition-validating inference that a fake Q is not a Q by building the privation rule into the meaning of 'fake'. This analysis gives little idea of how we will satisfy the range of finer grained empirical constraints I just mentioned—after all, meaning postulates are little more than a hat trick that sneaks the desired entailment into a constituent's meaning and allows us to yank it out at the point of inference. And this one does not even allow us to say, for example, that a fake *x* ought to at least resemble an *x* along some (contextually variable) axis. An elephant is not a fake gun, but since the semantics for 'fake' mandates only that fake *x*s not be *x*s, elephants should be in the extension. For this, what is required is a more detailed lexical semantics, i.e. a more detailed set of claims of the internal structure of the expression, projected into the functional representation. Accounts of privative adjectives that do so have been offered across linguistics and philosophy, most recently in Partee (2010a) and drawing on work in linguistics by Pustejovsky (1995), Del Pinal (2015). Advances on those finer grained constraints will occur within the Fregean program, at the level of modifications in function and argument.[6]

The point here is not to rehearse the semantics of privative adjectives, but rather to point out that it extends the Fregean program in a recognizable pattern. Drawing on a range of data and motivated by a range of theoretical desiderata, someone makes a proposal for function-argument

[5] For present purposes we can ignore the type description subscript and the type lowering operator $^{\vee}$ which simply serve to get the extension right.

[6] This is not something that Montague himself fretted about, since his aim was not to provide a empirically adequate theory of lexical meaning, but rather to use a formal language to capture semantic relations like entailment between sentence of natural language (cf. Janssen 2012, 244–246).

structure. Drawing on more data, or tightening the desiderata, someone else challenges or improves it. And this recipe is followed in many areas, including quantification, anaphora, conditionals, by someones who come from within both linguistics and philosophy.

Before moving on, let's pause to put together the results from the two programs and see where it gets us: compositionality and the Fregean program. These broad paradigms tell us that meanings are functions and their arguments that compose to build complex expression meaning. Applying the 'Lucky' entity to the 'is happy' function yields 1 just in case Lucky is happy and not otherwise. This leaves much open that philosophers and linguists have jointly explored and continued to explore, with points of good communication and even occasionally of agreement appearing as the successes that characterize the special relationship. The semantics of the further parts (the contribution of the copula, for example (e.g. Devitt 1990; Kratzer 1994)); the more detailed semantics of those parts covered (how do we understand names? (Kripke 1980; Montague 1973; Matushansky 2008) How do we understand properties like 'happiness'?; the justification of truth conditional interpretation, the right way to understand the process of composition, and the right way to understand what slice of linguistic reality the model represents. These further references indicate both the fertility of the programs and their enduring mutual interest.

2.3 Expanding the Fragment

What I have given so far is an impressionistic sketch of two programs that have herded both philosophers and linguists toward the same set of questions with the same set of framework assumptions and tools. The final program is headed by an ethos, inherited from the work of Richard Montague, that characterizes the way in which formal semantics has been carried out. So far I have introduced the guiding programs as largely indifferent between particular implementations: function compositions tracking to build disquotational truth conditions does not tell one what the underlying logical language looks like in which the functions are couched. Now I will fill in a little more detail about how this has in fact been fleshed out, in order to introduce the final program of expanding the fragment.

Formal semantic theory as we know it today arose as a product of the successful coordination of a range of fields, centrally: philosophy, linguistics and logic. Its core theoretical notions and basic framework arose from the fusion of multiple traditions that ran in parallel; predominantly those of Montague Grammar and Generative Semantics. Generative Semantics was, again broadly, the work that aimed to interpret the generative grammar of Chomsky with a semantic component, and though there was considerable dispute over exactly how to do that, I will sidestep these issues here in favor of introducing Montague's project.[7] Montague introduced the (now dominant) model-theoretic interpretation of sentences of natural language. In brief, constructing a model-theoretic interpretation of a language (logical or natural) requires first defining a model; this is a domain of individuals (sets and functions) and a function that assigns individuals from the domain to nonlogical constants of the target language. Rules must also be specified to recursively assign truth conditions to each expression generated by the syntax of the target language.[8] Semantical relations, such as entailment and contradiction, are then captured as relations between classes of models: entailment between sentences A and B, for example, is a relation in which all models in which A is true, B is true. To state the truth conditions of expressions of the object language, Montague's project used an intensional logic enriched with tools from set theory, and the lambda operator – a function that allows perspicuous expression of higher-order functions (in fact, the semantics for 'fake' above uses this operator).

Montague's work offers a natural setting in which to carry out both prior programs. It comes set up with a strong inbuilt form of compositionality known as rule-rule compositionality. Syntax and semantics walk in lock step: for every expression generated by the syntax there is an interpretation assigned by the model. It also offers a highly workable form of

[7] For the purposes of exegesis, I am recklessly charging over a considerable amount of history. The reader is heartily advised to consult Janssen (2012); Partee (2010b; 2016) for a fuller picture.

[8] For more details, cf. Abbott (1997), Janssen (2012). Note that this entails that all that was dependent on context was outside of the scope of semantics; until Kaplan () even pronominal elements were considered pragmatics (cf. Stalnaker 2014).

the Fregean Program; the elements of the domain that are assigned to nonlogical constants are sets and their characteristic functions: 'Lucky is happy' involves applying the interpretation of 'Lucky' (for Montague, this is the characteristic function of the set of 'Lucky' properties) to the interpretation 'happy' (the characteristic function of the set of happy properties in the model).

When Montague began specifying the truth conditions of natural language expressions in the typed lambda calculus, he began only with a fragment of English, namely a few intensional verbs and a few quantificational expressions. Quantifiers—expressions like 'most; some; all; partly; a bit of' and so on, concern generality, and allow us to make generalizations over cases, and to reason from the general to the particular. Their interest for Montague lay in their centrality to formalizing natural language inference. In PTQ Montague himself began with only a few of these—a small fragment of English; he anticipated, however, that his fragment would be gradually expanded as a reasonably adequate treatment of the truth conditions quantifiers was achieved. This is the program of *expanding the fragment*.[9]

Expanding the fragment was a program that many took up. Developing Montague's own insight, and integrating elements of the various ongoing semantic traditions to construct the discipline of formal semantics as we now know it was the work of both philosophers (including *inter anderem* David Lewis, Terence Parsons, Donald Davidson, Gilbert Harman, Jerry Fodor) and linguists (*inter anderem* Barbara Partee, Jerrold Katz, Paul Postal, Nino Cocciarella). From this union we inherit the groundwork of formal semantics as it is practiced today (though it would see updates to notation and logic); we also inherit a range of explanatory targets, some imported from the linguist's desire for close empirical fit, and some from a persisting logician's interest in tracking logical relations between sentences: a good semantic theory will account for ambiguity and synonymy, anomaly, entailment and contradiction.

---

[9] It should be noted here that I do not take this ethos to be inextricably tied to Montague-style model-theoretic semantics; at base it harbors a picture that the way to construct a semantic theory for a natural language is to select a representational medium of choice and extend it over a natural language.

3 Degenerating Research Programs?

The opening of this chapter cited a range of enthusiasms for the prospects of a close relationship between linguists and philosophers. Much enthusiasm stemmed and continues to stem from either success or from promise of success in the joint scope of the research programs above. If we have a guarantee that those programs will survive intact between disciplines, then there is reason to have considerable hope that the exchange will continue somewhat as it has. This is because shared framework assumptions will mean that even despite differences in the nature of the questions asked or the style and scope of the answers given, both linguists and philosophers will both count the same sort of thing as progress in semantics: a contribution that aids in expanding or adding detail to a compositional, post-Montagovian fragment.

This conditional optimism (granted, in an ultimately pessimistic paper) runs contrary to the line of another pessimistic missive: Cappelen (2017). Cappelen argues that philosophers ought to recognize a widening technical and empirical gap between themselves and linguists, and so ought to drop attempts to contribute to semantics.[10] He offers what he calls a 'division of labor' argument:

> [H]aving philosophers work on figuring out the semantics for a natural language is like having philosophers contribute to the literature on cancer treatment – we just don't have the training and it would be silly to have philosophers do that even if some of us have some half-baked ideas about how to do it (and took some summer courses on the side). (2017: 6)

I have general sympathy for this perspective if it is a claim about how things may go in the long term: just as the basic sciences specialized away from philosophy, so too many of its other offspring get their own department and (after some time moving back and forth) will entirely leave home. But as a claim about the present it is partially just false—linguists and philosophers can and do still expand the fragment together and often in much the same way; a cursory indication of this is that to determine

[10] See Nefdt (2019) for criticism.

the department of origin for much work in the major joint journals, a bit of Googling is required (Googling will probably fail to resolve the basic question of how to classify the work). And it partially rests on an assumption that useful contributions to semantics from philosophers ought to be exactly of the same kind as those made by linguists. It is not at all clear that this is what anyone wants or expects. Linguistically-minded philosophers and philosophically-minded linguists can offer novel logical tools (e.g. Kracht 2007; Fine 2017; Starr 2020), conceptual frameworks (e.g. Camp 2013; Ludlow 2014; Nickel 2016) and critical evaluation of work in linguistics (Collins 2007; Stanley 2000). And philosophically-minded linguistic semantics generates, operationalizes, and tests theories in a way that has wider implications for theories of the nature of mind and language (e.g. Asher 2011; Tonhauser et al. 2013; Potts 2010; Shan 2001).

My conditional optimism is not optimism, but it is not pessimism for Cappelen's reason. Cappelen's point was that the linguist and philosopher engaged in semantics are both trying to build the same house—only, one of them is not a builder but a speculative hobbyist and so is doing an embarrassing job. I have objected to that way of framing things (though I do not deny that all academic fields have well-meaning hobbyist neighbors). But I believe that we have real reason to be pessimistic if there is reason to believe that we are beginning to come apart at the level of the shared programs. A fissure at the level of the programs may spell the end for the special relationship because it will mean that philosophers and linguists will now have moved away from the idea of joint construction.

Why think that such a change might be in the offing? As Sarah Moss puts it:

> Semanticists are separated from philosophers of language in virtue of being alert to quite different adjacent fields: morphology, phonology and syntax, as opposed to metaphysics, epistemology and philosophy of mind, for instance. (2012, 10–11)

Moss believes that this means merely that philosophers and linguists will find different questions interesting, while remaining in broadly the same game, and Cappelen believes that it is already reason for philosophers to throw in the towel.[11] My position is intermediate between the two: I believe that it means that over time the fields will move in different directions. In linguistics, the early signs are perhaps more pronounced than in philosophy. Linguistics is, in general, under greater pressure to be able to integrate the results of one variety of linguistic analysis, with results from the relevant portions of other varieties. If a working linguist knows that, for example, conceptual structure, syntax, information structure will make a difference to interpretation then she is under pressure not to black box these fields out. Linguistic semantics routinely includes consideration of these areas. Philosophical semantics by contrast, centralizes related metaphysical and epistemological concerns.

So what is going wrong? I will outline a highly promising direction in which linguistics more broadly is moving namely that of statistical modeling. Routinized use of statistical tools can be observed in many subfields of linguistics, from phonology (cf. Kingston 2012), to pragmatics (cf. Schwarz 2017), and even in syntax (cf. Manning 2002). Models of processing, comprehension, acquisition, retrieval are very often statistical in nature, and this has been so since the nineties (cf. in particular Jurafsky 1996). And neither is this new; summaries of the widespread stochastic turn in linguistics date back to at least Abney (1996). At this point this direction is not dominant in formal semantics, and nor am I arguing that it will become so any time soon (though I do not score this out). I am merely pointing out that statistical methods have provided strong contenders for capturing and predicting linguistic behavior and suggesting that a stochastic turn within formal semantics will herald a destabilization of the special relationship. Here is the peal of doom: a wholesale shift will change the landscape of linguistic semantic theory sufficiently that bidirectional communication between philosophers and linguistics will see a significant drop in ease and fluency, with philosophy becoming restricted to the Montagovian corner. Philosophers will not be able to bootstrap into reading and contributing to a variety of trends in semantics merely from their training in intensional logic, and over time this gap will widen.

[11] "In order to understand natural language semantics you need the kind of training that linguists get. You need training in phonology, syntax, semantics, pragmatics, and forms of morphology, and all the other topics that you get a training in when you do a graduate program in linguistics" (Cappelen, 2017: 6).

As advances in linguistics and philosophy percolate down to general awareness and undergraduate-level teaching, newcomers to either linguistics or philosophy will fail to inherit many of the foundational tools to orient themselves in the world of the other. In what follows I will take two case studies: statistical methods in an area already familiar to and partially integrated with formal semantics, namely psycholinguistics, and a promising contender for the Montagovian throne in formal semantics, namely distributional semantics. So let's start with statistical methods in psycholinguistics.

3.1 Psycholinguistics

Psycholinguistics is the study of the psychology and neurobiology of interpretation; it can show us what the brain does when we interpret, and it can shed light on the nature of meaning indirectly, by relying on what we know about the brain and the cognitive system. Linguistics has always maintained a relationship with the output of cognitive science, and there has been a range of word meaning that stem directly from these, mostly entering at the level of pragmatics, with many enduring and standout contributions, including the prototype theory of Rosch (1975); the Relevance theory of Wilson and Sperber (1995); work on *ad hoc* scale construction by Hirschberg (1985). In recent years, however, experimental methods for testing and evaluating semantical hypotheses and statistical models for have been increasingly employed (and indeed solicited by mainstream joint journals) in semantics. I'll first consider experimental methods.

Formal semantics has seen a rise in the use of tools such as EEGs, Electrophysiology, self-paced reading studies, aphasia studies and more; most of these stemming from work in psycholinguistics. Needless to say, if the requirement that linguists *perform* experimental semantics gains influence then in Cappelen's words '[philosophers] don't have the training'. But then, many linguists will also be out of the game; access to the requisite practical training and equipment is not uniformly available for them either. But for our purposes the issue of interest is not the threat that semanticists should all join labs and philosophers will not be allowed in. Rather, it is that the increasing centrality of experimentation and it's interpretation lays the conceptual groundwork for linguists to leave the three programs behind entirely.

Consider the following. The majority of work coming from the psycholinguistic tradition as it concerns semantics moves from experimentation and the interpretation of results to semantic conclusions about meaning without adverting to a layer of compositional semantics (cf. e.g. Millburn et al. 2016; Zhang et al. 2018). The weight of disconfirming hypotheses about natural language rests entirely on experimentation and analysis. In a growing recognition of the significance of this work, courses in statistics, and statistics for linguists are increasingly common at graduate level. This is not at all to say that such psycholinguistic work has no downstream integration with compositional formal semantics—indeed much excellent work offers exactly this (Deo 2015). Merely that much self-standing work in semantics requires somewhat specialized statistical knowledge, including knowledge of relevant programming languages such as R and Matlab, and that this specialized knowledge serves a tradition on meaning that does not require the three programs.[12]

Again, the worry here is not Cappelen's point that philosophers mostly haven't taken the right courseload. It is certainly true that this discrepancy will present something of a threat to bidirectional literacy, as those out of the know will not be able to evaluate the connection between the results and the conclusion drawn as effectively as their trained counterparts. And without a layer of familiar semantic architecture, the philosopher's ability to develop or extend such work will be minimal. But let us not overstate: experimental semantics is rightly gaining importance and gaining ground but there is little reason to expect that it will overturn post-Montagovian semantics, because it does not itself provide a model of meaning that will rival the one offered by the three programs. In fact, it is often in one way or another borrowing assumptions from those programs, even if papers in this tradition have only unarticulated lambdas. I believe that a more genuine threat to the special relationship comes from deep familiarization of statistical methods in linguistic semantics.

[12] Indeed a popular textbook, Gries' *Statistics for Linguists* (2009) acknowledges and condenses much of this.

Montague will not fall from a bit of t-testing, but the psycholinguistic trend of omitting compositional semantics entirely or adding it in later raises a serious question about what it was contributing in the first place and whether anything else would be more useful. Irrelevance may beget irreverance, and open up an opportunity for an alternative program to enter. And there are alternatives. Indeed I will now turn to one candidate for replacement coming from Natural Language Processing (NLP): distributional semantics. Note that in what follows I do not claim that this is the only such approach that could displace the paradigms, but rather one of a family of approaches stemming from computational linguistics and NLP (cf. Ferrone and Zanzotto 2020 for overview).

3.2 Distributional Semantics

The fundamental insights of distributional semantics are present as early as Harris (1954) and Firth (1957), in his famous words 'you shall know a word by the company it keeps" (cf. Hermann and Blunsom 2013; Jurafsky and Martin 2019, ch. 6).[13] The basic claim is that the meaning of an expression can be represented by similarities between its contexts of occurrence, and that the more similar two expressions are, the more similar their contexts of occurrence will be. For our purposes it will suffice merely to lay out the basic character of the approach, and to show that its recent successes have marked it out as a promising contender to the formal semantic throne of the three programs (but for instructive detail see Jurafsky and Martin 2019). I do not claim, however, that it is the only such contender (cf. e.g. Pater 2019; Potts 2019).

Computational approaches to natural language semantics in linguistics are nothing new; they have been around since at least the 1980's, and persisted through the 90's with approaches to lexical meaning that were inspired by early Natural Language Processing; some of which tried to integrate its feature structures directly into the post-Montagovian framework and receive considerable and ongoing attention in philosophical

[13] For reasons of space I here omit discussion of the Fregean Context Principle, and Wittgensteinian use theory of meaning, but this connection has been noted (cf. Jurafsky & Martin 2019). See also Sahlgren 2008 for connections to structuralism.

and linguistic semantics (cf. Pustejovsky 1995, Copestake and Briscoe 2000). Insights from computational linguistics even drive the now mainstream formal semantic models for representing the semantic contribution of affect (Potts 2005, 2007). A more common, non-integrationist trend stemming from NLP, however, does away with the Montagovian framework and relies on a framework in which expressions and their semantic properties are represented by vectors (and perhaps complex expressions by their compositions (see Erk (this volume))). In brief, and skimming over much detail and variation, a vector is a sequence of numerical values. Different approaches build vectors in different ways, but the guiding intuition is that, for some target expression **w** (say, a word) and dataset **d** (say, a set of Wikipedia articles), these values will be based on counts for each expression in windows of consecutive expressions including **w** in **d** (cf. Jurafsky and Martin 2019 ch.6). This allows us to form a picture of the kinds of context in which **w** occurs: we might find that a target word 'dog', for example, frequently occurs in windows with 'bone' and 'eat' and not with 'apple' or 'antiquing'. A similarity measure on vectors for targets **w** and **w'** can be computed, and under the guiding intuition that like meanings occur in like contexts, its output is interpreted as similarity between the meanings of **w** and **w'** : we will find, for example, that 'dog' and 'doggy' have closer vectors than 'dog' and 'spoon' and correctly identify the former pair as more similar in meaning than the latter. Developments around this simple core allow us to build a very detailed, predictive picture of the way that meaning behaves in context.

Distributional semantics has several considerable advantages over a standard post-Montagovian semantics, though there is no space to detail them fully here. It has had considerable success in traditionally tricky areas such as in word sense disambiguation, and predicting novelty, e.g. novel compound availability. It has skills entirely absent in traditional Montagovian semantics but plausibly part of what one can do when one grasps a language, such as text summarisation, and ability to extract coarse-grained information about affect. It also has predictive power that outstrips anything possible within Montague semantics, with respect both to *diachronic* semantics (predicting and retrodicting meaning change) and with respect to predicting the probability of expression occurrence (what sub-expression is most likely to appear in the context of a larger complex) and expression context (in what contexts an expression is likely to occur).[14] Distributional semantics feeds on large sets of usage data, and so its empirical contact and ecological validity is far higher than a standard compositional semantics, which (when tested against usage data) is tested against small populations in experimental settings (e.g. via MTurk studies). It even appears to provide a better solution to a problem that historically provided abductive motivation for compositionality; namely learnability (cf. Pagin and Westerstahl 2010); Huebner and Willitz 2018).

Distributional semantics is not, at least not intrinsically, a compositional semantics, built around accounting for complex expression meaning in terms of part meaning. Composition was initially a vexing issue, in particular for vector-based semantics, although there have now been advances in showing how vectors compose to build representations for complex expressions (for overview see especially Mitchell and Lapata (2008), Le and Zuidema (2015), Potts (2019)). It does not represent expression meaning as function-argument structure and it is not couched in a model-theoretic framework. Contributing to it does not involve expanding the post-Fregean, Montagovian fragment. In short, it is not an extension of the programs above. But vector-based distributional semantics presents an exciting and increasingly popular proposal for modeling natural language semantics, and it is one that linguists are increasingly gaining exposure to (cf. Pater 2019; Potts 2019; Ferroni and Zanzotto 2020). It lacks a full account of a range of phenomena that compositional semantics has had success with: scopal behavior, anaphora, and the distinction between word- and world knowledge but these are open questions in a promising direction in the same way that those phenomena were open questions when Montague wrote PTQ.

[14] See Kutuzov et al. (2018) for overview.

4 Conclusion

I have presented three research programs that, I claim, have underpinned many of the success stories of mutual effort. Linguistics and philosophy have doubtless been held together in special relationship; one of often joint and often mutually fruitful inquiry, and while I have aimed to put a range of the successes of collaboration on display, I have also presented an emerging trend toward statistical modeling in linguistic semantics that threatens to destabilize the programs that held the disciplines together. What the future holds for philosophy and linguistics is unclear; over time perhaps the three programs will strengthen their grip, and no emergent trend will offer an alternative path to modeling expression meaning. Perhaps the truth-heavy, post-Montagovian fragment will be replaced by internet-fed algorithms, and linguists will abandon the philosophers' toolkit of semantic modeling.[15] Perhaps a synthesis will emerge. Or perhaps the relationship will survive drastic change, as philosophers of language and linguists make Herculean efforts to induct their students into the other field. For now, it remains to be seen. But it might be time for each to start thinking hard about what they really needed from the other one anyway.

References

Asher, Nicholas. 2011. *Lexical meaning in context: A web of words*. Cambridge: Cambridge University Press.

Barwise, Jon, and Robin Cooper. 1981. Generalized quantifiers and natural language. *Linguistics and Philosophy* 4 (2): 159–219.

Beaver, David. 2001. *Presupposition and assertion in dynamic semantics*. Stanford, CA: CSLI Publications.

Camp, Elisabeth. 2013. "Slurring perspectives". In *Analytic. Philosophy* 54 (3): 330–349.

[15] Arguably a similar progression can be seen in philosophy of science, from the Logical Positivists' search for a logic of induction to the stochastic shift to Bayesian factor analysis and Significance testing.

Cappelen, Herman. 2017. Why philosophers Shouldn't do semantics. *Review of Philosophy and Psychology.* 8: 743–762.

Chae, Hee Rahk. 2015. Idioms: Formally flexible but semantically non-transparent. In *Proceedings of the 29th Pacific Asia Conference on Language, Information and Computation*: 45–54.

Collins, John. 2007. Syntax, more or less. *Mind* 116 (464): 805–850.

Copestake, Ann, and Ted Briscoe. 2000. Semi-productive polysemy and sense extension. *Journal of Semantics* 12 (1): 15–67.

Cresswell, Max J. 1985. *Structured meanings*. Cambridge, MA: MIT Press.

Davidson, Donald. 1967. Truth and Meaing. *Synthese* 17 (1): 304–323.

Del Pinal, Guillermo. 2015. Dual content semantics, privative adjectives, and dynamic compositionality. *Semantics and Pragmatics.* 8 (7): 1–53.

Deo, Ashwini. 2015. Diachronic Semantics. *Annual Review of Linguistics* 1 (1): 179–197.

Dever, Josh. 1999. Compositionality as methodology. *Linguistics and Philosophy* 22 (3): 311–326.

Devitt, 1990. The Diachronic Development of Semantics in Copulas. Proceedings of the Sixteenth Annual Meeting of the Berkeley Linguistics Society, pp. 103–115.

Ferrone, Lorenzo and Fabio Massimo Zanzotto. 2020. Symbolic, Distributed and Distributional Representations for Natural Language Processing in the Era of Deep Learning: a Survey in: Frontiers in Robotics and AI https://arxiv.org/abs/1702.00764.

Fine, Kit. 2017. Truth maker semantics. In *A companion to the philosophy of language*, ed. Gilian Russell and Delia Graff Fara, 556–577. London: Routledge.

Firth, J. R. 1957. A synopsis of linguistic theory 1930–1955. In *Studies in linguistic analysis. Philological society*. Reprinted in Palmer, F. (ed.) 1968. Selected papers of J. *R. Firth*. Longman, Harlow.

Gries, Stephan Theodore. 2009. *Statistics for linguistics with R: A practical introduction*. (Trends in linguistics: Studies monographs) 1st Ed. Amsterdam: De Gruyter.

Gutzmann, Daniel. 2015. Use-conditional meaning. Studies in Semantics and Pragmatics. Oxford Studies in Semantics and Pragmatics 6. Oxford: Oxford University Press.

Harris, Zellig, and S. 1954. Distributional structure. *Word* 10: 146–162.

Heim, Irene, and Angelika Kratzer. 1998. *Semantics in generative grammar*. Malden: Wiley-Blackwell.

Hermann, Karl Moritz, and Phil Blunsom. 2013. The role of syntax in vector space models of compositional semantics. In *Proceedings of the 51st annual meeting of the association for computational linguistics*, Vol. 1, 894–904. Long Papers.
Higginbotham, James. 2002. On linguistics in philosophy, and philosophy in linguistics. *Linguistics and philosophy* 25 (5–6): 573–584.
Hirschberg, Julia. 1985. *A theory of scalar implicature*. Ph.D. thesis, University of Pennsylvania.
Horwich, Paul. 2005. *Reflections on meaning*. Oxford: Oxford University Press.
Huebner, Phillip A. and John A. Willitz. 2018. Structured Semantic Knowledge Can Emerge Automatically from Predicting Word Sequences in Child-Directed Speech. *Frontiers in Psychology.* Frontiers Media: Lausanne.
Janssen, Theo M.V. 2011. Montague semantics. In *The stanford encyclopedia of philosophy*, ed. Edward N. Zalta. Stanford: Stanford University. https://plato.stanford.edu/entries/montaguesemantics/. Accessed 10 June 2020.
———. 2012. "*Compositionality: its historic context*", in Werning et. al. (eds.), 2012, chapter 1, Oxford University Press. 19–46.
Jurafsky, Daniel. 1996. A probabilistic model of lexical and syntactic access and disambiguation. *Cognitive Science* 20: 137–194.
Jurafsky, Daniel, and James H. Martin. 2019. Vector semantics and embeddings. *Speech and language processing*. https://web.stanford.edu/~jurafsky/slp3/ed3book.pdf
Katz, Jerrold J. 1965. The relevance of linguistics to philosophy. *The Journal of Philosophy* 62 (20): 590–602.
King, Jeffrey C. 2009. Questions of Unity. *Proceedings of the Aristotelian Society*, Vol. CIX, Part 3, 257–277.
King, Jeffrey C., and Lewis, Karen S. 2016. Anaphora. In *The stanford encyclopedia of philosophy*, ed. E. N. Zalta. https://plato.stanford.edu/archives/sum2016/entries/anaphora/. Accessed 4 June 2020: https://plato.stanford.edu/entries/anaphora/.
Kracht, Marcus. 2007. Modal consequence relations. *Handbook of Modal Logic* 2007: 491–545.
Kratzer, Angelila. 1989. Stage and individual level predicates. Papers on Quantification. NSFGrant Report, Linguistics Department, University of Massachusetts, Amherst.
———. 1994. On External Arguments. In Functional Projections, eds. Elena Benedicto and Jeffrey Runner, 103-130. Amherst/Mass.: GLSA, UMass Amherst.
Kripke, Saul. 1980. Naming and Necessity. Harvard University Press.

Kutuzov, Andrey, Lilja Øvrelid, Terrence Szymanski, and Erik Velldal. 2018. Diachronic word embeddings and semantic shifts: A survey. In *Proceedings of the 27th international conference on computational linguistics*. Association for Computational Linguistics: 1384–1397.

Le, Phong and Willem Zuidema. 2015. Compositional Distributional Semantics with Long Term Short Term Memory. in: Proceedings of the Fourth Joint Conference on Lexical and Computational Semantics. Association for Computational Linguistics. Denver Colorado. pp.10–19.

Lepore, Ernest, and Jeffrey Pelletier. 2008. Linguistics and philosophy. In *Naturalism, reference, and ontology: Essays in honor of Roger F. Gibson*, ed. C. Wrenn, 183–215. New York: Peter Lang Publishing.

Lewis, David. 1975. Languages and Language. in Keith Gunderson (ed.), Minnesota Studies in the Philosophy of Science, Volume VII, Minneapolis: University of Minnesota Press, 3–35.

Ludlow, Peter. 2014. *Living words: Meaning underdetermination and the dynamic lexicon*. Oxford: Oxford University Press.

Matushansky, Ora. 2008. On the Linguistic Complexity of Proper Names. *Linguistics and Philosophy* 31 (5): 573–627.

McGrath, Mathew, and Devin Frank. 2018. Propositions. In *Stanford encyclopaedia of philosophy*, ed. E. N. Zalta. https://plato.stanford.edu/entries/propositions/. Accessed 3 June 2020.

Millburn, E., T. Warren, and M.W. Dickey. 2016. World knowledge affects prediction as quickly as selectional restrictions: Evidence from the visual world paradigm. *Language, Cognition, and Neuroscience* 31 (4): 536–548.

Mitchell, Jeff, and Mirella Lapata. 2008. Vector-based models of semantic composition. In *Proceedings of ACL-08: HLT*, 236–244. Columbus: Association for Computational Linguistics.

Montague, Richard. 1973. The proper treatment of quantification in ordinary English. in Approaches to Natural Language (Synthese Library 49). eds. K. J. J. Hintikka, J. M. E. Moravcsik, and P. Suppes. Dordrecht: Reidel, 221–242.

Moss, Sarah. 2012. The role of linguistics in the philosophy of language. In *The Routledge Companion to the Philosophy of Language*, ed. Delia Graff Fara and Gillian Russell, 513–525. London: Routledge.

Nefdt, Ryan. 2019. Why philosophers should do formal semantics (and a bit of syntax too): A reply to Cappelen. *Review of Philosophy and Psychology* 10 (1): 243–256.

Nickel, Bernard. 2016. *Between logic and the world: An integrated theory of generics*. Oxford: Oxford University Press.
Ninan, Dilip. 2010. De se attitudes: Ascription and communication. *Philosophy Compass* 5: 551–567.
Pagin, Peter, and Dag Westerstahl. 2010. Compositionality I: Definitions and variants. *Philosophy Compass* 5 (3): 250–264.
Partee, Barbara H. 1984. Compositionality. In *Varieties of formal semantics*, ed. F. Landman and F. Veltman, 281–312. Dordrecht: Foris. Reprinted in Partee, B.H. 2004. Compositionality. In *Formal Semantics: Selected Papers by Barbara H. Partee*, 153–181. Oxford: Blackwell Publishing.
———. 1996. The development of formal semantics in linguistic theory. In *The handbook of contemporary semantic theory*, ed. Shalom Lappin, 11–38. (Blackwell Handbooks in Linguistics 3). Oxford: Blackwell.
———. 2010a. Privative adjectives: Subsective plus coercion. In *Presuppositions and discourse: Essays offered to Hans Kamp*, ed. Rainer Bäuerle, Uwe Reyle, and Thomas Ede Zimmermann, 273–285. Bingley: Emerald Group Publishing.
———. 2010b. Formal Semantics: Origins, Issues, Early Impact. *The Baltic International Yearbook of Cognition, Logic and Communication* 6 (1).
———. 2016. Lexical Semantics in Formal Semantics, History and Challenges. ESSLLI Ref SemPlus Workshop, August 2016. https://esslli2016.unibz.it/wp-content/uploads/2015/10/ParteeESSLLI2016RefSemPlus.slides.pdf.
Pater, Joe. 2019. Generative linguistics and neural networks at 60: Foundation, friction, and fusion. *Language* 95(1): 41–74.
Piñango, Maria Mercedes, and Ashwini Deo. 2015. Reanalyzing the Complement Coercion Effect through a Generalized Lexical Semantics for Aspectual Verbs. *Journal of Semantics* 33 (2): 359–408.
Potts, Christopher. 2005. *The Logic of Conventional Implicature*. Oxford University Press.
———. 2007. The Expressive Dimension. *Theoretical Linguistics* 33 (2): 165–197.
———. 2019. A case for deep learning in semantics: Response to Pater. Language, Project MUSE. https://doi.org/10.1353/lan.2019.0003.
Pustejovsky, James. 1995. *The generative lexicon*. Cambridge: MIT press.
Quine, Willard Van Orman. 1960. *Word and object*. Cambridge, MA: MIT Press.
Rosch, Eleanor. 1975. Cognitive representations of semantic categories. *Journal of Experimental Psychology: General* 104 (3): 192–233.

Sahlgren, Magnus. 2008. The distributional hypothesis. *Rivista di Linguistica* 20 (1): 33–53.

Schiffer, Stephen. 2003. *The things we mean*. Oxford: Oxford University Press.

Shan, Chung-chieh. 2001. Monads for natural language semantics. In *Proceedings of the ESSLLI-2001 Student Session*, ed. Kristina Striegnitz, 285–298. 13th European Summer School in logic, language and information.

Sperber, Dan, and Deirdre Wilson. 1986. *Relevance, communication and cognition*. 2nd ed. Oxford: Basil Blackwell.

Stalnaker, Robert. 2014. *Context*. Oxford: Oxford University Press.

Stanley, Jason. 2000. Context and logical form. *Linguistics and Philosophy.* 23: 391–434.

Starr, Will. 2020. Conditional and counterfactual logic. In *The handbook of rationality*, ed. Alfred Mele and Piers Rawling. Cambridge, MA: MIT Press.

Szabó, Zoltán Gendler. 2000. *Problems of compositionality*. New York: Garland.

Tarski, Alfred. 1935. Der Wahrheitsbegriff in den formalisierten Sprachen. In *Studia Philosophica* 1: 261–405. Translation of Tarski 1933 by L. Blaustein, with a postscript added.

Tonhauser, Judith, David Beaver, Craige Roberts, and Mandy Simons. 2013. Toward a taxonomy of projective content. *Language* 89 (1): 66–109.

Vendler, Zeno. 1965. The relevance of linguistics to philosophy: Comments. *The Journal of Philosophy* 62 (20): 602–605.

Wilson, Dierdrie and Dan Sperber. 1995. Relevance: Communication and cognition (2nd ed.). Blackwell Publishing.

Yalcin, Seth. 2018. Semantics as model-based science. In *The Science of Meaning: Essays on the Metatheory of Natural Language Semantics*, ed. Derek Ball Brian Rabern, 334–360. Oxford: Oxford University Press.

Zadrozny, Wlodek. 1994. From compositional to systematic semantics. *Linguistics and Philosophy* 17 (4): 329–342.

Zhang, Muay, Maria Mercedes Piñango, and Ashwini Deo. 2018. Real time roots of meaning change: Electrophysiology reveals the contextual modulation processing basis of synchronic variation in the location-possession domain. In *Proceedings of the 40th Annual Conference of the Cognitive Science Society*, 2783–2788. Madison: Curan Associates.

Zimmermann, Thomas Ede, 2012, "*Compositionality Problems and How to Solve Them*", in Oxford Handbook of Compositionality. eds. Werning, Hinzen, & Machery 2012: 81–106.

Index[1]

[1] Note: Page numbers followed by 'n' refer to notes.

R. M. Nefdt et al. (eds.), *The Philosophy and Science of Language*, https://doi.org/10.1007/978-3-030-55438-5

G

H

I

J

K

L

M

Lightning Source UK Ltd.
Milton Keynes UK
UKHW010025210821
389204UK00001B/6